TENTH EDITION

Microbiology Experiments

A HEALTH SCIENCE PERSPECTIVE

John Kleyn

Anna Oller
University of Central Missouri

MICROBIOLOGY EXPERIMENTS: A HEALTH SCIENCE PERSPECTIVE, TENTH EDITION

Published by McGraw Hill LLC, 1325 Avenue of the Americas, New York, NY 10121. Copyright ©2022 by McGraw Hill LLC. All rights reserved. Printed in the United States of America. Previous editions ©2019, 2016, and 2012. No part of this publication may be reproduced or distributed in any form or by any means, or stored in a database or retrieval system, without the prior written consent of McGraw Hill LLC, including, but not limited to, in any network or other electronic storage or transmission, or broadcast for distance learning.

Some ancillaries, including electronic and print components, may not be available to customers outside the United States.

This book is printed on acid-free paper.

2 3 4 5 6 7 8 9 LKV 26 25 24 23 22

ISBN 978-1-264-34193-1 (bound edition)
MHID 1-264-34193-8 (bound edition)

Portfolio Manager: *Lauren Vondra*
Product Developer: *Erin DeHeck*
Marketing Manager: *Tami Hodge*
Content Project Manager: *Jeni McAtee*
Buyer: *Laura Fuller*
Designer: *David Hash*
Content Licensing Specialist: *Beth Cray*
Cover Image: *National Institute of Allergy and Infectious Diseases (NIAID)*
Compositor: *MPS Limited*

All credits appearing on page or at the end of the book are considered to be an extension of the copyright page.

Library of Congress Cataloging-in-Publication Data

Names: Kleyn, John G. | Oller, Anna, author.
  Title: Microbiology experiments : a health science perspective / John Kleyn,
  Anna Oller, University of Central Missouri.
  Description: Tenth edition. | New York : McGraw Hill LLC, [2022] | Includes index.
  Identifiers: LCCN 2020050147 | ISBN 9781264341931 (hardcover)
  Subjects: LCSH: Microbiology—Laboratory manuals. | Microbiology—Experiments.
  Classification: LCC QR63 .K54 2022 | DDC 579.078—dc23
  LC record available at https://lccn.loc.gov/2020050147

The Internet addresses listed in the text were accurate at the time of publication. The inclusion of a website does not indicate an endorsement by the authors or McGraw Hill LLC, and McGraw Hill LLC does not guarantee the accuracy of the information presented at these sites.

mheducation.com/highered

CONTENTS

PREFACE

To the Student

A microbiology laboratory is valuable because it actually gives you a chance to see and study microorganisms firsthand. In addition, it provides you with the opportunity to learn techniques used to study and identify these organisms. The ability to make observations, record data, and analyze results is useful throughout life.

It is very important to read the scheduled exercises before coming to class so that class time can be used efficiently. It is helpful to ask yourself the purpose of each step as you are reading and carrying out the steps of the experiment. Sometimes it will be necessary to read an exercise several times before it starts to make sense.

Conducting experiments in microbiology laboratories is particularly gratifying because the results can be seen in a day or two (as opposed, for instance, to plant genetics laboratories, which can take months). Opening the incubator door to see how your cultures grew and the results of the experiment is a pleasurable moment. We hope you enjoy your experience with microorganisms as well as acquire laboratory and critical thinking skills that will be valuable in the future.

To the Instructor

The manual includes a wide range of exercises—some more difficult and time-consuming than others. Usually more than one exercise can be done in a 2-hours laboratory period. In these classes, students can actually see the applications of the principles they have learned in the lectures and text. We have tried to integrate the manual with the text *Nester's Microbiology: A Human Perspective,* Tenth Edition, by Denise Anderson, et al.

The exercises were chosen to give students an opportunity to learn new techniques and to expose them to a variety of experiences and observations. We did not assume that the school or department has a large budget; thus, exercises were written to use as little expensive media and equipment as possible. The manual contains more exercises than can be done in one course so that instructors will have an opportunity to select the appropriate exercises for their particular students and class.

- An online Instructor's Manual is available from the publisher.
  - It lists equipment, cultures, media, and reagents needed for each exercise and has extensive information for storing cultures and making media.
- The Notes to the Instructor section gives suggestions for preparing and presenting the laboratory sessions.
- Additional questions that can be used to supplement those in the student manual are included.
- **We highly recommend that the instructors utilize the Instructor's Manual.** Contact your local McGraw Hill representative for the URL and password for this site.
- Revisions in the Tenth Edition include the following:
  - New exercise components:
    - Ex. 28 Bacteriophage Susceptibility was added to the titers.
    - Ex. 31 added Quorum Sensing to Epidemiology.
  - Changes to exercises:
    - Numerous Figures and Tables were added and updated throughout.
    - Numerous Lab Report tables were updated to guide students in recording lab results.
    - The Microscopy and Staining labs were moved to exercises 1–6, whereas Diversity and Aseptic Techniques were moved to exercises 7 and 8.
    - Additional tables were added to exercise 7 explaining media types.

- A more explicit explanation of dilution math was added to exercise 8.
- Exercises 16, 17, 18, and 19 were updated to minimize culturing pathogens from students.
- Demonstrations of representative microbes are included.
- Many graphics were updated to include color to differentiate and more clearly explain concepts.
- Measurement bars were added to most microscopy photos.
- Bacteriophage specificity was added to Prokaryotic Viruses in exercise 28.
- *Enterococcus* and MUG protocols were added to exercise 29 for those using it as the indicator organism.
- Using appropriate software to determine primers to the sequences was added to exercise 34.
- The culture media and reagents listing was updated to reflect the current exercises.
- An additional 11 color plates and many new figures were added to enhance student learning.
- Student objectives that meet suggested American Society for Microbiology (ASM) curriculum guidelines are designated by superscripted numbers.
- If a student has a special interest in microbiology and would like to do independent work, three projects are available online:
  ○ Methylotrophs
  ○ The UV-Resistant *Deinococcus*
  ○ Hydrocarbon-Degrading Bacteria

We hope these changes are helpful and that the manual contributes to the students' understanding of microbiology. We also hope both students and instructors enjoy their experience with a very interesting group of organisms.

Acknowledgments

We would like to thank Denise Anderson, Sarah Salm, and Deborah Allen for their text *Nester's Microbiology: A Human Perspective*. This text was the source of much of the basic conceptual material and figures for our laboratory manual.

We would also like to express our appreciation to Lauren Vondra, Erin DeHeck, Jeni McAtee, Beth Cray, David Hash, and all the others involved in publishing this manual.

LAB SAFETY RULES

1. Place backpacks, coats, etc., in designated areas. Do not place backpacks or coats on bench tops.
2. At the beginning and end of each lab period, wipe the bench tops with disinfectant and allow to air dry/proper contact time. Make sure the entire surface comes into contact with the solution.
3. Upon entering and leaving the lab, wash your hands thoroughly—30–45 seconds minimum (ABCs, etc.).
4. Proper attire is crucial for your safety.
   a. Safety goggles and a lab apron or coat MUST BE worn when you are in the laboratory. Tie back long hair. Your instructor may require you to wear gloves. Remove them before leaving lab and store your apron/coat and goggles in the lab until the end of the course.
   b. Hats, sandals, and open-toed shoes are NOT allowed. If you drop a slide or test tube, you may be injured.
5. Report any accidents to the instructor(s) immediately.
6. If you have an open wound, inform the instructor and ask for a band-aid to prevent infection.
7. If you are tardy, you miss imperative safety information. Thus, you may not be allowed to enter the lab.
8. Do not smoke, eat, drink, or CHEW GUM. Do not apply cosmetics (chapstick), or insert contact lenses.
9. Do not chew on pens/pencils. Leave a pen/pencil in your lab drawer to avoid taking contamination home.
10. ALWAYS flame loops and needles BEFORE and AFTER obtaining ORGANISMS.
11. Do not place CONTAMINATED loops, needles, pipettes, etc. on bench tops or in lab drawers.
12. DO NOT LICK labels needing moistened. Use a water dropper/bottle to moisten. Do not pipette by mouth.
13. Do not work on top of the lab manual. If a spill occurs, it cannot be easily disinfected.
14. Loose papers are NOT permitted on the bench top. Secure in binders, use clips, etc., to prevent fires.
15. Use test tube racks or designated containers to carry and store cultures.
16. Hold the entire tube when handling cultures. Kimcaps can (and often do) drop and break on the bench top or floor when held by only the cap. Mix cultures gently to avoid creating aerosols.
17. Do not take lab items (plates, slides, etc.) outside of the lab, unless instructed to do so by your instructor.
18. Place contaminated slides, used staining reagents, broken glassware, etc., in the proper container(s).
19. You may only use items from your assigned area, regardless of how many students are in the class.
20. At the end of the lab period, place cultures, tubes, plates, etc., in designated areas. Push stools in fully.
21. Fill up water dropper bottles using distilled water. Restock labels, lens paper, etc., before leaving.
22. IF A CULTURE IS SPILLED/DROPPED:
    a. **Notify the instructor or lab technician immediately.**
    b. If you are NOT contaminated, move any lab books, pencils, sharpies, etc. out of the way.
    c. If you ARE contaminated, ask someone else to help move books, etc.
    d. Disinfect the bench area and any contaminated tubes or racks with disinfectant.
    e. Wash your hands immediately if they come into contact with any microbes or chemicals. Most chemicals need approximately 15 minutes of thorough washing to be completely removed.

23. When lighting Bunsen burners, open the gas jet and use a flint to light the burner. Do not leave the gas on for PROLONGED PERIODS OF TIME before lighting the flame. Turn off the burners (gas) when you are not using them. Make sure that paper, alcohol, the gas hose, and your microscope are not close to the flame.

24. You will be given an unknown culture. It is your responsibility to transfer it and place it in proper containers to ensure its survival. If you do not transfer your unknowns, your results may be questionable.

25. Make sure electronic devices (if allowed by your instructor) are enclosed in disposable/disinfected plastic covers.

26. When the laboratory is in session, keep the doors and windows shut. Post a sign on the door indicating that a microbiology laboratory is in session.

27. If you are immunocompromised (including pregnancy), consult a physician before taking this class.

28. You are expected to stay the entire lab time. Ask the instructor if you are truly done with the day's lab.

29. Comply with all federal and state guidelines regarding microbial disposal and disinfection, and do not use microbes in any manner that could be threatening to others.

30. Improper behavior WILL NOT be tolerated, and you will be asked to leave. Improper behavior may include attitude, verbiage, clothing/dressing, and actions, which are safety violations.

I understand the above safety rules and regulations and agree to abide by them. I have also read the course syllabus and understand its contents. I also know where the following lab safety equipment is located and how to properly use it:

Safety Devices: Sink(s):   Eyewash(es):

Fire Extinguisher(s):   Fire Blanket(s):

MSDS Sheets:

_____   _____

                                     Signature   Date

INTRODUCTION to Microbiology Stains & Culturing

When you take a microbiology class, you have an opportunity to explore an extremely small biological world that exists unseen in our own ordinary world. Fortunately, we were born after the microscope was perfected, so we can observe these extremely small organisms. We now also understand how the cell transfers the information in its genes to the operation of its cellular mechanisms so that we can study not only what bacteria do but also how they do it.

A few of these many and varied organisms are pathogens (capable of causing disease). Special techniques have been developed to isolate and identify them as well as to control or prevent their growth. The exercises in this manual will emphasize medical applications. The goal is to teach you basic techniques and concepts that will be useful to you now or can be used as a foundation for additional courses. In addition, these exercises are also designed to help you understand basic biological concepts that are the foundation for applications in all life science fields.

As you study microbiology, it is also important to appreciate the essential contributions of microorganisms as well as their ability to cause disease. Most organisms play indispensable roles in breaking down dead plant and animal material into basic substances that can be used by other growing plants and animals. Photosynthetic bacteria are an important source of the earth's supply of oxygen. Microorganisms also make major contributions in the fields of antibiotic production, food and beverage production, food preservation, and DNA technologies such as CRISPR. The principles and techniques demonstrated here can be applied to these fields as well as to medical laboratory science, nursing, or patient care. This course is an introduction to the microbial world, and we hope you will find it useful and interesting.

In the next few exercises, you will be introduced to several basic procedures: the use of the microscope and aseptic and pure culture techniques. These are skills that you will use not only throughout the course but in any microbiology laboratory work you do in the future. The exercises that meet the recommended ASM guidelines for laboratory skills have been designated in each lab. These are needed skills for future employment.

Note: The use of pathogenic organisms has been avoided whenever possible, and non-pathogens have been used to illustrate the kinds of tests and procedures that are actually carried out in clinical laboratories. In some cases, however, it is difficult to find a substitute, and organisms of low pathogenicity are used. These exercises will have an additional safety precaution.

Staining

Bacteria are difficult to observe in a broth or wet mount because there is very little contrast between them and the liquid in which they are suspended. This problem is solved by staining bacteria with dyes. Although staining kills bacteria so their motility cannot be observed, the stained organisms contrast with the surrounding background and are easier to see. The determination of the shape, size, and arrangement of the cells after dividing is useful in identifying an organism. These can be demonstrated best by preparing a smear on a glass slide from a broth culture, or a colony from a plate, then staining the smear with a suitable dye.

Examining a stained preparation is one of the first steps in identifying an organism. Staining procedures can be classified into two types: the simple stain and the multiple (or differential) stain. In the **simple stain**, a single stain such as methylene blue or crystal violet is used to dye the bacteria. The most commonly used simple stains are basic dyes, which are positively charged and make it possible to stain negatively charged bacteria.

The shape and arrangement of organisms can be determined, but all organisms (for the most part) are stained the same color. Another kind of simple stain is the negative stain. In this procedure, the organisms are

Table I.1.1   A Summary of Stains and Their Characteristics

Stain	Characteristic
Simple Stains	A basic dye is used to stain cells. Easy way to increase the contrast between otherwise colorless cells and a colorless background. Allows visualization of shape and size. Examples include crystal violet, safranin, and methylene blue.
Differential Stains	A multistep procedure is used to stain cells and distinguish one group of microorganisms from another.
Gram stain	Used to separate bacteria into two major groups: Gram-positive and Gram-negative. The staining characteristics of these groups reflect a fundamental difference in the chemical structure of their cell walls. This is by far the most widely used staining procedure.
Acid-fast stain	Used to detect organisms that do not easily take up stains, due to mycolic acid on the cell wall, particularly members of the genus *Mycobacterium*.
Special Stains	A staining procedure used to detect specific cell structures.
Capsule stain	The common method darkens the background and a counterstain stains the bacterial cell, so the capsule stands out as a clear area surrounding the cell. Capsules are produced by *Klebsiella* and *Enterobacter* species.
Endospore stain	Stains endospores, which do not readily take up stains due to the presence of dipicolic acid and calcium. Members of the genera *Bacillus* and *Clostridium* are among the few species that produce endospores.
Flagella stain	The staining agent adheres to and coats the otherwise thin flagella, making them visible with the light microscope.
Fluorescent Dyes and Tags	Some fluorescent dyes bind to compounds found in all cells; others bind to compounds specific to only certain types of cells. Antibodies to which a fluorescent molecule has been attached are used to tag specific molecules. Flagella are seen on many bacteria, including *Escherichia coli* and *Spirillum volutans*.

mixed with a dye on a slide and the mixture is permitted to air dry. When the stained slide is viewed under the microscope, the organisms are clear against a dark background. The types of stains are shown in **table I.1.1**.

The multiple stain involves more than one dye. The most widely used example is the differential Gram stain. After staining, some organisms appear purple and others pink, depending on the structure of their cell wall, which are depicted in **table I.1.2**.

Special stains are used to observe specific structures of bacteria. Compared with eukaryotic organisms, prokaryotic organisms have relatively few morphological differences. Several of these structures, such as endospores, capsules, acid-fast cell walls, storage granules, and flagella, can be seen with differential stains. You will have an opportunity to stain bacteria with a variety of staining procedures and observe these structures over the next few labs.

Table I.1.2   Comparison of Features of Gram-Positive and Gram-Negative Bacteria

Gram-Positive / Gram-Negative

	Gram-Positive	Gram-Negative
Color of Gram-Stained Cell	Purple	Pink
Representative Genera	*Bacillus, Staphylococcus, Streptococcus*	*Escherichia, Neisseria, Pseudomonas*
Distinguishing Structures/Components		
Peptidoglycan	Thick layer	Thin layer
Teichoic acids	Present	Absent
Outer membrane	Absent	Present
Lipopolysaccharide (endotoxin)	Absent	Present
Porin proteins	Absent (unnecessary because there is no outer membrane)	Present; allow molecules to pass through outer membrane
General Characteristics		
Sensitivity to penicillin	Generally more susceptible (with notable exceptions)	Generally less susceptible (with notable exceptions)
Sensitivity to lysozyme	Yes	No

EXERCISE 1

Introduction to the Compound Light Microscope

Definitions

Compound microscope. A microscope with two lenses, often called a bright-field compound microscope.

Condenser. A structure located below the microscope stage that contains a lens for focusing the light rays on the specimen as well as an iris diaphragm.

Iris diaphragm. An adjustable opening that regulates the amount of light illuminating the specimen.

Magnification. The microscope's ability to optically increase the specimen size.

Objective lens. The rotating lenses that magnify the items from the slide, such as a 10× or 40×.

Ocular lens. The eyepiece looked through to view microscope specimens; usually has a 10× magnification.

Parfocal. Lenses that are on one plane to allow for fine adjustment between objective lenses of different magnifications.

Resolution. The smallest separation that two structural forms—two adjacent cilia, for example—must have in order to be distinguished optically as separate.

Viable. Alive.

Wet mount. A laboratory technique in which a microscopic specimen in liquid is added to the surface of a slide and covered with a coverslip.

Objectives

1. Describe the parts of the microscope and why they are important.[ASM 1]

2. Describe stained and unstained materials.

3. Explain the use and proper care of the microscope.[ASM 1]

Pre-lab Questions

1. Why should you carry a microscope in the upright position?

2. What is the only material that can be used to wipe an objective lens?

3. Why must you wipe off the ocular lens before storing the microscope?

Getting Started

Microbiology is the study of organisms too small to be seen with the naked eye, including a vast array of bacteria, viruses, protozoa, fungi, and algae. Van Leeuwenhoek, a Dutch merchant in the late seventeenth century whose hobby was lens making, was the first to see these previously unknown creatures. His microscope consisted of one simple lens, but it was enough to observe some of these tiny living things (**figure 1.1**). Although he made drawings of some of these organisms, he did not suspect that they were essential for the existence of our world, or that a small percentage was responsible for contagious diseases. It was only after the improvement of the compound microscope almost 200 years later that it was possible to understand the role of microorganisms in disease. After the

Figure 1.1 Model of a van Leeuwenhoek microscope. The original was made in 1673 and could magnify the object being viewed almost 300×. The object being viewed is brought into focus with the adjusting screws. This replica was made according to the directions given in the *American Biology Teacher* 30:537, 1958. Note its small size. © Tetra Images/Alamy Stock Photo.

microscope was perfected in the 1870s and 1880s, real progress was made in determining the actual cause of disease (see Appendix 6 for a history of the development of the microscope). In 1877, Robert Koch saw bacteria in the blood from an animal with anthrax. In combination with his postulates, Koch could see that bacteria caused the disease anthrax, and the disease was not caused by swamp gas or evil spirits.

The modern **compound light microscope** consists of two lens systems. The first is the **objective** lens, which is closest to the material on the slide, and the second is the eyepiece, or **ocular lens**, which magnifies the image formed by the objective lens. The total magnification is found by multiplying the ocular lens magnification by the objective lens magnification. For example, if the ocular lens magnification is 10×, and the objective lens magnification is 45×, the total magnification is 450 diameters.

Although it is possible to put additional magnifying glasses on top of the ocular lenses of the microscope, they would not improve the ability to see more detail. The reason is that the actual limiting factor of the light microscope is the **resolution**. This is the ability to distinguish two close objects as distinct from one another rather than as one round, hazy object. The resolution of a lens is limited by two factors: the angle of the lens and the wavelength of light entering the microscope. When using an objective lens of **magnification** 100× and light is optimal, a compound light microscope has a maximum magnification of about 1,000×.

A magnification of 1,000× is sufficient to easily visualize single-celled organisms such as algae and fungi. Bacteria, however, still appear very small

(about the size of the letter *l* on a printed page), and their appendages, such as flagella, cannot be observed without special stains. Viruses are also usually too small to be seen in a light microscope.

To surmount the limitations of light and lens, the electron microscope, which uses electrons instead of light, was developed in the 1930s. It magnifies objects 100,000×, permitting the visualization of viruses and structures within cells. Because the electron microscope is a very large piece of equipment requiring specialized techniques, it is usually found only in universities or research facilities. More recently, many other microscopes have been developed but are usually for specific research applications.

In this exercise, you will have an opportunity to become familiar with a compound light microscope and learn how to use and care for it. You will prepare **wet mounts** of unstained organisms and learn to examine previously stained and unstained organisms. The microscope is an expensive and complex piece of equipment. Treat it with great care.

The Parts of the Microscope[ASM 1]

The eyepieces are the lenses at the top of the microscope. They usually have a magnification of 10× (see **figure 1.2**).

Objective lenses. Most microscopes have at least three objective lenses, and some have a fourth—4×. Each lens is color coded with a band around the objective. They include:

Low power (yellow band)	10×
High power (blue band)	40×
Oil immersion (white or red band)	100×

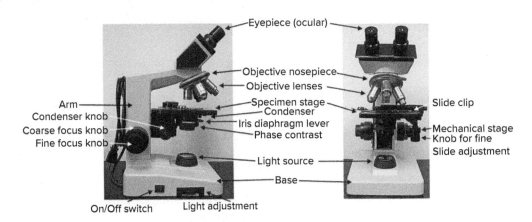

Figure 1.2 Modern bright-field compound microscope. Courtesy of Anna Oller, University of Central Missouri.

These lenses can be rotated and the desired lens clicked into place. The high power is often called high-dry, because it is considered the highest power without using oil (hence, dry).

Stage. The stage is below the objective lens. A slide clip keeps the microscope slide in place. It usually has a device called a *mechanical stage* for holding the slide, as well as knobs that permit the slide to be moved smoothly while viewing.

Condenser. Below the stage is the **condenser**, which focuses the light on the slide. If it is lowered, the amount of light is reduced, but the resolution is also lowered. For our purposes, the condenser should remain at its highest position under the stage.

Diaphragm. A lever on top of the condenser but under the stage controls the **iris diaphragm**. The diaphragm is important for adjusting the amount of light illuminating the slide. The higher the magnification, the greater the amount of light that is needed.

Coarse and Fine Focus Knob. The large course focus knob controls the large distance between the slide and the objective. It is used to bring a specimen into view on 10× or 40×. The fine focus brings a specimen into clear, or fine, focus using smaller increments than the coarse focus. The fine focus is important to see microbes viewed at 100×.

Light Source. The light source is at the base of most microscopes. Usually, the light source is set at maximum, and the amount of light on the slide is adjusted with the iris diaphragm. There is often a blue (or other color) filter held below the diaphragm to help correct the color that you see.

Precautions for Proper Use and Care of the Microscope[ASM 1]

1. Carry the microscope with both hands by the arm and base. Keep it upright. If the microscope is inverted, the eyepieces (oculars) may fall out.

2. Do not remove the objective or ocular lenses for any purpose.

3. ****Use only <u>one</u> hand to turn a course or fine focus knob.**** Using a hand on both sides of the focusing knobs can twist the knobs improperly, stripping the gears inside. This makes the microscope unable to properly focus, and costs money to repair.

4. If something seems stuck or you have problems making adjustments, do not apply force. Consult the instructor.

5. Never touch or wipe the objective or eyepiece lenses with anything but lens paper. Clean the lens by gently drawing a flat piece of lens paper across it. The presence of foreign particles can be determined by rotating the ocular lenses manually as you look through the microscope. A pattern that rotates is evidence of dirt. If wiping the lenses with lens paper does not remove the dirt, consult the instructor. It may be on the inside surface of the lens.

6. Before storing the microscope:
   If the microscope has an adjustable tube, rack it down so that the microscope can be stored more easily. Make sure the lowest objective (a blank, a 4×, or 10×) is clicked into place.

   Make certain the eyepiece lenses are clean. Sweat deposits from your eyes can etch the glass.

   Important: If you have used the oil immersion lens, be sure to wipe off all the oil. If not removed, it can leak into the lens and cause severe damage.

Materials

Prepared slides of a printed letter *e* with coverslips (optional)

Prepared stained slides of protozoa, baker's yeast, or other large cells

Suspension of protozoa and/or algae

Suspension of baker's yeast, *Saccharomyces cerevisiae*

PROCEDURE

Important note: In this introduction to the microscope, the 100× oil objective lens (usually labeled with a white band) will not be used. It is the most expensive lens, and its use with immersion oil will be explained in exercise 2. Be particularly careful not to hit this lens on the stage.

1. Place the microscope on a clear space on your bench, away from any flame or heat source. Identify the different parts with the aid of figure 1.2 and the Getting Started section, "The Parts of the Microscope."

2. Before using the microscope, be sure to read the Getting Started section, "Precautions for Proper Use and Care of the Microscope."

3. Obtain either a prepared slide with the printed letter *e* covered with a coverslip or a large stained specimen of protozoa, fungi, baker's yeast, or algae.

4. Place the slide on the mechanical stage, coverslip/specimen side up, and turn on the light at the base of the microscope.

   Tip: If a slide is placed upside down on the stage, the specimen will always appear fuzzy.

5. Move the ocular lenses apart, and then while looking through the lenses, push them together until you see one circle. This is the ocular width of your eyes. If you note the number on the scale between the lenses, you can set the microscope to this number each time you use it.

6. Rotate the low-power objective lens (10×) into place. When looking from the side, bring the lens as close to the slide as possible (or the slide to the lens depending on the microscope). Then when looking through the microscope, use the large coarse-adjustment knob, to raise the lens (or lower the stage) until the object is in focus. Increase and decrease the amount of light with the iris diaphragm lever to determine the optimal amount of light. Continue to focus until the specimen is in sharp view using the smaller fine-adjustment knob. Remember that the 10× objective lens should never touch the surface of the slide or coverslip.

7. Move the slide back and forth. When viewing objects through the microscope, the image moves in the opposite direction than the slide is actually moving. It takes a while to become accustomed to this phenomenon, but later it feels normal.

8. View the specimen at a higher magnification. Rotate the 40× objective lens into place (sometimes called the high–dry objective lens). Be very careful not to hit the slide or the stage with the oil immersion lens. Notice the change in the amount of light needed. Also note that the specimen is almost in focus. Use only the fine focus adjustment to bring the specimen into focus. Most microscopes are **parfocal**, meaning that the objective lens can be rotated to another lens and the slide remains in focus. View several prepared slides until you become comfortable with the microscope. Always start focusing with the low-power objective lens and then move to the high–dry objective lens.

Wet Mounts

1. Prepare a wet mount (**figure 1.3**). Clean a glass slide with a mild cleansing powder such as Bon Ami or as directed by your instructor.

2. Place one drop of suspension on the slide using a dropper. Carefully place a coverslip on edge next to the drop and slowly lower it so that it covers the drop. Try to avoid air bubbles. If the drop is so big it leaks out from under the coverslip, either (1) add a second coverslip to wick away the excess liquid, or (2) discard the slide into a container designated by the instructor and prepare another wet mount.

3. Examine the wet mount using the low-power objective lens and then switch to the high-power objective lens. When viewing unstained material, it is necessary to reduce the amount

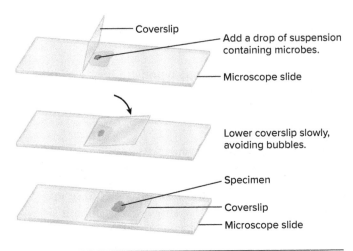

Coverslip

Add a drop of suspension containing microbes.

Microscope slide

Lower coverslip slowly, avoiding bubbles.

Specimen

Coverslip

Microscope slide

Figure 1.3 Preparation of a wet mount.

of light to increase the contrast between the cells and the liquid. If you are having difficulty focusing on the material, try to focus on the edge of the coverslip and then move the slide into view. Do not try to view the wet mount on the oil immersion lens, as the lens will push the liquid out from the coverslip and contaminate the microscope.

4. Draw the slides you viewed in the report. Indicate the **total** magnification.

5. Dispose your slides as directed by your instructor. The material on the slide is still **viable**, so the slides should be boiled, placed in bleach, or autoclaved.

6. Wipe the ocular lens (eyepiece) with lens paper. After clicking the proper objective into place, turn the microscope off. Then center the mechanical stage (so it does not stick out), rack the microscope down, and return your microscope carefully to its storage space.

EXERCISE

1

Laboratory Report: Introduction to the Compound Light Microscope

Results

1. Draw four fields of the specimens you observed. Label the objective power used and calculate the total magnification. If possible, show the same field or material at two different magnifications.

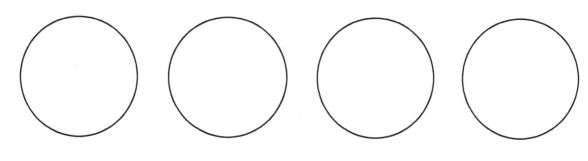

Specimen   _____   _____   _____   _____

Objective lens   _____   _____   _____   _____

Total magnification _____   _____   _____   _____

2. On your microscope, what is the magnification of

   a. the ocular lenses?   _____

   b. the low-power lens? _____

   c. the high–dry lens?   _____

Questions

1. When you increase the magnification, is it necessary to increase or decrease the amount of light? Why?

2. When looking at unstained material, do you need more or less light than what is needed to view a stained preparation? Explain why.

3. Why is it convenient to have a parfocal microscope?

4. Why couldn't you see a virus with your microscope even if you increased the ocular lens magnification to 100×?

5. Why was Koch's observation of bacteria in blood so significant?

6. When observing a specimen through the microscope, how do you calculate the total magnification?

EXERCISE

2

The Oil Immersion Lens ASM 1

Definitions

Brownian movement. A hovering back and forth motion seen when a liquid hits a solid object, causing a microbe to appear as moving when it is not truly motile.

Immersion oil. Oil placed on a slide to minimize refraction of the light entering the lens.

Refraction. The bending of light as it passes from one medium to another.

Refractive index. The ratio of the velocity of light in the first of two media to its velocity in the second medium as it passes from one medium into another medium.

Objectives

1. Explain the use of the oil immersion lens when viewing stained slides.
2. Explain the use of the oil immersion lens to observe motile bacteria in a wet mount preparation (optional).

Pre-lab Questions

1. What is the total magnification of your microscope when you are observing stained bacteria with the oil immersion lens?
2. How can the refraction of light be reduced as it goes through the glass slide and into the lens?
3. What is Brownian movement?

Getting Started

To observe stained bacteria, it is necessary to use the oil immersion lens. The magnification of this lens is 100× and when viewing a specimen through an eyepiece of 10×, the total magnification is 1,000×. This lens is the longest and usually has a white colored band around it. Extra care must be taken when using it. Not only is it the most valuable, but it is also possible to scratch the lens or hit and break the slide when attempting to focus with it because of its length. Also, care must be taken not to hit the lens on the stage when rotating a lens into place.

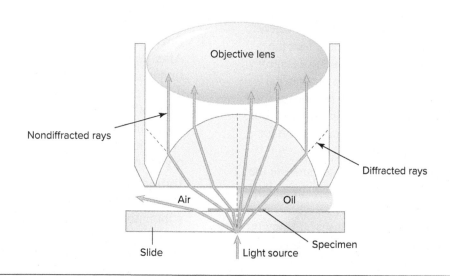

Figure 2.1 Oil immersion lens. Oil is required to change the light ray direction so more light rays are nondiffracted and enter the objective lens. This allows the specimen to be viewed in focus.

The reason this lens is called the oil immersion lens is that **immersion oil** is added to the top of a stained slide and the lens is then carefully rotated into the oil (**figure 2.1**). The purpose of the oil is to prevent **refraction** of the light entering the lens. Light bends each time it goes through a different medium. (This can be seen in the appearance of a spoon in a glass of water.) The **refractive index** of oil is the same as that of glass. The oil prevents diffracted rays of light so the specimen appears sharp and in focus. Without the addition of oil, the specimen would appear blurry. This can be observed if your oil dispensing bottle has a glass dropper. The dropper will disappear when you immerse it in the oil because the light does not bend between the immersion oil and the glass (see Appendix 6 for additional information).

In this exercise, the oil immersion lens will be used to examine stained slides. Observing wet mounts to determine whether bacteria are motile is optional, because this is a fairly difficult procedure and can be done later when you have had more experience using the microscope.

Materials

Prepared stained slides

PROCEDURE

The Oil Immersion Lens[ASM 1]

1. Hold your slide up into the air and look where organisms are located on the slide. Place the slide label side up onto the mechanical stage and into the stage clip. Using the mechanical stage knobs, move the area with organisms over the hole where light is coming through the condenser. If you cannot see the microbe, aim for the coverslip center.

2. Click the low-power 10× into place and use the coarse focus (big knob) to move the stage all the way up close to the objective.

3. Now look through the ocular lenses and slowly move the stage down using the course adjustment, until color is seen. (The field may be fuzzy.) By starting at the top, if you turn too quickly past the organism, the stage can easily be reset at the top and started over.

4. Once color is seen, use the fine-adjustment knob (smaller outer knob) (see exercise 1) to bring the object into sharp focus. Increase or decrease the light with the iris diaphragm for optimal viewing.

5. Once the object in sharp focus on 10×, simply rotate the 40× objective into place and you should just need to do a slight fine focus adjustment to bring the object into sharp focus.

6. Without moving the stage, rotate the objective to be on the lowest power. Place a drop of immersion oil directly on top of the slide where you see light coming through the slide.

7. Carefully rotate the oil immersion lens into the oil and click into place. The lens is retractable and will not hit the slide as long as the correct plane was originally in focus. A slight fine focus adjustment (three or four fine focus knob rotations) to bring the object into sharp focus should be all that is needed.

8. While looking through the ocular lens, focus the microscope using the fine-adjustment knob. If the microscope is parfocal, it should take very little adjustment. Never rotate the oil immersion lens toward the slide while looking through the microscope. If you are on the wrong plane, the lens will be too long and hit the slide with the lens. This will either scratch or break the lens, or crack the slide.

9. Notice that you must increase the amount of light as you increase the magnification.

10. Examine the stained slides to become familiar with the appearance of stained bacteria. This will be useful when you prepare your own slides.

11. Draw examples of several fields you observed. After you feel comfortable using the oil immersion lens, rotate it away from the slide, remove the slide, and take the oil off the slide with a piece of lens paper or as directed by your instructor. If the slide does not have a coverslip, do not wipe the slide as it will remove the stained organisms.

12. **Important:** Before storing the microscope, remove the oil from the lens by carefully wiping the lens with several flat pieces of lens paper. If the oil is not removed, it can seep around the cement holding the lens in place, damaging the microscope. Do not crumple the paper, but slide it over the lens. Also carefully wipe the eyepiece with clean lens paper.

Determining Motility with a Wet Mount (Optional)[ASM 2]

Some bacteria can move through liquid using flagella (singular, *flagellum*). These flagella are much smaller than those seen on protozoa, and can only be visualized in the light microscope if special stains are used.

Although bacterial cell motility is usually determined by the semisolid agar stab inoculation method, it is sometimes determined by direct microscopic examination of wet mounts. This has the advantage of immediate results and also gives some additional information about the shape and size of the organism. Sometimes it is possible to see how the cells are arranged, as in tetrads or chains. However, these groupings are easily broken up in the process of making the wet mount, so the designations are not reliable.

One disadvantage to determining motility from a wet mount is that some types of media inhibit motility, so microbes not actively multiplying are no longer motile.

When observing bacteria in wet mounts, it is important to distinguish true motility from other kinds of motion. Evaporation at the edge of the coverslip causes convection currents to form. The suspended cells appear to be moving along in a stream flowing to the edge of the coverslip. If cells are truly motile, they swim in random directions.

Brownian movement is a form of motion caused by molecules in liquid striking a solid object—in this case, a bacterial cell—causing the cell to slightly bounce back and forth. Bacterial cells appear to jiggle in liquid, but if the cell is actually motile it moves from one point to another.

Materials

Glass slides
Coverslips
Cultures of the following:
    Overnight cultures in nutrient broth of the following:
        Spirillum volutans
        Bacillus subtilis
        Alcaligenes faecalis
    Overnight culture in trypticase soy broth of
        Staphylococcus epidermidis
Vaseline (optional)
Toothpicks (optional)

PROCEDURE

1. Using aseptic technique, add one drop of culture to a slide using a dropper or loop. If using a loop, sterilize it in the Bunsen burner, permit it to cool, obtain a loopful of broth, and place the drop on a clean slide. The drop should be about the size of a pencil eraser. Remember to sterilize the loop before putting it down. It is easier to use a separate slide for each culture when beginning.

2. Carefully place a coverslip on one edge next to the drop and lower it down (see **figure 1.3**). If some of the drop seeps out from under the coverslip, it may contaminate the lens. You can either add a second coverslip and see if it wicks enough liquid away, or dispose of the slide as directed by the instructor—it cannot be simply washed off because it contains viable microorganisms.

3. Focus the slide as usual, first with the low-power lens and then with the high–dry lens. You will have to greatly reduce the amount of light using the iris diaphragm because there is very little contrast between the broth and the unstained organisms.

4. Place a drop of immersion oil on the coverslip of the wet mount and carefully click the oil immersion lens into place. Refocus the microscope, being very careful not to touch the slide with the lens. You should only need to use the fine-adjustment knob.

5. Observe the bacteria to determine if they are motile. Distinguish between Brownian movement, convection currents, and true motility (see the "Getting Started" section).

6. If you want to have more time to observe your wet mounts, you can coat the edge of the coverslip with vaseline using a toothpick before covering the drop on the slide. This will prevent the specimen from evaporating as quickly.

7. Dispose your slides as directed by your instructor.

8. Record your results.

EXERCISE

2

Laboratory Report: The Oil Immersion Lens ASM 1

Results

Make drawings of the slides as observed under the oil immersion lens.

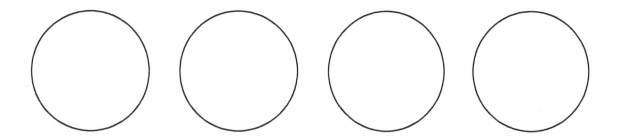

Specimen viewed _____   _____   _____   _____

Questions

1. How do you adjust the amount of light when viewing a slide through the microscope?

2. Why is it important to remember that the 100× lens is longer than the lower power lens?

3. When increasing the magnification while observing a stained slide, do you increase or decrease the amount of light?

4. What **must** you do before storing your microscope after using the oil immersion lens?

5. How can you distinguish true motility from other movement?

6. How do cells appear when Brownian movement is causing their motion?

7. How do cells appear when convection currents are causing their motion?

EXERCISE

3

Simple Stains: Positive and Negative Stains

Definitions

Bacillus. A rod-shaped bacterial cell.

Capsule. A distinct, thick gelatinous material that surrounds some microorganisms.

Chain. In bacterial arrangement, a single bacterium attached to the end of another bacterium to form a chain.

Coccus. A round or spherical bacterial shape.

Differential stain. A stain that is used to distinguish one group of bacteria from another.

Diplococcus. In bacterial arrangement, two cocci cells attached to each other; often observed in Streptococci.

Inclusion bodies. Microscopically visible structures, often granules used for storage, such as sulfur, iron, or phosphate.

Micrometer (abbreviated μm). The metric unit used to measure bacteria. It is 10^{-6} m (meter) and 10^{-3} mm (millimeter). One meter = 1,000 mm.

Negative stain. A staining technique that uses negatively charged acidic dye to stain the background against which colorless cells can be seen.

Positive stain. A staining technique that uses a positively charged, basic dye to stain microbes.

Parfocal. All the objective lenses of the microscope are in the same plane, so that it is not necessary to refocus when changing lenses.

Sarcina. In bacterial arrangement, cocci remain attached as a set of eight cells, resembling a domino or cube pattern.

Simple stain. A staining technique that uses a basic dye to add color to bacterial cells.

Smear. In a staining procedure, the film obtained when a drop of microbe-containing liquid is placed on a slide and allowed to dry.

Spirillum. A rigidly coiled curvy cell.

Spirochete. A flexible, twisted spiral cell containing a special structure called an axial filament for motility.

Tetrad. In bacterial arrangement, two cocci remain attached to two other cocci, making a square of four cells.

Vibrio. A short, curved rod.

Objectives

1. Describe the preparation and staining of a bacterial smear using a positive stain.[ASM 2, 6, & 7]

2. Describe the preparation of a negative stain.

3. Describe the morphologies and arrangements of bacteria observed in a simple stain.

Pre-lab Questions

1. What is the dried mixture of cells and water on a slide called?

2. How big is the average bacterial cell?

3. Name two kinds of information you can learn from a simple stain.

Getting Started

Two kinds of simple stains will be done in this exercise: the positive stain and the negative stain. The **positive stain** uses a basic, or positively charged, dye to adhere to the negatively charged cell wall (**figure 3.1**). Many dyes such as crystal violet, methylene blue, or safranin can be used. The **negative stain** uses an acidic dye that is negatively charged so it does not adhere to the negatively charged cell wall. Rather, it stains the background of the slide and the microbes remain clear. Microbiologists most frequently stain organisms with the Gram stain, but in this exercise a **simple stain** using one dye will be used to give you practice staining and observing bacteria before doing the more complicated **differential stains**.

Another kind of simple stain is the **negative stain** (color plate 1). Although it is not used very often, it is advantageous in some situations. Organisms are mixed into a drop of nigrosin or India ink on a glass slide. After drying, the organisms can be observed under the microscope as clear areas in a black background. This technique is sometimes used to observe **capsules** or **inclusion bodies**. It also prevents eyestrain when many fields must be scanned.

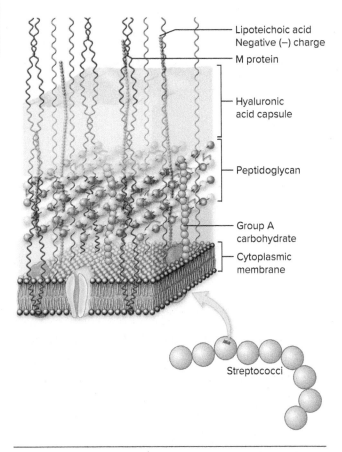

Lipoteichoic acid
Negative (−) charge

M protein

Hyaluronic
acid capsule

Peptidoglycan

Group A
carbohydrate

Cytoplasmic
membrane

Streptococci

Figure 3.1 The structure of a *Streptococcus* cell showing a negative cell wall charge due to the lipoteichoic acid.

The dye tends to shrink away from the organisms, causing cells to appear larger than they really are.

In both of these simple stains, you can determine the shape of the bacteria and the arrangement after cell division. Some organisms tend to stick together after dividing and form chains or clumps. Other organisms are most often observed as individual cells and do not exhibit an arrangement. However, arrangement formation depends on how the organisms are grown. *Streptococcus* bacteria form long, fragile **chains** in broth, but if grown on plates as colonies, it is difficult to make a smear with intact chains. *Staphylococcus aureus* appears as grape-like clusters, *Micrococcus* often appears in groups of four as tetrads, and Sarcina appear as a set of eight as **sarcina**. Some *Bacillus* species present as chains.

After you have stained your bacterial **smears**, you can examine them with the oil immersion lens, which will allow you to distinguish the morphology of different organisms. The typical bacteria you will see are about 0.5–1.0 **micrometers** (μm)

in width to about 2–7 μm long and are usually rods, **cocci**, or spiral-shaped. Sometimes, rods are referred to as **bacilli**, but since that term is also a genus name *(Bacillus)* for a particular organism, the term *rod* may be preferred. Other shapes include a **spirillum**, a rigidly coiled curvy cell containing flagella, of which *Spirillum* is an example. A **spirochete** is a flexible, twisted spiral cell composed of an axial filament for motility. An example of a spirochete is *Treponema pallidum*, the bacterium which causes the sexually transmitted disease syphilis. Other shapes include a **vibrio**, which is a short, curved rod and includes the genus *Vibrio*.

Materials

Cultures of the following:
   Bacillus subtilis
   Staphylococcus epidermidis
   Streptococcus mutans
   Micrococcus luteus
Staining bottles with
   Crystal violet
   Methylene blue
   Safranin
Glass slides
Waterproof marking pen or wax marking pen
Tap water in small dropper bottle (optional)
Inoculating loop
Slide labels (optional)
Alcohol sand bottle (a small screw cap bottle half full of sand and about three-quarters full of 95% alcohol) (optional)
Disposable gloves (optional)

PROCEDURE

Positive Stain

1. Clean a glass slide by rubbing it with slightly moistened cleansing powder such as Boraxo or Bon Ami. Rinse well and dry with a paper towel. New slides should be washed because sometimes they are covered with a protective coating.

2. Draw one or two circles with a waterproof pen or wax pencil on the **underside** of the slide.

Slide label or frosted portion

Water

Figure 3.2 Slide with two drops of water. Two different bacteria can be stained on one slide.

If the slide has a frosted portion, you can also write on it with a pencil. This is useful because it is easy to forget the order in which you placed the organisms on the slide and you can list them, for instance, from left to right (**figure 3.2**).

3. Add a drop of water to the slide on top of the circles. Use your loop to transfer tap water or use water from a dropper bottle. This water does not need to be sterile. Although some organisms are in municipal water systems, they are non-pathogens and are too few to be seen.

If you are preparing a smear from a broth culture, as you will do in the future, add only the broth to the slide. Broth cultures are relatively dilute, so no additional water is added.

Tip: It is helpful if you always place the microbes in the same location on the slide so you know where to look for them after staining.

4. Sterilize a loop by holding it at an angle in the flame of the Bunsen burner. Heat the entire wire red hot but avoid putting your hand directly over the flame or heating the handle itself (**figure 3.3a**).

5. Allow the loop to cool a few seconds and then remove a small amount of a bacterial culture and suspend it in one of the drops of water on the slide (see figure 3.3b). Continue to mix in bacteria until the drop becomes slightly cloudy. If your preparation is too thick, it will stain unevenly, and if it is too thin, you will have a difficult time finding organisms under the microscope. In the beginning, it may be better to err on the side of having a slightly too cloudy preparation—at least you will be able to see organisms and you will learn from experience how dense to make the suspension.

6. Heat the loop red hot again. It is important to burn off the remaining organisms so that you will not contaminate your benchtop. If you then rest your loop on the side of your Bunsen burner, it can cool without burning anything on the bench.

Sometimes the cell material remaining on the loop spatters when heated. To prevent this, some laboratories remove bacterial cell material from the loop by dipping the loop in an alcohol sand bottle. Then the loop is heated red hot in the Bunsen burner.

7. Permit the slide to dry. Do not heat it in any way to hasten the process, because the cells will become distorted. Place the slide off to the side of the bench so that you can proceed with other work. In some labs, a hot plate at 50°C may be used.

8. When the slide is dry (in about 5–10 minutes), heat-fix the organisms to the slide by *quickly* passing it, **sample side up**, through a Bunsen burner flame two or three times so that the bottom of the slide is barely warm. This step causes the cells to adhere to the glass so they will not wash off in the staining process (figure 3.3e).

9. Place the slide on a staining loop over a sink or pan. Alternatively, hold the slide over the sink with a forceps or clothespin. Cover the specimen with a stain of your choice—crystal violet is probably the easiest to see (**figures** 3.3f and **3.4a**).

10. After about 20 seconds, pour off the stain by tilting the slide and rinse with tap water (figure 3.4b).

11. Carefully blot the smear dry with a paper towel or bibulous paper. Do not rub the slide from side to side as this will remove the organisms. Permit the slide to air dry until you are sure the slide is completely dry (figure 3.4c).

12. Observe the slide under the microscope. Because you are looking at bacteria, you must use the oil immersion lens in order to see them. Focus the slide on low power and then again on high power. Add a drop of immersion oil to the slide at the spot where you see light coming through and move the immersion lens into place. Use only the fine focus adjustment. If your microscope is **parfocal**, it should be very close to being in focus. Note that no coverslip is used when looking at stained organisms.

Remember, never move the immersion lens toward the slide while looking through the microscope. You may hit the slide with the lens and damage the lens. When you have a

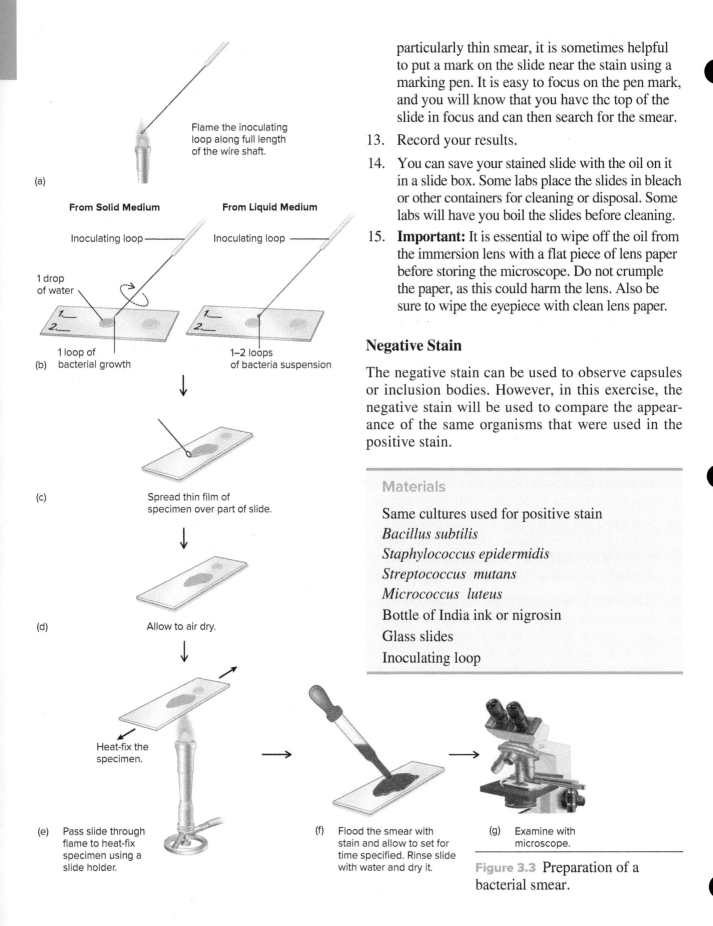

(a) Flame the inoculating loop along full length of the wire shaft.

From Solid Medium

Inoculating loop

1 drop of water

1. ___
2. ___

(b) 1 loop of bacterial growth

From Liquid Medium

Inoculating loop

1. ___
2. ___

1–2 loops of bacteria suspension

(c) Spread thin film of specimen over part of slide.

(d) Allow to air dry.

(e) Pass slide through flame to heat-fix specimen using a slide holder.

Heat-fix the specimen.

(f) Flood the smear with stain and allow to set for time specified. Rinse slide with water and dry it.

(g) Examine with microscope.

particularly thin smear, it is sometimes helpful to put a mark on the slide near the stain using a marking pen. It is easy to focus on the pen mark, and you will know that you have the top of the slide in focus and can then search for the smear.

13. Record your results.

14. You can save your stained slide with the oil on it in a slide box. Some labs place the slides in bleach or other containers for cleaning or disposal. Some labs will have you boil the slides before cleaning.

15. **Important:** It is essential to wipe off the oil from the immersion lens with a flat piece of lens paper before storing the microscope. Do not crumple the paper, as this could harm the lens. Also be sure to wipe the eyepiece with clean lens paper.

Negative Stain

The negative stain can be used to observe capsules or inclusion bodies. However, in this exercise, the negative stain will be used to compare the appearance of the same organisms that were used in the positive stain.

Materials

Same cultures used for positive stain
Bacillus subtilis
Staphylococcus epidermidis
Streptococcus mutans
Micrococcus luteus
Bottle of India ink or nigrosin
Glass slides
Inoculating loop

Figure 3.3 Preparation of a bacterial smear.

1. Add a drop if India ink onto a clean slide.
2. Using a sterile loop, add a loopful of bacteria to the India ink on the slide, being sure to spread the mixture out across the slide. It may be helpful to use a push slide to thin the smear. Discard the push slide in bleach.
3. Sterilize the loop. Do NOT heat-fix the slide.
4. Let the slide dry and examine under the microscope. Bacteria can be seen as clear areas on a black background.
5. Record your results.

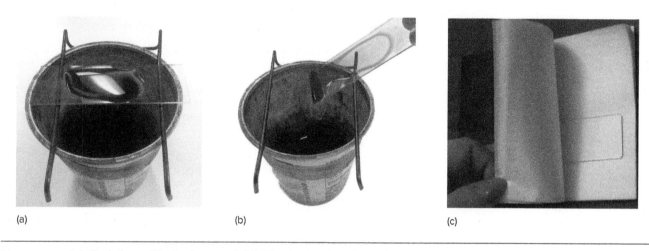

(a) (b) (c)

Figure 3.4 (*a*) Staining, (*b*) rinsing, and (*c*) blotting a simple stain. Courtesy of Anna Oller, University of Central (*a–c*) Courtesy of Anna Oller, University of Central Missouri.

EXERCISE

3

Laboratory Report: Simple Stains:
Positive and Negative Stains

Results

1. Positive stain

	Staphylococcus	*Bacillus*	*Micrococcus*	*Streptococcus*
Draw shape and arrangement				
Dye used				

2. Negative stain

	Staphylococcus	*Bacillus*	*Micrococcus*	*Streptococcus*

Questions

1. What three basic shapes of bacteria can be seen in a simple stain?

2. How does the appearance of the negative stain compare to the appearance of the positive stain?

3. Why do you gently heat the slide before staining?

4. What might happen if you heat-fix a slide before it is dry?

5. When blotting dry a stained slide, what will happen if you rub it from side to side?

6. How many μm are in a millimeter (mm)? _____
   How many μm are in a meter (m)? _____

7. (To ponder) If a large dill pickle is 100 mm long (or 10 cm) and you magnified it 1,000×, how long would it appear? (Assume a yard is approximately a meter.)

   a. As long as a table tennis table (~3 m)

   b. As long as a swimming pool (25 m)

   c. As long as a football field (100 m)

Hint: 10 cm is 0.1 m.

EXERCISE

4

Differential and Other Special Stains ASM 2, 7

Definitions

Amphitrichous. A flagellum or flagella seen at both ends of a bacterium, such as *Alcaligenes faecalis*.

Basal body. In flagella, rings that attach a flagellum and anchor it into the cell membrane.

Cords. A bacterial arrangement in which cells remain together in parallel and are formed by the genus *Mycobacterium*.

Counterstain. In a differential stain, the stain applied to give a contrasting color to bacteria that do not retain the primary stain.

Endospore. A resting structure containing dipicolonic acid and calcium that is produced when *Bacillus* and *Clostridium* encounter an unfavorable environment, such as heat or lack of nutrients. Endospores become vegetative cells when favorable environmental conditions return.

Filament. The long, helical length of a flagellum that extends from the cell surface, and is composed of the protein flagellin.

Hook. In flagella, a flexible joint or sheath that connects the filament to the basal body to allow motion to occur.

Inclusion bodies. An accumulation of an abundant nutrient within a cell, like phosphate or sulfur, for future use by the cell. They are also known as storage granules.

Lophotrichous. A tuft of flagella at one end of an organism, such as *Spirillum*.

Monotrichous. A flagellum at one polar end of a bacterial cell, such as *Vibrio cholerae*.

Mordant. A substance that increases the adherence of a dye.

Peritrichous. Flagella surrounding a bacterial cell, such as *Escherichia coli*.

Peptidoglycan. The macromolecule found only in bacteria that provides strength to the bacterial cell wall. The basic structure has NAM (*N*-acetylmuramic acid) and NAG (*N*-acetylglucosamine) crosslinked by peptide bridges.

Primary stain. The first dye applied in a multistep differential staining procedure.

Vegetative cell. An actively replicating cell.

Virulence factor. A characteristic that enhances the ability of a microbe to cause disease in a host by changing an immune response or adhering to cells. Examples include adhesins, capsules, enzymes, proteins, toxins, etc.

Objectives

1. Describe the Gram-stain procedure.ASM 2 & 7
2. Differentiate between Gram-positive organisms and Gram-negative organisms.
3. Describe other special stains and explain in what situation the particular stain would be used.

Pre-lab Questions

1. Which structure of the bacterial cell determines whether the organism is Gram positive or Gram-negative ?
2. What is the composition of both Gram-positive and Gram-negative cell walls?
3. What color are Gram-positive cells after Gram-staining?

A Differential Stain: The Gram Stain

Getting Started

Differential stains usually involve at least two dyes and are used to distinguish one group of organisms from another. For example, the Gram stain determines whether organisms are Gram positive or Gram negative.

The Gram stain is especially useful as one of the first procedures employed in identifying organisms. It reveals not only the morphology and the arrangement of the cells but also information about the cell walls.

In the late 1800s, Christian Gram devised the staining procedure when trying to stain bacteria so that they contrasted with the tissue sections he was observing. Many years later, it was found that

Table 4.1 Appearance of the Cells After Each Procedure

	Gram +	Gram −
Crystal violet	Purple	Purple
Iodine	Purple	Purple
Alcohol	Purple	Colorless
Safranin	Purple	Pink

purple (Gram positive) bacteria had thick cell walls of **peptidoglycan**, whereas pink (Gram negative) bacteria had much thinner cell walls of peptidoglycan surrounded by an additional lipid membrane (color plate 2). The thick cell wall retains the purple dye in the procedure, but the thin wall does not (**table 4.1**). The *Escherichia coli* and *Alcaligenes faecalis* used in this lab serve as a Gram-negative control, and *Bacillus, Staphylococcus, Streptococcus*, and *Micrococcus* serve as a Gram-positive control.

In the Gram stain, a bacterial smear is dried and then heat-fixed to adhere bacteria to the glass slide (as in the simple stain). It is then stained with crystal violet dye, the **primary stain** (**figure 4.1**), which is rinsed off and replaced with an iodine solution. The iodine acts as a **mordant**—that is, it binds the dye to the cell. The smear is then decolorized with alcohol and **counterstained** with safranin. In Gram-positive organisms, the purple crystal violet dye, complexed with the iodine solution, is not removed by the alcohol, and thus the organisms remain purple. On the other hand, the purple stain is removed from Gram-negative organisms by the alcohol, and the colorless cells take up the red color of the safranin counterstain.

Note: In the past, many clinical laboratories used a 50/50 mixture of alcohol and acetone because it destains faster than 95% alcohol. However, most labs now use 95% alcohol, which is just as effective, but the stain must be decolorized longer.

Special Notes to Improve Your Gram Stains

1. Gram-positive organisms can lose their ability to retain the crystal violet complex when the culture is old or has been incubating 18 hours or longer. The genus *Bacillus* is especially apt to become Gram-negative. Use young, overnight cultures whenever possible. Gram-positive organisms can appear Gram-negative, but Gram-negative organisms rarely appear Gram-positive.

2. Gram-positive organisms can also appear falsely Gram-negative by over-decolorizing in the Gram-stain procedure. If excessive amount of alcohol is used, almost any Gram-positive organism will lose the crystal violet dye and appear Gram-negative.

3. If you are staining a very thick smear, it may be difficult for the dyes to penetrate the organisms properly. This is not a problem with broth cultures, which are quite dilute, but be careful not to make the suspension from a colony in a drop of water too thick (**figure 4.2**).

4. When possible, avoid making smears from inhibitory media such as eosin methylene blue (EMB) because the bacteria frequently give variable staining results and can show atypical morphology.

5. The use of safranin in the Gram stain is not essential. It is simply used as a way of dying the colorless cells so they contrast with the purple. If you are color-blind and have difficulty distinguishing pink from purple, try other dyes as a counterstain and visit with your instructor.

PROCEDURE

Procedure for Gram Stain

1. Label the slide with your name, date, and section. Then add 2 circles to the underside of the slide. Add a "U" for the unknown

Figure 4.2 A slide containing multiple organisms can be stained simultaneously to gauge proper staining. Add the Unknown to the first circle. Record the bacterial names used for the Control to the slide label. Add one Gram-positive and one Gram-negative organism to the second Control circle.

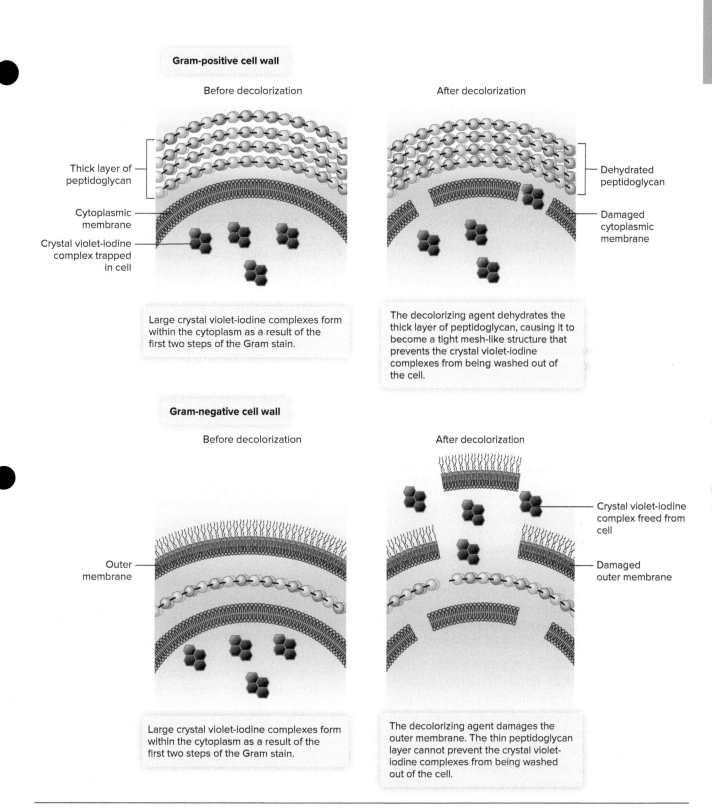

Gram-positive cell wall

Before decolorization

After decolorization

Thick layer of peptidoglycan

Cytoplasmic membrane

Crystal violet-iodine complex trapped in cell

Dehydrated peptidoglycan

Damaged cytoplasmic membrane

Large crystal violet-iodine complexes form within the cytoplasm as a result of the first two steps of the Gram stain.

The decolorizing agent dehydrates the thick layer of peptidoglycan, causing it to become a tight mesh-like structure that prevents the crystal violet-iodine complexes from being washed out of the cell.

Gram-negative cell wall

Before decolorization

After decolorization

Outer membrane

Crystal violet-iodine complex freed from cell

Damaged outer membrane

Large crystal violet-iodine complexes form within the cytoplasm as a result of the first two steps of the Gram stain.

The decolorizing agent damages the outer membrane. The thin peptidoglycan layer cannot prevent the crystal violet-iodine complexes from being washed out of the cell.

Figure 4.1 The Gram stain.

Materials

Staining bottles of the following:

   Crystal violet

   Iodine

   95% alcohol or acetone

   Safranin

Slides

Clothespin or forceps (slide holders)

Staining bars or staining rack

Metal or glass

Disposable gloves

Water bottle or dropper

Sterile swabs (optional)

Slide labels (optional)

Overnight cultures growing on trypticase soy agar slants of the following:

   Escherichia coli or *Alcaligenes faecalis* (Gram-negative)

   Bacillus subtilis

   Staphylococcus epidermidis

   Streptococcus mutans

   Micrococcus luteus

   Culture labeled as "Unknown"

and a "C" for the control. Be sure to include at least abbreviations for the two microbes used in the control so slides can be compared.

2. Add two separated drops of water to a clean slide (**figure 4.2**). In the first drop, make a suspension of the unknown organism to be stained just as you did for a simple stain (see exercise 3 for instructions on preparing a smear). In the second drop, add a loopful of one of the Gram-negative organisms. Properly flame your loop, cool, and then add a loopful of one of the Gram-positive organisms. This mixture is a control to ensure that your Gram-stain procedure (**figure 4.3**) gives the proper results. Heat-fix the slide after the slide is dry.

3. Place the slide on a staining bar or rack across a sink (or can). Alternatively, hold the slide with a clothespin or forceps over a sink.

4. Cover the smear with crystal violet. Leave the stain for 30 seconds and then discard. The timing is not critical. Tilt the slide and rinse with water from a water bottle or dropper. (figure 4.3*a* and *b*).

5. Cover the smear with iodine for about 60 seconds and then tilt the slide and rinse with water (figure 4.3*c* and *d*).

6. Tilt the slide at 45° angle and carefully drip 95% ethanol, the decolorizer, over it until no more purple dye runs off. Immediately rinse the slide with water. Thicker smears may take longer than thinner ones, but ethanol should usually be added for about 20 seconds. Timing is critical in this step (figure 4.3*e* and *f*).

7. Cover the smear with safranin and leave it for 60 seconds—timing is not important. Tilt the slide and rinse with water. Safranin is a counterstain because it stains the cells that have lost the purple dye (figure 4.3*g* and *h*).

8. If your instructor uses different times than those listed, fill in the following blanks:

   a. Add crystal violet for _____ seconds.

   b. Add iodine for _____ seconds.

   c. Add _____ drops of alcohol/acetone for _____ seconds.

   d. Add safranin for _____ seconds.

9. Blot the slide carefully with bibulous paper or paper towel to remove the water, but do not rub from side to side. When it is completely dry, observe the slide under the microscope. Remember that you must use the oil immersion lens to observe bacteria. Compare your "Unknown" stain to the control mixture on the same slide and with color plate 2.

10. Describe the appearance of your stained bacteria in the Results section of the Laboratory Report.

11. Be sure to carefully wipe off the immersion oil from the objective and ocular lenses with lens paper before storing the microscope.

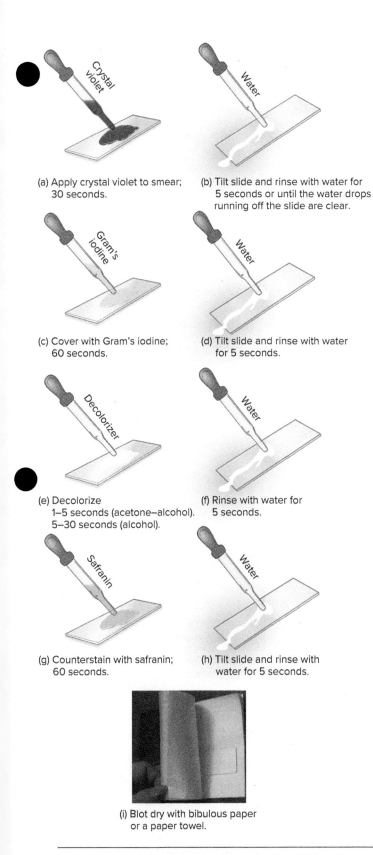

(a) Apply crystal violet to smear; 30 seconds.

(b) Tilt slide and rinse with water for 5 seconds or until the water drops running off the slide are clear.

(c) Cover with Gram's iodine; 60 seconds.

(d) Tilt slide and rinse with water for 5 seconds.

(e) Decolorize 1–5 seconds (acetone–alcohol). 5–30 seconds (alcohol).

(f) Rinse with water for 5 seconds.

(g) Counterstain with safranin; 60 seconds.

(h) Tilt slide and rinse with water for 5 seconds.

(i) Blot dry with bibulous paper or a paper towel.

Figure 4.3 Gram-stain procedure.
(*i*) Courtesy of Anna Oller, University of Central Missouri.

A Differential Stain: Acid-Fast Stain (Optional)

Getting Started

The acid-fast stain is useful for identifying bacteria with a waxy lipid cell wall. Most of these organisms belong to an order of bacteria called Actinomycetales, and includes the mycobacteria. Although many harmless bacteria are in this group, it does include the pathogen *Mycobacterium tuberculosis,* which causes tuberculosis in humans. These organisms have a Gram-positive cell wall structure, but the lipid in the cell wall prevents proper staining with Gram-stain dyes. Further, the mycobacteria exhibit an arrangement called **cords**, where the rods remain in parallel to other cells and tend to appear in groups rather than individual cells.

In the Ziehl–Neelsen acid-fast stain procedure, the dye carbolfuchsin stains the waxy cell wall when the slide is heated. Once the lipid-covered cell has been dyed, it cannot be decolorized easily—even with alcohol containing hydrochloric acid (HCl) (called acid–alcohol). Non-mycobacteria are also dyed with the carbolfuchsin but are decolorized by acid–alcohol. These colorless organisms are then counterstained with methylene blue, so they contrast with the pink acid-fast bacteria that were not decolorized. The Kinyoun acid-fast staining methods do not use heat. The Kinyoun method uses HCl, whereas the modified Kinyoun method uses alcohol containing sulfuric acid.

The acid-fast stain is important because it is the initial method of diagnosing *Mycobacterium tuberculosis* in a patient's sputum. A bacterial colony takes about a month to appear on special agar. (Sputum is a substance that is coughed up from the lungs and contains puslike material.) Tuberculosis is a very serious disease worldwide that is re-emerging due to antibiotic-resistant strains. Because the process of finding acid-fast organisms in sputum can be difficult and time-consuming, this test is usually performed in state health laboratories or reference labs. Your instructor may have you view the specialized stains on already prepared slides using the oil immersion lens. Remember to wipe off the oil from the microscope before putting your microscope away.

Objectives

1. Interpret the acid-fast stain.
2. Examine acid-fast organisms.

Materials

Slides

Broth culture of the following:

  Mycobacterium smegmatis or *M. phlei*
  Staphylococcus epidermidis

Staining bottles of the following:

  Carbolfuchsin

  Acid–alcohol

  Methylene blue

Water bottle or dropper

Staining bars or staining rack

Procedure for Acid-Fast Stain (Kinyoun Modification)

1. Prepare a smear of the material by adding both the *Mycobacterium* and *Staphylococcus* to one area on a slide, similar to the control you made in figure 4.2, heat-fix (see exercise 3).

2. Cover the smear with carbolfuchsin and stain for 5 minutes (**figure 4.4**).

3. Tilt the slide and rinse well with water.

4. Decolorize with acid–alcohol for 1 minute, or until the carbolfuchsin stops running out of the cells.

5. Tilt the slide and rinse well with water.

6. Counterstain with methylene blue for 1 minute.

7. Tilt the slide and rinse well with water.

8. Blot dry carefully and examine under the oil immersion lens.

9. The acid-fast *Mycobacterium* will appear as pink rods, and the *Staphylococcus* will appear as blue cocci (color plate 3).

10. Record results.

Special Stains: Capsule, Endospore, Inclusion Bodies, and Flagella

Getting Started

Although bacteria have few cell structures observable by light microscopy when compared to other organisms, some have capsules, endospores, inclusion bodies (storage granules), or flagella. The following procedures will enable you to see these structures.

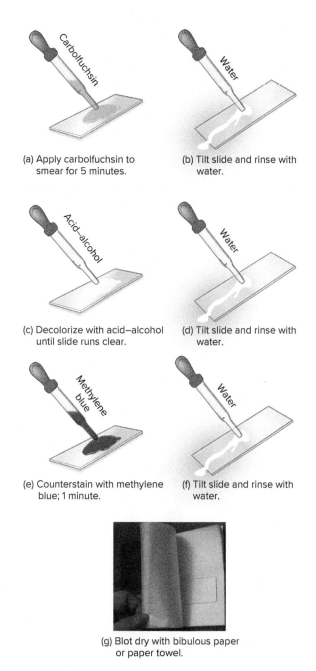

(a) Apply carbolfuchsin to smear for 5 minutes.

(b) Tilt slide and rinse with water.

(c) Decolorize with acid–alcohol until slide runs clear.

(d) Tilt slide and rinse with water.

(e) Counterstain with methylene blue; 1 minute.

(f) Tilt slide and rinse with water.

(g) Blot dry with bibulous paper or paper towel.

Figure 4.4 Acid-fast stain procedure. (*g*) Courtesy of Anna Oller, University of Central Missouri.

Objectives

1. Examine various structures and storage products of bacteria.

2. Interpret staining procedures for these structures.

Capsule Stain (Optional)

A capsule is composed most commonly of polysaccharides surrounding a bacterial cell. However, some bacteria produce capsules that consist of amino acids or carbohydrates instead. A capsule can protect a bacterium from engulfment by white blood cells like macrophages, and help the bacterium attach to host cells, so it is considered a **virulence factor**. The ability to produce a capsule frequently depends on the availability of certain sugars. *Streptococcus mutans*, for example, produces a capsule when growing on sucrose but not when growing on glucose. Capsules can also be destroyed by heat or water.

The capsule stain is similar to a negative stain, except a counterstain is added to stain the bacterium. You will need to be careful to avoid contamination with this stain since the microbes will still be alive.

Materials

Culture of the following:

Klebsiella or other organism with a capsule growing on a slant

India ink or nigrosin

Safranin

Slides

Coverslips

Procedure for Capsule Stain

1. Add a small drop of India ink or nigrosin to a clean slide.
2. Add a suspension of organisms to the India ink or nigrosin drop.
3. Either procedure A or B will be used:

   A. Add a small drop of safranin to the nigrosin drop, and then carefully lower a cover slip over the drops. You may need a large coverslip or two small coverslips so that all of the liquid is under a coverslip.

   B. Add a small drop of safranin to the nigrosin drop. Then use a second push slide to make a thin smear and allow the slide to dry. The push slide should be placed in bleach or other discard container.

4. Examine the slide under the microscope on oil immersion and find a field where you can see the cells surrounded by a halo in a black background. The halo is the capsule surrounding the cell (color plate 4).
5. After viewing, place slides in bleach or in a container of boiling water and boil for a few minutes before cleaning. This is necessary because the bacteria are not killed during the staining process.
6. Record results.

Endospore Stain (Optional)

Some organisms such as *Bacillus* and *Clostridium* can form a resting stage called an **endospore** (or sometimes called a *spore*), which protects them from heat, chemicals, and starvation. When the cell determines that conditions are becoming unfavorable due to a lack of nutrients or moisture, it forms an endospore. When conditions become favorable again, the spore can germinate and the cell can continue to divide. The endospore is resistant to most stains so special staining procedures are needed.

Materials

Slides
Culture of the following:

Bacillus cereus or *B. subtilis* on a nutrient agar slant after 3–4 days' incubation at 30°C

Staining bottles of the following:

Malachite green

Safranin

Metal or glass staining bars

Water bottle or dropper

Beaker or can (if boiling water)

Slide warmer set at 55°C (optional)

Procedure for Endospore Stain

Note: Due to potential fumes given off during the heating process, performing the heating under a proper hood is recommended.

(a) Saturate paper with malachite green, and steam slide on heat source for a total of 5 minutes. Add time for removing slide from heat.

(b) Remove paper, cool, tilt, and rinse completely with water.

(c) Counterstain with safranin for 30 seconds to 1 minute.

(d) Tilt and rinse with water until slide runs clear.

(e) Blot dry with bibulous paper or paper towel.

Figure 4.5 Endospore stain procedure. (*e*) Courtesy of Anna Oller, University of Central Missouri.

1. Make a suspension of *Bacillus* in a drop of water on a clean slide, allow the slide to air dry, and heat-fix.

2. Add about an inch of water to a beaker and bring it to boil.

3. If using a slide warmer, place the slide directly on top of the flat surface of the unit.

4. Place two short staining bars over the beaker and place the slide on them.

5. Tear a piece of paper towel, a little smaller than the slide, and lay on top of the smear. The paper prevents the dye from running off the slide and keeps the dye in contact with the cells.

6. Saturate the paper towel with malachite green and steam the slide for a <u>continuous</u> 5 minutes.

Continue to add stain to prevent the dye from drying on the slide (**figure 4.5**). Be sure to let the slide cool before adding cold dye otherwise the slide may break. If you remove the slide from the heat source to add more stain, you must add more steaming time.

7. Remove the paper towel from the slide using forceps and allow the slide to cool.

8. Tilt the slide and rinse <u>completely</u> with water until the water runs clear. This step often requires almost an entire small dropper bottle to properly rinse the malachite green off. The **vegetative cells** (dividing cells) lose the dye, but the endospores retain the dye.

9. Counterstain the slide with safranin for 30 seconds to 1 minute. Then rinse the slide until the water runs clear. Blot dry carefully.

10. Observe with the oil immersion lens. The endospores appear green and the vegetative cells appear pink (color plate 5). Sometimes the endospore is seen within the cell, and its shape and appearance can be helpful in identifying the organism. Frequently, endospores are outside of the bacterial cells, and can appear anywhere in the smear because the cells around them have disintegrated (**figure 4.6**).

11. Record results.

12. Prepare and observe a Gram stain of the same culture (optional).

Note: When bacteria containing endospores are Gram stained, the endospores do not stain and the cells appear to have holes in them (figure 4.6).

Inclusion Body Stain (Optional)

Many organisms can accumulate currently abundant materials as storage granules, often called **inclusion bodies**, from their environment for future use. Storage granules can contain phosphate, sulfur, iron, nitrate, lipids, etc. depending on the microbe. For example, phosphate can be stored as metachromatic granules (also called volutin granules). When organisms containing these granules are stained with methylene blue, the phosphate granules are stained a darker blue. Further, microbes like *Bacillus cereus* store polyhydroxybutyrate (PHB), a type of lipid that can be stained using Sudan Black B, which stains the lipids in the bacterium black.

Spore Stain of *Bacillus* with Malachite Green

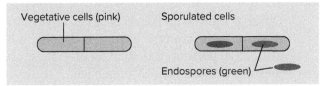

Gram Stain of *Bacillus*

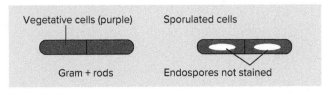

Figure 4.6 Appearance of endospores stained with a spore stain and Gram stain. **Note:** *Bacillus* species frequently lose their ability to stain Gram-positive, which is why determining bacterial size is important.

Materials

Slides

Culture of the following:

> *Spirillum, Bacillus cereus*, or *B. megaterium* grown in nutrient broth

Methylene blue

Water bottle or dropper

Sudan Black B

Coverslips

Procedure for Inclusion Body Stain

1. **Metachromatic Granules:** Prepare a smear from the broth. Add a loopful of microbes to a clean slide, allow to dry, and then heat fix the slide.

2. Flood the smear with methylene blue for 30 seconds to 1 minute. Rinse with water and blot dry.

3. **Lipid Inclusions:** Add one drop of Sudan Black B to a clean slide. Add one drop of safranin to the drop. Add a loopful of *Bacillus* to the Sudan Black B drop. Gently lower a coverslip (depending on the drop size you may need two coverslips) onto the slide.

4. Observe the slides using the oil immersion lens. Metachromatic granules should appear

as small, dark blue bodies within the cells, and lipids will appear as black bodies within the cells (color plate 6).

5. The lipid inclusion slide contains live organisms, so dispose the slide in the designated container.

6. Record results.

Flagella Stain (Optional)

Some bacteria have flagella (singular, *flagellum*) for motility. Their width is below the resolving power of the microscope so they cannot be seen in a light microscope (the flagella seen at each end of *Spirillum* in a wet mount are actually a tuft of flagella). Flagella can be visualized if they are dyed with a special stain that precipitates on them, making them appear much thicker (color plate 7). The arrangement of the flagella can aid in identification as a particular genus or species will consistently exhibit the same arrangement. There are four basic kinds of arrangement of flagella: **peritrichous, Monotrichous** (or Polar), **amphitrichous**, and **lophotrichous**. A bacterium may exhibit one of the four arrangements, assuming the microbe is motile. A flagellum is comprised of a **basal body**, which attaches the flagellum into the cell, a **hook** that anchors the filament to the basal body, and a **filament**, which makes up the length of the flagellum (**figure 4.7**).

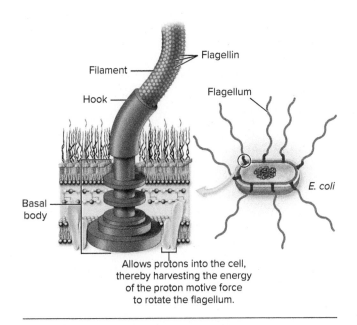

Allows protons into the cell, thereby harvesting the energy of the proton motive force to rotate the flagellum.

Figure 4.7 The flagellum structure.

Stained demonstration slides of *Escherichia coli*, *Spirillum volutans*, *Alcaligenes faecalis*, or *Pseudomonas*

Procedure for Flagella Stain

1. Observe slides with stained flagella of several organisms and note the pattern of flagella. It is difficult to perform this staining procedure, so prestained slides are recommended.

2. Record results.

EXERCISE

4

Laboratory Report: Differential
and Other Special Stains [ASM 2, 7]

Results

Draw the organisms that you see on oil immersion and record the colors you see on the slide. Circle the gram reaction seen. If a microbe has an arrangement, list the arrangement seen.

Gram-Stain	Gram Reaction	Arrangement (Sketch)	Optional Stains	Organism(s) Used	Appearance
E. coli	+ −		Acid-fast		
Color seen					
B. subtilis	+ −		Capsule		
Color seen					
S. epidermidis	+ −		Endospore		
Color seen					
Streptococcus mutans	+ −		Metachromatic granules		
Color seen					
M. luteus	+ −		Lipid granules		
Color seen					
Unknown	+ −		Flagella		
Color seen					

Questions

1. List the Gram-stain reagents in order and explain the function of each reagent. Indicate the step that the timing is critical.

    a.

    b.

c.

d.

2. Give two reasons Gram-positive organisms can appear Gram negative.
   1.

   2.

3. What is the purpose of using a control in the Gram stain?

4. What is a capsule?

5. What are inclusion bodies, and why are they important to the cell?

6. How does an endospore appear (draw and indicate color)
   a. when Gram stained?

   b. when spore stained?

7. Why can't you Gram stain an acid-fast organism?

8. Why do you need a special staining procedure for flagella?

INTRODUCTION to Microbial Growth

The ability to grow bacteria is very important for studying and identifying them. Organisms in the laboratory are frequently grown either in a broth culture or on a solid agar medium. A broth culture is useful for growing large numbers of organisms. Agar medium is used in a Petri dish when a large surface area is important, as in a streak plate. On the other hand, agar medium in tubes (called slants) is useful for storage because the small surface area is not as easily contaminated and the tubes do not dry out as fast as plates. You will be able to practice using media in all these forms (**figure I.5.1**).

You will also use different kinds of media in this section. Most media are formulated so that they will support the maximum growth of various organisms, but other media have been designed to permit the growth of desired organisms and inhibit others (selective). Still other media have been formulated to change color or in some other way distinguish one bacterial colony from another (differential). These media can be very useful when trying to identify an organism.

It is also important to know how to count bacteria. You will have an opportunity to learn about several techniques and their advantages and disadvantages.

In these exercises, no pathogenic organisms are used, but it is very important to treat these cultures as if they were harmful because then you will be prepared to work safely with actual pathogens. Also, almost any organism can cause disease if there are large numbers in the wrong place.

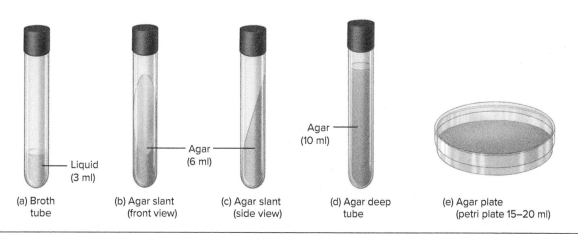

(a) Broth tube — Liquid (3 ml)
(b) Agar slant (front view) — Agar (6 ml)
(c) Agar slant (side view)
(d) Agar deep tube — Agar (10 ml)
(e) Agar plate (petri plate 15–20 ml)

Figure I.5.1 Various media in different forms.

Notes:

5 Diversity of Microorganisms

Definitions

Agar. A carbohydrate derived from seaweed used to solidify a liquid medium.

Colony. A visible population of microorganisms growing on a solid medium.

Contamination. The growth of unwanted microbes.

Inoculate. To transfer organisms to a medium to initiate growth.

Media (singular, *medium*). The broth or agar containing nutrients that support the growth of microorganisms.

Mesophile. A microorganism that grows at 37°C, body temperature.

Microbiome. All of the microorganisms present in a location, such as the intestinal microbiome.

Normal microbiota. The organisms usually found associated with parts of the body (previously termed normal flora).

Pathogen. An organism capable of causing disease.

Psychrophiles. Microorganisms that prefer to grow at refrigerator temperature, 4°C.

Sterile. Free of known viable microorganisms, including viruses.

Ubiquity. The existence of organisms in many locations at the same time.

Objectives

1. Explain why some organisms are **ubiquitous.**
2. Explain how organisms are grown on laboratory culture **media.**
3. Explain the lab safety protocols you will use for each microbiology lab (see page vi).[ASM 11&12]

Pre-lab Questions

1. Why won't one kind of agar medium support the growth of all microorganisms?

2. We require oxygen to live. Is this also true of all bacteria?

3. What is the microbial advantage of being able to grow at body temperature?

Getting Started

Microorganisms are everywhere—in the air, soil, and water; on plant and rock surfaces; and even in unlike places such as hot springs and Antarctic ice. Millions of microorganisms are also found living within or on animals; for example, the mouth, the skin, and the intestine (all exterior to our actual tissues) support huge populations of bacteria. In fact, the interior of healthy plant and animal tissues is one of the few places free of microorganisms. In this exercise, you will obtain samples from the environment and your body to determine which organisms are present that will grow on laboratory **media.**

An important point to remember, as you try to grow organisms, is that no one condition or **medium** permits the growth of all microorganisms. The trypticase soy **agar** used in this exercise is a rich medium (a digest of meat and soy products, similar to a beef and vegetable broth) and will support the growth of many diverse organisms, but bacteria growing in a freshwater lake, that is very low in organic compounds, would find it too rich (similar to a goldfish in vegetable soup). However, organisms that are accustomed to living in our nutrient-rich throat might find the same medium lacking substances they require.

Temperature is also important. Organisms associated with endothermic (warm-blooded) animals usually prefer temperatures close to 37°C, which is approximately the body temperature of most animals. These microbes are termed **mesophiles.** Soil organisms generally prefer a cooler temperature of 30°C. Organisms growing on glaciers would find room temperature (about 25°C) too warm and would probably grow better in the refrigerator. Microorganisms that prefer to grow at 4°C are termed **psychrophiles.**

Microorganisms also need the correct atmosphere. Many bacteria require oxygen, whereas other organisms find it extremely toxic and will only grow in the absence of air. Therefore, the organisms you see growing on the plates will be a small sample of the organisms originally present.

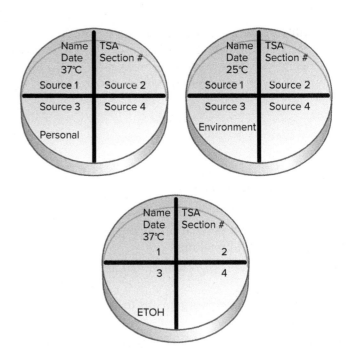

Figure 5.1 Plates labeled on the bottom for ubiquity exercise.

Materials

Per team of two (or each individual, depending on number of plates available):

Isopropyl alcohol pads, or paper towel pieces saturated with alcohol, 2

Trypticase soy agar plates, 3

Sterile swabs as needed

Sterile water (about 1 ml/tube) as needed

Waterproof marking pen or wax pencil

Bleach container or autoclave bag for swab disposal

Alternatively, your instructor may have prepared a tube containing a mixed culture of common microbes to use for the inoculations.

PROCEDURE

First Session

1. Each pair of students should obtain three Petri plates of trypticase soy agar. Notice that the lid of a Petri plate fits loosely over the bottom half, which contains a smooth layer of agar. Keep the lid closed as much as possible to prevent unwanted **contamination**.

2. Using a wax pencil or waterproof marker, label the bottom of the plates with your name, lab section, media type (TSA), date, and activity. Labeling the bottom of the plates prevents accidental switching of lids, and improper results.

3. Divide each plate in quarters with two lines on the bottom of the petri plate. Label the two plates 37°C and the other 25°C (**figure 5.1**).

4. **Inoculate** the 37°C personal plate with samples from your body. For example, moisten a sterile swab with sterile water and rub it on your skin and then gently roll and press the swab onto one of the quadrants (be careful not to gouge the agar). **Sterile** water is water containing no living organisms and prevents contamination of your results from the water. Try placing a hair on the plate. Try whatever interests you. (Be sure to place all used swabs into an autoclave container or a bucket of disinfectant after use.) Label each quadrant well.

5. For the ETOH plate, gently touch your index finger to the 1st quadrant. Place the same finger to the 2nd quadrant. Next, use an alcohol pad or create one using a small piece of paper towel and adding ethanol to saturate the paper towel. Rub the same finger used previously with ethanol for 5 seconds. Let dry. Gently press your finger to the 3rd quadrant. Obtain a new pad or paper and saturate as before. Rub the same finger previously used for 10 seconds and let dry. Gently press your finger to the 4th quadrant.

6. Inoculate the plate labeled 25°C (room temperature) with environmental samples from

the room. It is easier to pick up a sample if the swab is moistened in sterile water first. **Sterile** water is water containing no living organisms and prevents contamination of your results from the water. Try sampling the bottom of your shoe or some dust, or press a coin or other objects lightly on the agar. Be sure to label each quadrant well so that you will know what you sampled. Place swabs in an autoclave or disinfectant container.

7. Incubate the plates at the temperature written on the plate. Place the plates in the incubator or basket upside down. This is important because it prevents condensation from forming on the lid and dripping onto the agar below. The added moisture would permit colonies of microorganisms to run together.

Second Session

Important: Handle all plates with colonies as if they were potential **pathogens,** disease-causing organisms. Follow your instructor's directions carefully. Your plates may have a clear film called Parafilm keeping the plate lid and base shut tight. The plates should not be opened. Do not remove the parafilm!

Note: For best results, the plates incubated at 37°C should be observed after 2 days, but the plates incubated at 25°C will be more interesting after 5–7 days. If possible, place the 37°C plates either in the refrigerator or at room temperature after 2 days so that all plates can be observed at the same time.

1. Examine the plates you prepared in the first session and record your observations on the report sheet for this exercise. There will be basically four kinds of **colonies:** fungi (molds), yeasts, bacteria, and actinomycetes (filamentous bacteria). Mold colonies are usually large and fluffy, like the type seen on moldy bread. Bacterial colonies are usually soft and glistening and tend to be cream colored or yellow. Yeast colonies like *Candida spp.* may appear similar to bacterial colonies, except yeast colonies may appear matte or dull with small hyphae digging into the agar. Actinomycetes often have a textured, raised appearance but are not fluffy. Compare your colonies with color plates 8 and 9.

2. When describing the colonies, include (**figure 5.2**):
   a. relative size as compared to other colonies
   b. shape
   c. color
   d. margin (entire, undulate, etc.)
   e. elevation (flat, raised, etc.)
   f. surface (dull or shiny)
   g. texture (smooth or rough)
   h. optical (opaque, transparent, etc.)
   i. consistency (dry, moist, or mucoid)

3. Expect surprising results. If you pressed your fingers to the agar before and after washing, you may find more organisms on the plate after you washed your hands. The explanation is that your skin has **normal microbiota** (organisms that are always found growing on your skin). When you wash or rub your hands, you remove the organisms you picked up from your surroundings as well as a few layers of skin cells. This exposes more of your normal microbiota; therefore, you may see different colonies of bacteria or yeast before you wash or rub your hands than afterward. Your microbiota are important in preventing undesirable organisms from growing on your skin. Hand washing is an excellent method for removing organisms that are not part of your normal microbiota.

Note: In some labs, plates with molds are opened as little as possible and immediately discarded in an autoclave container to prevent contaminating the lab with mold spores. Some mold spores are toxic. Plates may be sealed by stretching a plastic strip called Parafilm over where the lid meets the bottom. Parafilm is usually applied after incubation to prevent opening the plates or to prevent the media from dehydrating.

4. Follow the instructor's directions for discarding plates. All agar plates are autoclaved before washing or discarding in the municipal garbage system.

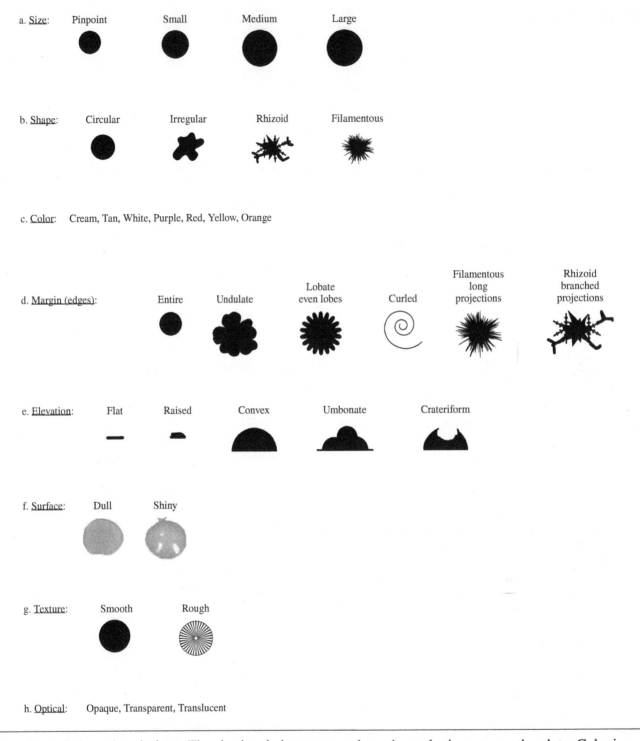

a. Size: Pinpoint   Small   Medium   Large

b. Shape: Circular   Irregular   Rhizoid   Filamentous

c. Color: Cream, Tan, White, Purple, Red, Yellow, Orange

d. Margin (edges):   Entire   Undulate   Lobate even lobes   Curled   Filamentous long projections   Rhizoid branched projections

e. Elevation: Flat   Raised   Convex   Umbonate   Crateriform

f. Surface: Dull   Shiny

g. Texture: Smooth   Rough

h. Optical: Opaque, Transparent, Translucent

Figure 5.2 Colony descriptions. The size is relative compared to other colonies seen on the plate. Colonies can be seen in color on color plate 8. Courtesy of Anna Oller, University of Central Missouri.

EXERCISE

5

Laboratory Report: Diversity of Microorganisms

Results

Room Temperature (about 25°C) Environmental Plate

	Plate Quadrant			
	1	**2**	**3**	**4**
Source/Location				
Number of colonies Number of colony types				

Choose 1 colony in each quadrant to describe:
Colony appearance
  Size
  Shape
  Color
  Margin
  Elevation
  Surface
  Texture
  Optical				

37°C Personal Plate

	Plate Quadrant			
	1	**2**	**3**	**4**
Source/Location				
Number of colonies Number of colony types				

Choose 1 colony in each quadrant to describe:
Colony appearance
  Size
  Shape
  Color
  Margin
  Elevation
  Surface
  Texture
  Optical				

37°C Personal Plate

	Plate Quadrant			
	1	**2**	**3**	**4**
Index finger				
Number of Colonies Number of Colony Types				
Choose 1 colony to describe, if growth was observed				
Move colony appearance up				
Margin Elevation Surface Texture Optical				

Questions

1. Give three reasons why all the organisms you placed on the trypticase soy agar plates might not grow.

2. Why were some agar plates incubated at 37°C and others at room temperature?

3. Why do you invert agar plates when placing them in the incubator?

4. Name one place that might be free of microorganisms. _____

5. Why must you wipe down the bench top and wash your hands before and after the lab?

6. Why can't you apply cosmetics, chapstick, or place your pencil in your mouth?

EXERCISE

6

Pure Culture and Aseptic Technique ^{ASM 3, 4, 6, 11, 12}

Aseptic Technique and Streak Plate Technique

Definitions

Aseptic. Free of contamination.

Aseptic Technique. Procedures that help prevent the introduction of unwanted microbes, or contamination.

Contamination. The growth of unwanted microbes.

Control. A known positive and negative test done concurrently with an unknown test to ensure test reagents and ingredients reacted properly.

Flaming. The process of heating a loop or needle until bright red in order to kill any remaining environmental microbes.

Incineration. The process of heating microbes to 1800°C by flaming a loop or needle in order to kill any remaining microbes.

Incubate. Store cultures under conditions suitable for microbial growth, often in an incubator at a specified temperature.

Inoculate. To transfer organisms to a medium to initiate growth.

Pure culture. A population of cells descending from the growth of a single cell and free from contamination.

Sterile. Free of viable microorganisms, including endospore-formers and viruses.

Streak plate. A technique for isolating pure cultures by spreading organisms on an agar plate. The streak plate is often called an isolation streak.

Objectives

1. Describe aseptic technique procedures.^{ASM 3, 6, 11, 12}

2. Describe the isolation of separate colonies using the streak plate technique.^{ASM 4}

Pre-lab Questions

1. What is the purpose of transferring sterile broth from one tube to another tube of sterile broth?

2. How will you determine if bacteria entered the broth during the transfers?

3. How do you ensure no organisms remain on your wire loop?

Note: Some laboratories use electric incinerators instead of Bunsen burners (**figure 6.1**). Others have neither heat source and perform all transfers with disposable sterile loops. Follow the directions of your instructor for appropriate procedures.

Getting Started

The two goals of **aseptic technique** are to prevent contamination of your culture with organisms from the environment and to prevent the culture from contaminating you or others. In this exercise you will learn three important procedures using aseptic technique: transferring material with a sterile **flamed** loop, transferring liquid with a sterile pipet, and isolating a **pure culture**.

First you will transfer sterile broth from one tube to another using a sterile flamed loop. Flaming the loop is also called **incineration**. The goal is to transfer the liquid in such a way that no environmental organism enters the tubes. Then you will use a sterile pipet to **aseptically** transfer broth between two broth tubes. After you have practiced transferring the broth, you will **incubate** the broth tubes for a few days to determine if they remain **sterile**. If you used good technique, the broth will remain clear; if environmental organisms entered the broth tubes, the broth will be cloudy from microbial growth. If you can successfully transfer sterile broth aseptically, you can use the same technique to transfer a pure culture without **contaminating** it or the environment.

A **control** broth inoculated with *Escherichia coli* will also be incubated with the practice tubes. This control broth tube shows that the broth supports microbial growth. Without this control you would not know if the practice tubes were clear after

Figure 6.1 An electric incinerator used to sterilize loops and needles. Courtesy of Anna Oller, University of Central Missouri.

incubation because of excellent technique or because the organism could not grow in the broth. All experiments have a control to ensure the test was performed properly.

The isolation streak, or **streak plate**, is the third essential technique used to obtain a pure culture. It permits you to isolate a colony formed by a single cell deposited onto a plate from a mixture containing millions of cells. You will start with a broth culture containing two different organisms and separate them into two different colony types. As bacteria are spread out, the cells are able to form individual colonies separated from other individual colonies (color plate 10). Thus, the goal is to pull bacteria from one plate section to another to thin the bacteria out. Performed properly, you eventually have well isolated colonies.

Aseptic Technique

Broth-to-Broth Transfer with a Wire Loop

Materials

Per student

    Tubes of trypticase soy broth, 3

    Inoculating loop

    Broth culture of *Escherichia coli* or the yeast *Saccharomyces cerevisiae*

PROCEDURE

1. Always properly label tubes before adding anything to them. In this exercise, you will transfer sterile broth from one tube to another, so two tubes can be labeled "P," for practice. Label the third tube "C," for control. Labels should include your name, date, lab section, medium, temperature, and the microbe inoculated.

2. Grip the loop as you hold a pencil and flame the wire portion red hot. Hold it at an angle so that you do not burn your hand (**figure 6.2**).

3. After the loop has cooled, pick up a practice tube in the other hand and remove the cap of the tube with the little finger (or the fourth and little fingers) of the hand holding the loop.

4. Flame the mouth of the tube by passing it quickly through a Bunsen burner flame, and then use the sterile loop to obtain a loopful of liquid from the tube. Flame the mouth of the tube briefly again and replace the cap. If you have trouble picking up a loopful of material, check that your loop is a complete circle without a gap.

   Note: If you flame the mouth of the tube too long, it may melt or burn you. One brief passage through the flame is sufficient.

5. Set down the first tube and pick up the second practice tube. Remove the cap, briefly flame the tube, and deposit a loopful of material into the liquid of the second tube. Withdraw the loop, flame the tube, and then replace the cap. Be sure to flame the loop before setting it on the bench (your loop would normally be contaminated with the bacteria you were inoculating).

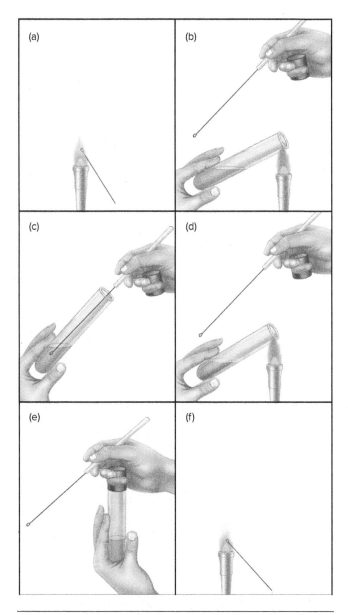

Figure 6.2 Aseptic technique for removing a loopful of broth culture. (*a*) Hold the culture tube in your left hand at 45° angle and the loop in your right hand (reverse if you are left-handed). Flame the loop until bright red along the length of the loop to sterilize it, and allow to cool. (*b*) Remove the culture tube cap, and **briefly** flame the mouth of the culture tube. (*c*) Insert the sterile loop into the culture tube. (*d*) Remove the loopful of inoculum, and flame the mouth of the culture tube again. (*e*) Replace the culture tube cap. Place the culture tube in a test tube rack and perform the inoculation. (*f*) Reflame the loop.

6. Using the flamed and cooled loop as before, inoculate the control tube with the broth culture provided by the instructor. Incubate the "P" tubes at 37°C and the "C" at 37°C for *Escherichia coli* or 25°C for *Saccharomyces cerevisiae*.

7. When learning aseptic technique, it is better to hold one tube at a time; later, you will be able to hold multiple tubes at the same time.

    Note: Many laboratories no longer flame the mouth of the tube, especially if the tubes are plastic. However some laboratories still follow this procedure as warm air creates convection currents that carry air contaminants away from the tube opening.

Broth-to-Broth Transfer with a Pipet

Note: Sterile pipets are used when it is necessary to transfer known amount of material. Some laboratories use plastic disposable pipets, and others use reusable glass pipets. Do not lay a pipet on the benchtop once it has been removed from the container or after transferring microbes. Be sure to follow the instructor's directions for proper disposal after use (**never put a used pipet on your benchtop**). Mouth pipetting is dangerous and is not permitted. A variety of bulbs or devices are used to draw the liquid up into the pipet, and your laboratory instructor will demonstrate their use (**figure 6.3**).

Materials

Per student
   Trypticase soy broth tubes, 3
   1 ml sterile pipet
   Bulb or other device to fit on end of the pipet

PROCEDURE

First Session

1. Label the tubes as P-P1, P-P2, and P-C for pipet tube 1, pipet tube 2, and pipet negative control to distinguish them from the tubes in the previous exercise. For the P-C tube, do not open the tube (do nothing to it but label it) so it serves as a true negative control and incubate it at 37°C.

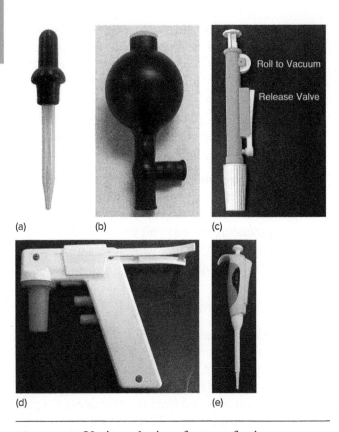

(a)   (b)   (c)

(d)   (e)

Figure 6.3 Various devices for transferring liquids. (*a*) Small rubber bulb. (*b*) Bulb and valve pipet filler. (*c*) Pi-Pump® pipettors. Pipettor color indicates liquid capacity. Green pipets hold up to 10 ml (shown) and blue pipets hold 2 ml. (*d*) Automatic pipet aid. (*e*) Micropipettor.
(*a*): © Ken Karp/McGraw-Hill Education; (*b-e*): Courtesy of Anna Oller, University of Central Missouri.

2. Remove a sterile pipet from the package or canister by only touching the top of the pipet. Do not touch the pipet tip (or the last few inches by the tip that will be submerged in liquid). Insert the larger, round end of the pipet into a bulb. Grip the pipet as you hold a pencil. The pipet is plugged with cotton to filter the air going into it. Discard the pipet if liquid inadvertently wets the plug—air will no longer enter the pipet and the measured liquid will not flow out. **Notify your instructor if the bulb is contaminated**.

3. Pick up a tube with your other hand and remove the cap with the little finger of the hand holding the pipet. Flame the tube entrance. Expel air from the rubber bulb and insert the pipet tip into the liquid. Note that the liquid must be drawn to the 0 mark for 1 ml when using a 1 ml pipet. Draw up 1 ml of broth, remove the pipet from the tube, briefly flame the tube opening, replace the cap, and then place the tube back into the rack.

4. Pick up the second tube and repeat the steps used with the first tube except that the liquid in the pipet is expelled into the tube. Be sure to disinfect any drops from the pipet tips that may have landed on the benchtop during the transfer.

5. Repeat the above steps with the same tubes until you feel comfortable with the procedure.

6. Dispose of the pipet as directed.

7. Incubate the tubes at 37°C until the next lab.

Second Session

1. Observe the control broth with *Escherichia* or *Saccharomyces* and note if it is cloudy. This cloudiness indicates growth and demonstrates that the broth can support the growth of organisms.

2. Observe the practice tubes of broth for cloudiness. You may need to **gently** swirl the tubes to ensure no microbes fell to the bottom of the tub. Compare them to the uninoculated tube of broth. If they are cloudy, organisms contaminated the broth during your practice and grew during incubation. With a little more practice, you will have better technique. If the broths are clear, there was probably no contamination, and you transferred the broth without permitting the entry of any organisms into the tubes.

3. Record your results on the report sheet.

Streak Plate Technique

Materials

Per student

Trypticase soy agar plates, 2

Broth culture containing a mixture of two organisms, such as *Micrococcus luteus* or *M. roseus,* and *Staphylococcus epidermidis*

First Session

1. Label the agar plate on the bottom with your name, date, media type, temperature, microbes, and lab section.

2. Divide the plate into three sections with a *T* as diagrammed (**figure 6.4.**)

   Note: Some instructors prefer four sections.

3. Sterilize the loop in the flame by heating the whole length of the wire red hot. Hold it at an angle so you do not heat the handle or roast your hand.

4. Gently swirl the culture to ensure organisms are suspended. Once the loop has cooled, aseptically remove a loopful of organisms and, holding the loop as you would a pencil, use a gentle pressure to spread the bacteria on section 1 of the plate by streaking back and forth several times. The more streaks, the better chance of an isolated colony. As you work, keep the lid on the Petri dish as much as possible to minimize organisms falling on the plate from the air. Use a gliding motion and avoid digging into the agar. Don't press the loop into the surface. If your loop is not smooth or does not form a complete circle, it can gouge the agar, and colonies may run together. Note that you can see the streak marks on the agar surface if you hold the petri dish at an angle in the air.

5. Burn off all the bacteria from your loop by heating it red hot. This is very important because it eliminates the bacteria on your loop. Wait until the loop has cooled.

   Tip: If using a metal loop and shaft, you can slowly move your hands up the metal shaft as it cools. Once the beginning of the shaft has cooled (where the wire attaches), wait a few seconds to ensure the end of the loop has sufficiently cooled.

6. Without going into the broth again, streak section 2 (see figure 6.4) of the Petri plate. Go into section 1 with about three streaks and spread by filling section 2 with closely spaced streaks. Be very careful not to touch the streaks in section 1 again.

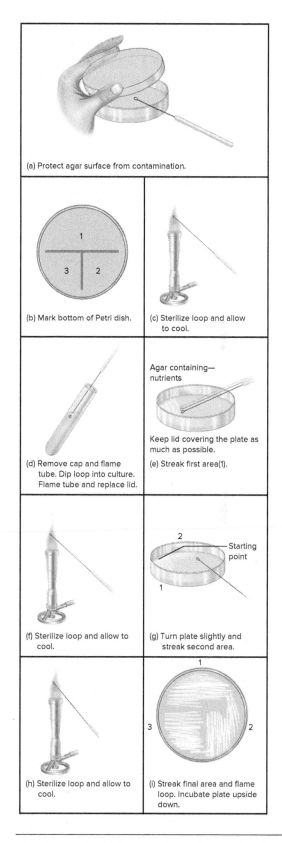

(a) Protect agar surface from contamination.

(b) Mark bottom of Petri dish.

(c) Sterilize loop and allow to cool.

(d) Remove cap and flame tube. Dip loop into culture. Flame tube and replace lid.

Agar containing—nutrients

Keep lid covering the plate as much as possible.

(e) Streak first area(1).

(f) Sterilize loop and allow to cool.

2
Starting point
1

(g) Turn plate slightly and streak second area.

(h) Sterilize loop and allow to cool.

(i) Streak final area and flame loop. Incubate plate upside down.

Figure 6.4 *(a–i)* Preparation of a streak plate. (color plate 11)

7. Again heat the loop red hot and allow to cool. Go into section 2 with one or two streaks and spread by filling section 3 with streaks. The more streaks you are able to make, the more likely you will obtain isolated colonies.

8. Heat the loop red hot before placing it on the benchtop. Usually you can rest the loop on the bottom, metal part of the Bunsen burner so it can cool without burning anything.

9. Repeat the procedure with a second plate for additional practice.

10. Incubate the plates, inverted, in the 37°C incubator.

Second Session

Observe your streak plates and record the results. Since you will probably have a lot of growth in the 1st and 2nd sections, you can semi-quantitate plates by using a 1+, 2+, or 3+. If the microbe only grew in the 1st section, it is a 1+. If the microbe grew in the 2nd section, it is a 2+; 3rd section is a 3+.

EXERCISE

6 **Laboratory Report:** Pure Culture and Aseptic Technique ASM 3, 4, 6, 11, 12

Results

Make a check mark or X under the appropriate column.

Loop Transfer	Clear	Turbid
Control: Microbe used _____		
Tube 1		
Tube 2		

Pipetting	Clear	Turbid
Control: P-C		
Tube 1: P-P1		
Tube 2: P-P2		

Streak Plate	Growth in Sections 1, 2 or 3?			Semi-quantitate the growth
	1	2	3	
Micrococcus				
Staphylococcus				

After observing the broth tubes, were you able to transfer the broth aseptically with the loop? With the pipet?

After observing the control broth, were microbes able to grow in the broth?

Did you obtain isolated colonies of each culture? If not, explain one potential reason why.

Questions

1. What is the definition of a *pure culture*?

2. Provide two reasons why sterile technique is important.

3. What did the control broth inoculated with *Escherichia coli* or *Saccharomyces cerevisiae* demonstrate?

4. What is the purpose of a streak plate?

5. Why is it important to avoid digging into the agar with the loop?

7

Chemically Defined, Complex, Selective,
and Differential Media

Definitions

Enteric. Associated with the intestine.

Fastidious. Organisms that require many growth factors in order to grow.

Growth factor. A compound such as a vitamin or an amino acid that is required by an organism for growth because it cannot synthesize it.

Objectives

1. Interpret the growth of organisms on a chemically defined medium and a complex medium.

2. Interpret the growth of organisms on a selective medium and a differential medium.

3. Explain the relationship between the growth of an organism and the composition of the medium.

Pre-lab Questions

1. How does a complex medium differ from a chemically defined medium?

2. What is the purpose of a selective medium?

3. What is the purpose of a differential medium?

Getting Started

Microbiologists have developed different types of culture media to grow a variety of microbes (**table 7.1**). Normal biota often contaminates samples, so a stool culture may be mixed with *Escherichia coli* normally found on the intestines and other microbes. Thus, isolating or even detecting pathogens from the non-pathogens present can be a challenge.

One way to overcome this is to inoculate the sample onto an Enriched medium, which contains pre-formed nutrients so organisms don't use up energy and materials to synthesize nutrients for growth. Some of the nutrients utilized by bacteria are shown in **table 7.2**. Many bacteria, like the intestinal pathogen *Campylobacter jejuni*, only grow on enriched media as they cannot synthesize necessary

Table 7.1   Types of Culture Media Used to Grow Microorganisms

Media Type	Description	Example
Chemically Defined (also called Synthetic)	A medium containing known amounts of specific chemicals, such as 0.5 grams of glucose.	Glucose mineral
Complex	A medium containing unknown amounts of some chemicals; it contains extracts (like yeast extract) or enzymatic digests of protein that can vary between batches or lots.	Trypticase soy Nutrient broth/agar Brain heart infusion
Enriched	A medium, usually complex, containing growth factors like serum or hemoglobin that support the growth of one, desired organism over others. Microbes that grow on enriched agars require special nutrients in order to grow.	Chocolate agar
Differential	A medium containing ingredients that grow many microbes, but visually differentiate them based on metabolic traits.	Blood agar (shows hemolysis of red blood cells)
Selective	A medium containing ingredients like antibiotics, dyes, or salts that grow one type of microorganism, like Gram positives or Gram negatives, but inhibits the growth of other microbes.	Mannitol salt (salt inhibits Gram-negatives)
Some medium types are considered both selective and differential.		Eosin methylene blue MacConkey Xylose Lysine Deoxycholate Mannitol salt (phenol red is a pH indicator)

Table 7.2 Chemicals Used by Microorganisms

Chemical	Function
Carbon, oxygen, and hydrogen	Component of amino acids, lipids, nucleic acids, and sugars
Nitrogen	Component of amino acids and nucleic acids
Sulfur	Component of some amino acids
Phosphorus	Component of nucleic acids, membrane lipids, and ATP
Potassium, magnesium, and calcium	Required for the functioning of certain enzymes; additional functions as well
Iron	Part of certain enzymes

Table 7.3 Energy and Carbon Sources Used by Different Groups of Microorganisms

Type	Energy Source	Carbon Source
Photoautotroph	Sunlight	CO_2
Photoheterotroph	Sunlight	Organic compounds
Chemolithoautotroph	Inorganic chemicals (H_2, NH_3, NO_2^-, Fe^{2+}, H_2S)	CO_2
Chemoorganoheterotroph	Organic compounds (sugars, amino acids, etc.)	Organic compounds

components needed for growth, and must be provided pre-formed amino acids, vitamins, and other **growth factors**. **Fastidious** microorganisms require specific vitamins or amino acids in order to grow (think picky eaters); thus, they usually need to be cultured on an enriched medium. Chemoorganoheterotrophs (**table 7.3**) need these nutrients supplied in the growth media.

Another way to grow organisms from the sample is to inoculate it onto a differential medium so a microbe of interest (like *Escherichia coli* O157:H7) can be identified by observing the colony color growing on the agar. One can determine characteristics like if the microbe can ferment specific sugars. The sample may also be inoculated onto a selective medium, which permits one kind of bacteria to grow (like *Salmonella enteriditis*) but not others, so the one kind can be isolated, even if they only make up a small percentage of the population.

In this exercise, you will observe the growth of organisms on three different media.

Glucose Mineral Agar This chemically defined medium will grow organisms that make all their cellular components from glucose and inorganic salts.

Trypticase Soy Agar This rich, complex medium made from an enzymatic digest of protein and soy product will grow organisms that require vitamins or other growth factors.

Eosin Methylene Blue (EMB) Agar This selective medium promotes Gram-negative **enteric** rod growth and inhibits Gram-positive microorganisms. The sugar lactose makes this medium differential as organisms that can ferment lactose produce dark pink or purple colonies (like *Escherichia coli*), but non-lactose fermenters appear as clear or very light pink colonies (color plate 45). *Escherichia coli* can grow as colonies with a distinctive metallic green sheen due to the large amounts of acid produced. The colonies of *Enterobacter*, also a lactose fermenter, usually are mucoid with pink to purple centers (Mucoid colonies have a slimy appearance.).

Materials

Cultures growing in trypticase soy broth of the following:

  Escherichia coli

  Staphylococcus epidermidis

  Alcaligenes faecalis

  Enterobacter aerogenes

Media per team

  Trypticase soy agar plate (TSA)

  Glucose mineral agar plate (GM)

  Eosin methylene blue agar plate (EMB)

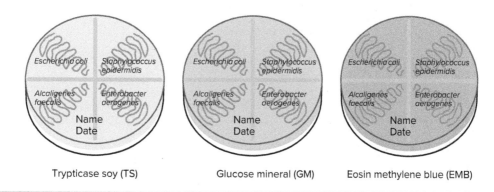

Trypticase soy (TS)   Glucose mineral (GM)   Eosin methylene blue (EMB)

Figure 7.1 A diagram of the labeled media plates. See step 1 for all labels to include.

PROCEDURE

First Session

1. With a marking pen, divide the bottom of the Petri plates into quadrants. Label around the plates with your name, date, media type, temperature, and lab section. Label each quadrant with the organism as shown in **figure 7.1**.

2. Inoculate each quadrant of the plate with a loopful of the culture using a wavy line or S streak.

3. Invert the plate and incubate at 37°C for 48 hours.

Second Session

1. Observe and compare the growth on the three plates.

2. Record the results.

EXERCISE 7

Laboratory Report: Chemically Defined, Complex, Selective, and Differential Media

Results

Compare the bacterial growth on the three plates. Describe the colony color and appearance on each plate. For each organism on the plate, indicate TFTC (Too Few to Count) if no colonies are seen. If there is some growth, indicate with a +. If the streak is a lawn of growth (continuous growth), indicate with a +++. For the EMB plate, indicate if lactose was or was not fermented.

	Glucose Mineral	TS		EMB
Escherichia coli				
Amount of growth				lac–/lac+
Colony appearance				
Staphylococcus epidermidis				
Amount of growth				lac–/lac+
Colony appearance				
Alcaligenes faecalis				
Amount of growth				lac–/lac+
Colony appearance				
Enterobacter aerogenes				
Amount of growth				lac–/lac+
Colony appearance				

Lac – = no lactose fermentation on EMB
Lac + = lactose fermentation on EMB (color plate 45)

Questions

1. Which tested organisms could not grow on the glucose mineral medium?

   Which tested organisms could grow on it?

2. Which organisms tested did not require any growth factors?

3. Of the organisms that grew on both trypticase soy agar and glucose mineral agar, did some organisms grow better on the trypticase soy agar than on the glucose mineral? Can you propose a reason?

4. Which organisms grew on the EMB agar?

5. Which organisms fermented lactose?

6. In general, EMB agar selects for what kinds of organisms?

7. What kinds of organisms does EMB agar differentiate?

EXERCISE

8

Quantification of Microorganisms:
Serial Dilutions ASM 5

Definitions

Colony-forming units (CFUs). A single cell or multiple cells attached to one another that give rise to one colony.

Diluent. A liquid used to dilute a sample: usually water, saline, or broth.

McFarland standards. A set of nine tubes with a known turbidity and bacterial concentration that are used to adjust bacterial concentrations for inoculations such as antibiotic sensitivities.

Optical density (O.D.). A measure of the amount of turbidity. Frequently called *absorbance* as light rays are absorbed by cells and not transmitted through a solution.

Pour plate. An inoculation method in which a sample is first diluted, then pipetted into the bottom of an empty, sterile Petri dish and lastly, agar is poured into the Petri dish and allowed to solidify.

Serial dilution. Preparing a dilution in steps instead of one dilution.

Spread plate. An inoculation technique in which a diluted sample is pipetted on top of the agar and spread out using a spreader.

Turbidity. Cloudiness in a liquid.

Viable (bacteria). Capable of growing and dividing.

Objectives

1. Describe how bacteria are counted.
2. Explain the plate count method of counting bacteria.
3. Explain the use of dilutions. ASM 5
4. Calculate the number of organisms that can be present in a clear liquid and in a turbid liquid.

Pre-lab Questions

1. Name three ways bacteria are counted.
2. Discuss one advantage and disadvantage of using a plate count for counting bacteria.
3. Explain how to perform a serial dilution.

Note: When serially transferring concentrated suspensions to less concentrated (more diluted) suspensions, a new pipette must be used for each transfer to prevent carryover of organisms. (In this exercise, the instructor may use each pipette twice to conserve materials.) However, when transferring less concentrated to more concentrated suspensions, the same pipette can be used without significant error because only a few organisms will be added to the much larger sample.

Getting Started

Counting bacteria is important in numerous applications, ranging from determining the number of bacteria in a sample of raw chicken, or the number of bacteria per milliliter of pond water. Several techniques can be used to enumerate bacteria, each with advantages and disadvantages. Three common methods are discussed: plate counts, direct microscopic counts, and turbidity.

Plate Count This counting method assumes that one **viable** bacterium produces one colony when grown on an agar plate. This method also assumes the bacterium can grow on the specific agar it was inoculated onto. Plate counts can be further divided into three types: a streak plate, a pour plate, and an isolation streak. Isolation streak plates are discussed in exercise 6, and a **pour plate** and **spread plate** are shown in figure 8.2(*d*) and (*e*).

In plate counts, a known amount of sample or material to be counted is suspended in liquid (also referred to as diluted) using a **diluent** such as saline or broth, and pipetted onto a Petri plate. After incubation, each organism produces a colony in the agar that can then be counted (color plate 12). Plate counts are used frequently to determine the concentration of microorganisms, and has advantages and disadvantages to consider prior to use, including:

1. Plate counts are normally used only if a sample contains at least 1×10^2 cells/ml. Otherwise,

not enough organisms are present to ensure a colony would form. On the other hand, bacteria can be present in large numbers (an overnight broth culture of *Escherichia coli* can easily contain 1 billion cells/ml), so samples must be diluted enough so distinct colonies will form that can be counted. The minimum number of colonies is usually set at 30 and the maximum number of colonies that can be accurately counted is 300 per plate and are designated as **colony-forming units (CFUs)**. Because it is difficult to know exactly how diluted to make a sample to obtain a countable plate, several different dilutions must be plated.

2. Some bacteria will stick together, so sometimes two or more bacterial cells give rise to one colony. This results in a lower number of calculated cells than are actually present.

3. If a sample contains many different kinds of bacteria, one medium or growth condition would most likely not support the growth of all the different bacteria present. For example, soil may contain organisms that require a temperature above 50°C; in contrast, other organisms are inhibited at 50°C. These conditions must be considered when a sample of mixed bacteria is enumerated.

4. The tubes or plates the sample is added to must be swirled sufficiently in order to have countable colonies, whereas a spread plate requires sufficient back and forth streaks in order for colonies to be counted.

5. Microbes take at least 24 hours to grow to obtain results of a plate count.

6. Two advantages of the plate count method over other methods include: (1) only viable organisms grow and are counted, which are usually the microbes of interest (versus dead or inactive cells), and (2) samples containing insufficient numbers for counting by other methods can be enumerated.

Direct Microscopic Count This counting method adds a suspension of cells to a special slide, called a counting chamber, hemocytometer, or McMaster's slide, that has been ruled into squares and holds a specific volume of liquid. By counting the cells that appear within the grids, the number of total organisms (both viable and non-viable) in the sample can

be calculated. Counting cells is faster than growing colonies on plates, but it does have drawbacks. First, about 1×10^7 organisms/ml must be present in order to be seen, and both viable and non-viable organisms appear identical under a microscope.

Turbidity Turbidity can be used to semi-quantitate or quantitate bacterial cultures. One semi-quantitative method uses **McFarland's standards**, which are a series of nine tubes designated as 0.5–8 that have a specific turbidity and are used to visibly adjust bacterial concentrations. A 0.5 McFarland standard is about 1.5×10^8 cells/ml, whereas the 8 equals about 24×10^8 cells/ml. If a 0.5 McFarland tube was measured spectrophotometrically at 625 nm, it should give an absorbance reading between 0.08 and 0.1, whereas an 8 should read 0.94–0.98. McFarland standards are commonly used to ensure proper bacterial inoculums are added to automated identification cards, wells, or tubes, as well as to consistently inoculate bacteria onto agar plates to determine antibiotic sensitivities.

Another method semi-quantitates by visually observing the cloudiness of a microbial suspension. Just as plate sections can be quantitated using a 1+, 2+, 3+ (exercise 6), tubes can also be semi-quantitated for clinical situations, such as the cloudiness of urine or spinal fluid. You can view printed text directly behind the tubes, or visualize the test tube rack separation line behind the tube as a reference point (color plate 13). Gently swirl the tube to make sure any cells that fell to the bottom of the tube are suspended. If you do not see any cells in the suspension and the rack line or text is clear, then it is a 0. If a suspension is slightly cloudy with more microbes than you initially added, but the rack or text is easily seen, it is a 1+.

If you can see some text or the rack line, but you cannot read the words, then the growth is a 3+. If the rack line or text is completely obscured and no longer visible, then the growth is a 4+. Thus, it is important to pay close attention to the amount of bacteria you added to a tube with a loop to distinguish between a 0 and a 1+.

In this counting method, a spectrophotometer is used to measure the **turbidity** or **optical density (O.D.)** of cells in a liquid. For bacteria, the more bacteria present, the cloudier the broth, and higher the O.D. In this method, plate counts must first be correlated with O.D. readings taken at 600 nm of each bacterial strain investigated because organisms are of different sizes. For example, an O.D. reading of

0.2 for a broth culture of one *E. coli* strain is equal to 1×10^8 cells/ml. The same cell density of another organism would have a different O.D. Once the correlation between plate counts and O.D. are determined, turbidity is a convenient method of determining number of organisms. This method is used to determine the generation time in exercise 10, and water quality is analyzed in exercise 29.

The plate count method is used in this exercise to count the number of organisms in two broth cultures: one cloudy (sample A), and the other with no visible cloudiness (sample B). You will also determine the turbidity of the suspensions so you will be able to compare the dilution sample O.D. to the plate counts.

Alternate Spread Plate Procedure

There are two methods of preparing plate counts: **pour plates** and **spread plates**. For pour plates, a known amount of sample is mixed with melted agar in a Petri plate and colonies appearing in and on the agar are counted. For spread plates, a known amount of sample is placed on the surface of the agar plate and spread with a bent glass rod or spreader. All of the colonies appear on the agar surface. The spread plate directions for this exercise is found in Appendix 4. It is widely used but can be a little more difficult.

Materials

Per team using sample A (cloudy suspension):

9.9 ml trypticase soy broth blanks, 3

9.0 ml trypticase soy broth blanks, 3

Spectrophotometer tubes, 4
Spectrophotometer tube containing trypticase soy broth (blank), 1

1 ml pipettes, 9 (14 if dilution tubes will not fit in the spectrophometer)

Pipette bulb (figure 6.3)

Trypticase soy agar deeps, 4 or (optional) 1 spreader

Sterile Petri dishes, 4

Spectrophotometer for the class to use
Suspension A: an overnight trypticase soy broth culture (without shaking) of *E. coli* or *Saccharomyces cerevisiae* diluted 1:1 with trypticase soy broth

First Session

1. Observe and record whether the suspension is cloudy or clear.

2. Label all blanks and plates with the dilution, as shown in **figure 8.1**. A plate should be lined up with each tube dilution. Label the Petri dishes with your name, organism, date, media type (TSA), temperature, concentration, and lab section.

3. Melt 4 trypticase soy agar deeps and hold at 50°C. It is very important not to let the deeps cool much lower than 50°C because the agar will harden and will need to be heated to boiling (100°C) before it will melt again.

4. Make **serial dilutions** of the bacterial suspension.

   a. Mix **bacterial suspension A** by rotating between the hands or gently pipetting up and down a few times. Transfer 0.1 ml of suspension to the 9.9 ml tube labeled 10^{-2}. Discard pipette into the proper container.

   b. Mix the tube well and transfer 0.1 ml of the 10^{-2} dilution to the 9.9 ml tube labeled 10^{-4}. Discard pipette into the proper container.

   c. Mix the tube well and transfer 0.1 ml of the 10^{-4} dilution to the 9.9 ml tube labeled 10^{-6}. Discard pipette into the proper container.

   d. Mix the tube well and transfer <u>1.0 ml</u> of the 10^{-6} dilution to the 9.0 ml tube labeled 10^{-7}. (**Note change from 0.1 to 1.0 ml.**) Discard pipette into the proper container.

   e. Mix the tube well and transfer 1.0 ml of the 10^{-7} dilution to the 9.0 ml tube labeled 10^{-8}. Discard pipette into the proper container.

   f. Mix the tube well and transfer 1.0 ml of the 10^{-8} dilution to the 9.0 ml tube labeled 10^{-9}. Discard pipette into the proper container.

5. Transfer 1.0 ml from dilution tubes 10^{-9}, 10^{-8}, 10^{-7}, and 10^{-6} into the corresponding sterile, labeled Petri plates as follows:

   a. Mix the 10^{-9} dilution, and transfer 1.0 ml into the sterile Petri plate labeled 10^{-9}. Do not discard the pipette.

Team A

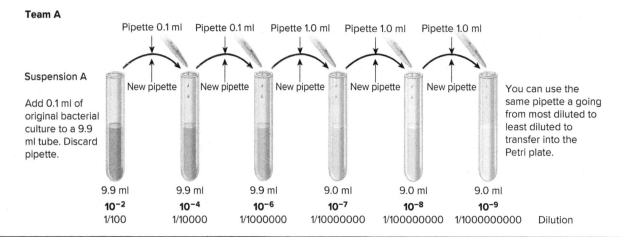

Figure 8.1 Dilution scheme for Team A.

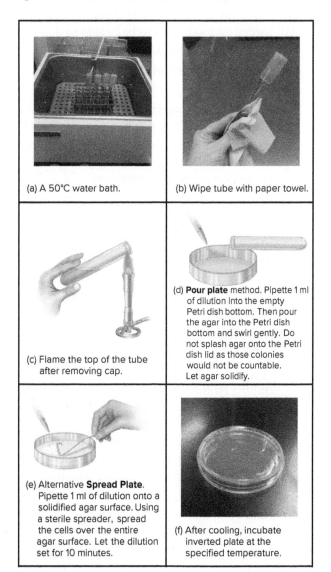

(a) A 50°C water bath.

(b) Wipe tube with paper towel.

(c) Flame the top of the tube after removing cap.

(d) **Pour plate** method. Pipette 1 ml of dilution into the empty Petri dish bottom. Then pour the agar into the Petri dish bottom and swirl gently. Do not splash agar onto the Petri dish lid as those colonies would not be countable. Let agar solidify.

(e) Alternative **Spread Plate**. Pipette 1 ml of dilution onto a solidified agar surface. Using a sterile spreader, spread the cells over the entire agar surface. Let the dilution set for 10 minutes.

(f) After cooling, incubate inverted plate at the specified temperature.

Figure 8.2 The pour plate and spread plate technique.
(a,b,f): Courtesy of Anna Oller, University of Central Missouri.

b. Mix the 10^{-8} dilution, and with the same pipette, transfer 1.0 ml into the Petri plate labeled 10^{-8}. Do not discard the pipette.

c. Mix the 10^{-7} dilution, and with the same pipette, transfer 1.0 ml into the Petri plate labeled 10^{-7}. Do not discard the pipette.

d. Mix the 10^{-6} dilution, and with the same pipette, transfer 1.0 ml into the Petri plate labeled 10^{-6}. Discard the pipette into the proper container.

6. Obtain the melted agar once you are ready to pour plates.

a. Add a tube of melted agar (wipe off the outside of the tube and flame the opening before pouring) to each plate (**figure 8.2**) and swirl the plate gently by moving the plate in a figure-eight pattern on the bench. Do not splash agar onto the lid or colonies may not be countable.

b. Place the plates right (lid) side up out of the way so they solidify. Do not move plates around or they may end up with an uneven surface. You should immediately rinse out excess melted agar left at the bottom of the tubes with water.

7. Label dilution factors on the spectrophotometer tubes and line them up from 10^{-9}, 10^{-8}, 10^{-7}, 10^{-6}, and C for the control trypticase soy broth blank.

a. Start with the 10^{-9} dilution (with microbes), pipette 1 ml from the suspension into the spectrophotometer tube labeled 10^{-9}. Repeat the pipetting for the remaining dilutions into corresponding spec tubes. Discard the pipette into the proper container.

8. Wipe fingerprints off the trypticase soy blank using a Kimwipe and place spectrophotometer tubes into a spectrophotometer set to 600 nm. Press the blank button.

9. Wipe off fingerprints from each tube and then read the O.D. of each tube. Record the O.D. results in the Report.

10. Invert the plates (lid side down) after you are sure the agar has hardened (about 10 minutes), and incubate at 37°C for *Escherichia coli*, or at 25°C for *Saccharomyces cerevisiae*.

Materials

Per team using sample B (a clear suspension)

9.9 ml trypticase soy blanks, 2

9.0 ml trypticase soy blanks, 3

1 ml pipettes, 9

Pipette bulb (figure 6.3)

Spectrophotometer tubes (fitting the spectrophotometer), 4

Spectrophotometer tube containing trypticase soy broth (blank), 1

Trypticase soy agar deeps, 4 or (optional) spreader

Sterile Petri dishes, 4

Suspension B: a non-turbid suspension of microorganisms

PROCEDURE FOR TEAM B

First Session

1. Observe whether the suspension is cloudy or clear.

2. Label all tubes and plates with the dilution (**figure 8.3**). Label the Petri dishes with your name, date, media type (TSA), temperature, concentration, and lab section.

3. Melt 4 trypticase soy agar deeps and hold at 50°C. It is very important not to let the agar deeps cool much lower than 50°C

because the agar will harden and will need to be heated to boiling (100°C) before it will melt again.

4. Make serial dilutions of the bacterial suspension.

   a. Mix **bacterial suspension B** and transfer 0.1 ml to the 9.9 ml tube labeled 10^{-2}. Discard pipette into the proper container.

   b. Mix the tube well and transfer 0.1 ml of the 10^{-2} dilution to the 9.9 ml tube labeled 10^{-4}. Discard pipette into the proper container.

   c. Mix the tube well and transfer 1.0 ml of the 10^{-4} dilution to the 9.0 ml tube labeled 10^{-5}. (**Note change from 0.1 ml to 1.0 ml.**) Discard pipette into the proper container.

   d. Mix the tube well and transfer 1.0 ml of the 10^{-5} dilution to the 9.0 ml tube labeled 10^{-6}. Discard pipette into the proper container.

   e. Mix the tube well and transfer 1.0 ml of the 10^{-6} dilution to the 9.0 ml tube labeled 10^{-7}. Discard pipette into the proper container.

5. Transfer 1.0 ml from dilution tubes 10^{-7}, 10^{-6}, 10^{-5}, and 10^{-4} into the corresponding sterile, labeled Petri plates as follows:

   a. Mix the 10^{-7} dilution, and transfer 1.0 ml into the sterile Petri plate labeled 10^{-7}. Do not discard the pipette.

   b. Repeat for the 10^{-6}, 10^{-5}, and 10^{-4} dilutions. Discard the pipette into the proper container.

6. Obtain the melted agar once you are ready to pour plates.

   a. Add a tube of melted agar (wipe off the outside of the tube and flame the opening before pouring) to each plate (figure 8.2) and swirl the plate gently by moving the plate in a figure-eight pattern on the bench. Do not splash agar onto the lid or colonies may not be countable.

   b. Place the plates out of the way right (lid) side up so they will solidify. Do not move plates around or they may develop an uneven surface. Your instructor may have you immediately rinse out excess melted agar left at the bottom of the tubes with water.

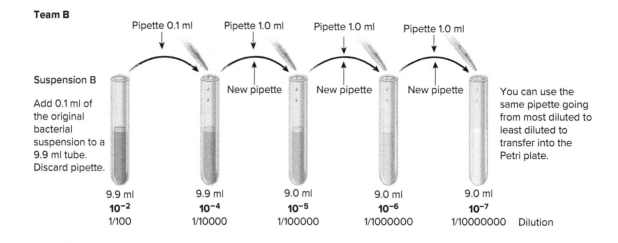

Figure 8.3 Dilution scheme for Team B.

7. Invert the plates (lid side down) after you are sure the agar has hardened (about 10 minutes), and incubate at 37°C for *Escherichia coli*, or at 25°C for *Saccharomyces cerevisiae*.

8. Label the dilution factor on the spectrophotometer tubes and line them up from 10^{-7}, 10^{-6}, 10^{-5}, 10^{-4}, and C for the control. Beginning with the 10^{-7} dilution containing the microorganism, pipette 1 ml from the suspension into the spectrophotometer tube labeled 10^{-7}. Repeat pipetting the remaining dilutions into corresponding spectrophotometer tubes. Discard the pipette into the proper container.

Second Session for Both A and B Teams

1. Count the colonies on the agar plates. Use a marking pen on the bottom of the plate to dot the colonies as you count them. Colonies growing in the agar tend to be lens-shaped and smaller than those growing on the surface, although all are counted equally. If there are more than 300 colonies on the plate, label it TNTC—**t**oo **n**umerous **t**o **c**ount.

2. Choose the plate that has between 30 and 300 colonies (less than 30 gives results with a high sample error), being sure to record the dilution. Calculate the number of organisms/ml using the following formula:

   number of organisms on the plate × 1/sample volume × 1/dilution = the number of organisms/ml in the original suspension

   Note: In this exercise, the volume of all the samples is 1.0 ml.

3. Record your results. Post your results on the blackboard so that average numbers of organisms/ml for suspension A and suspension B can be calculated.

Understanding Dilutions

(See also Appendix 3.)

1. To make a dilution, use the following formula:

$$\frac{\text{Sample}}{(\text{diluent} + \text{sample})} = \text{the dilution}$$

Example 1: How much is a sample diluted if 1 ml is added to 9.0 ml of water? (The water is sometimes called a diluent.)

$$\frac{1}{(1+9)} = \frac{1}{10} \text{ or } 10^{-1}$$

Example 2: How much is a sample diluted if 0.1 ml is added to 9.9 ml of water?

$$\frac{0.1}{(0.1 + 9.9)} = \frac{0.1}{10} = 1:100 \text{ or } 10^{-2}$$

2. When a sample is serially diluted, multiply each dilution together for the final dilution. The final dilution in tube B is 1:100 or 10^{-2}.

3. To calculate the number of organisms in the original sample, use this formula:

$$\frac{\text{\# of organisms}}{\text{ml in original sample}} = \text{\# of colonies on}$$

$$\text{plate} \times \frac{1}{\text{sample volume}} \times \frac{1}{\text{dilution}}$$

Example 3: Suppose you counted 120 organisms on a plate diluted 10^{-2}. The sample size was 0.1 ml.

Solution: $120 \times \dfrac{1}{0.1} \times \dfrac{1}{10^{-2}} = 120 \times 10 \times 100$

$$= 120.0 \times 10^{3}$$

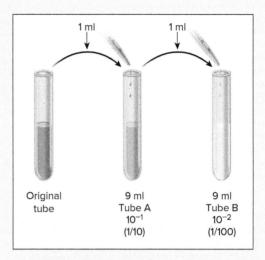

1 ml   1 ml

Original tube | 9 ml Tube A 10^{-1} (1/10) | 9 ml Tube B 10^{-2} (1/100)

Negative exponents become positive when moved above the division line. The 120 needs a decimal moved from after the 0 to between 1 and 2.

Since the decimal is moved 2 places, a 2 to be needs added to the 3 from above.

$$120.0 \times 10^{3+2=5} \quad \boxed{= 1.2 \times 10^{5} \text{ organism/ml}}$$

Example 4: Suppose you counted 73 colonies on the plate marked 10^{-6}. If the sample size was 1.0 ml, then

$$73 \times \frac{1}{1.0} \times \frac{1}{10^{-6}} = 73 \times 1 \times 10^{6} = 73.0 \times 10^{6}$$

Decimal needs to be moved from behind 3 to between 7 and 3.

Since the decimal is moved 1 place, 1 needs to be added to the 6.

$$10^{6+1} = 7.3 \times 10^{7} \text{ per ml}$$

(It is important to label the answer "per ml.")

EXERCISE

8

Laboratory Report: Quantification
of Microorganisms: Serial Dilutions ASM 5

Results

1. How many organisms/ml were in the original suspension of flask A or B that you counted? Calculate the number using the number of colonies and the dilutions of the plate with 30–300 colonies. Show calculations. Record in chart and on the blackboard.

O.D. taken at _____ nm

Appearance of Broth (circle one)		Control	10^{-4}	10^{-5}	10^{-6}	10^{-7}	10^{-8}	10^{-9}
Suspension A Cloudy/Clear	O.D. Reading =							
	Plate Counts =							
	Concentration =							
Suspension B Cloudy/Clear	O.D. Reading =							
	Plate Counts =							
	Concentration =							

Which plate contained between 30 and 300 colonies? _____

Concentration Determination:

Number of organisms/ml (class average): _____

Number of organisms/ml (your data): _____

Record TFTC (Too few to count) if plate count is 0.
Record TNTC (Too numerous to count) if plate count is 300+.

Questions

1. From these results obtained in class, about how many organisms/ml can be in a cloudy broth? (Show calculations.)

2. From these results obtained in class, about how many organisms/ml can be in a clear broth without showing any sign of turbidity? (Show calculations.)

3. What are two sources of error in this procedure?

4. If you serially dilute a sample with three 1:10 dilutions, what is the dilution of the last tube?

5. If you add 1.0 ml to 99 ml of water, what is the dilution of the sample?

6. If you had a solution containing 6,000 organisms/ml, how could you dilute and plate a sample so that you had a countable plate?

EXERCISE

9

Aerobic and Anaerobic Growth

Definitions

Aerotolerant. Microbes that can live in the presence or absence of oxygen, but will not use oxygen if present.

Agar deep. A test tube containing solidified agar.

Anaerobic. In the absence of oxygen.

Capnophile. Requires a higher CO_2 concentration than normal atmospheric CO_2 concentrations of 0.03%.

Facultative anaerobe. Grows best if O_2 is present, but can also grow without it.

Microaerophilic. Requires small amounts of (5–10%) of O_2, but higher concentrations are inhibitory.

Obligate aerobe. Requires O_2 to survive. Air contains about 20% oxygen. Sometimes called a strict aerobe.

Obligate anaerobe. Cannot grow in the presence of O_2. Sometimes called a strict anaerobe.

Objectives

1. Explain how *aerobes, anaerobes*, and *facultative anaerobes* differ in their oxygen requirements.
2. Explain the role of each item in an anaerobe container in creating an anaerobic environment.

Pre-lab Questions

1. How does an obligate aerobe differ from an obligate anaerobe?
2. What are organisms termed that can grow either in the presence or in the absence of oxygen?
3. How are you going to create anaerobic conditions to grow obligate anaerobes?

An organism cannot grow and divide unless it is in a favorable environment. Environmental conditions that influence how an organism grows include temperature, availability of nutrients, moisture, oxygen, osmotic pressure, and the presence of toxic products.

Each bacterial species has its own set of optimal conditions that allows maximum growth. For example, organisms found in a hot spring in Yellowstone National Park require a much higher temperature for growth in the laboratory than organisms isolated from the throat of humans.

Getting Started

All the animals we are familiar with, including humans, have an absolute requirement for oxygen. It is important to note that ambient air is composed of approximately 21% oxygen and 0.03% carbon dioxide. It seems rather surprising then that there are groups of organisms that cannot grow or are even killed in the presence of oxygen. Still other kinds of organisms can grow either with or without oxygen. The groups of microbes associated with their respective oxygen requirements are shown in **table 9.1**. Determining a microorganism's oxygen preference is

Table 9.1   Oxygen Requirement Terms

Oxygen (O_2) availability	Oxygen (O_2) requirement/tolerance reflects the organism's energy-harvesting mechanisms and its ability to inactivate reactive oxygen species.
Obligate aerobe	Requires O_2. *Mycobacterium*, a soil bacterium, is an obligate aerobe (color plate 14).
Facultative anaerobe	Grows best if O_2 is present, but can also grow without it. *Escherichia coli*, an intestinal bacterium, is a facultative anaerobe (color plate 14).
Obligate anaerobe	Cannot grow in the presence of O_2 but oxygen sensitivity varies. Methanogens found in swamps are killed by a few molecules, whereas *Clostridium* can then survive oxygen exposure but cannot grow until conditions become anaerobic (color plate 14).
Microaerophile	Microaerophile requires small amounts of O_2, but higher concentrations are inhibitory. *Micrococcus*, a skin bacterium (color plate 14), is microaerophilic.
Aerotolerant anaerobe (obligate fermenter)	Indifferent to O_2 and cannot use O_2 in metabolic pathways. *Propionibacterium shermanii*, a food bacterium, is aerotolerant.

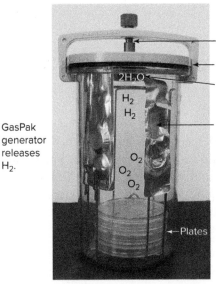

Spring to hold lid

Lid with gasket

Palladium pellets catalyze H and O to bind to form H₂O.

Methylene blue indicator turns white in anaerobic conditions.

$2H_2O$

H_2
H_2

O_2
O_2
O_2

GasPak generator releases H_2.

Plates

Figure 9.1 An anaerobe container showing the components required to obtain an anaerobic state. Other systems utilize different shapes or sealed plastic bags. Courtesy of Anna Oller, University of Central Missouri.

important not only to help in identification, but also to find potential treatments for infections.

One method of creating an anaerobic state is by using an anaerobe container (**figure 9.1**). Inoculated Brewer's anaerobe plates are placed in a container. A methylene blue or other indicator is added before the container is sealed to visually ensure that an anaerobic state is achieved. Methylene blue becomes white in the absence of oxygen and turns blue again when oxygen is detected. Different packets can be purchased to create carbon dioxide (CO_2) and hydrogen ions (H^+) once placed into the container. Palladium pellets added to the container catalyze the H^+ binding to the O_2 present to form nontoxic water (H_2O).

Another method of determining oxygen requirements utilizes test tubes that have the oxygen driven off during autoclaving. A thioglycollate tube has a pink layer showing how deeply oxygen has penetrated the tube. In order to truly determine a microbe's oxygen requirements, both a Brewer's anaerobe plate and thioglycollate tube should be inoculated.

There are several other categories of organisms, such as **microaerophiles**, which grow best in reduced amounts (5–10%) of oxygen (color plate 14), and those that prefer more CO_2 than the amount found in the atmosphere, called **capnophiles**. In this exercise, we examine the oxygen requirements of an

obligate aerobe, an obligate anaerobe, and a facultative anaerobe. Try to identify them by growing each culture **aerobically** and **anaerobically**.

Materials

Per team

Trypticase soy or brain heart infusion broth Cultures labeled A, B, C, and D of the following:

Escherichia coli

Mycobacterium smegmatis or *M. phlei* in brain heart infusion broth

Micrococcus

Clostridium sporogenes

Brewer's anaerobic plates, 4

Anaerobe system for plate incubation

Methylene blue or other anaerobic detection strip or tablet

Gas pak generator
Thioglycollate tubes, 4

PROCEDURE

First Session

1. Label the thioglycollate tubes and Brewer's anaerobe plates with the letter inoculated, name, date, media type, temperature, and laboratory section.

2. Flame your loop and allow it to cool. Inoculate a thioglycollate broth tube with a loopful of culture A.

3. Flame your loop and allow it to cool. Inoculate a Brewer's anaerobic plate with culture A, making an isolation streak (see exercise 6).

4. Repeat with cultures B, C, and D.

5. Place all Brewer's plates in the anaerobic system to be sealed up.

6. Inoculate all tubes and plates at least 48 hours at room temperature (25°C).

Second Session

1. Observe the surface of the Brewer's anaerobic plates and record growth.

2. Observe the thioglycollate broth tubes and record where the growth occurred (top, bottom, throughout, etc.).

3. Determine which cultures are the obligate aerobes, facultative anaerobes, and obligate anaerobes.

Alternative Method

Materials

Per team

Trypticase soy or brain heart infusion broth cultures labeled A, B, C, and D of the following:

Escherichia coli

Mycobacterium smegamatis or *M. phlei* in brain heart infusion broth

Micrococcus

Clostridium sporogenes

Trypticase soy + 0.5% glucose agar deeps, 4

Trypticase soy agar slants, 4

First Session

1. Put the agar deeps in a beaker or can filled with water to the height of the agar. Boil the deeps for 10 minutes. This will not only melt the agar but also drive off all the dissolved oxygen.

   After the agar hardens, air gradually diffuses into the tube so that approximately the top several millimeters of the agar become aerobic, but the remainder of the tube is anaerobic.

2. Cool the agar in a 50°C water bath for about 10 minutes. (Check the temperature of the water with a thermometer.) Be careful so that the agar does not cool much lower than 50°C or the agar will solidify and will need to be boiled to melt it again. The tube will feel hot, but you will be able to hold it, or use a tube holder or hot hand.

3. Label the tubes and slants.

4. Inoculate a melted agar deep with a loopful of culture A (**figure 9.2a**) and mix by rolling between the hands. Permit the agar to harden. This technique is often called a *shake tube*.

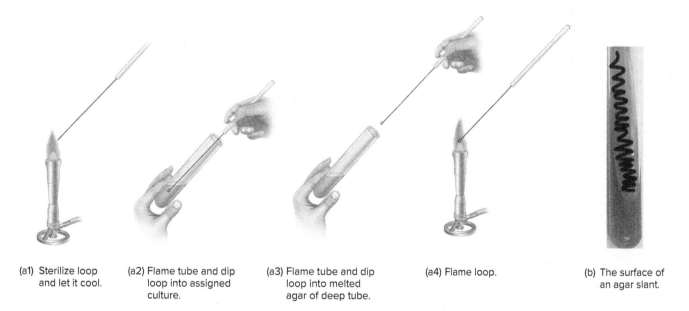

(a1) Sterilize loop and let it cool.

(a2) Flame tube and dip loop into assigned culture.

(a3) Flame tube and dip loop into melted agar of deep tube.

(a4) Flame loop.

(b) The surface of an agar slant.

Figure 9.2 Inoculating (*a*) a melted agar deep and (*b*) an agar slant. (*b*) Courtesy of Anna Oller, University of Central Missouri.

5. Inoculate a slant with a loopful of culture A by placing a loopful of broth on the bottom of the slant and making a wiggly line on the surface to the top of the slant (**figure 9.2b**).

6. Repeat with cultures B, C, and D.

7. Incubate all slants and deeps at least 48 hours at room temperature (25°C). Some cultures grow so vigorously at 37°C that the gas produced blows apart the agar.

Second Session

1. Observe the surface of the slants and the deeps and record the growth. Compare to **figure 9.3**.

   Note: Sometimes the anaerobes seep down and grow between the agar slant and the walls of the glass tube, where conditions are anaerobic, but not on the surface of the slant, which is aerobic.

2. Determine which cultures are the obligate aerobes, the facultative anaerobes, and the obligate anaerobes.

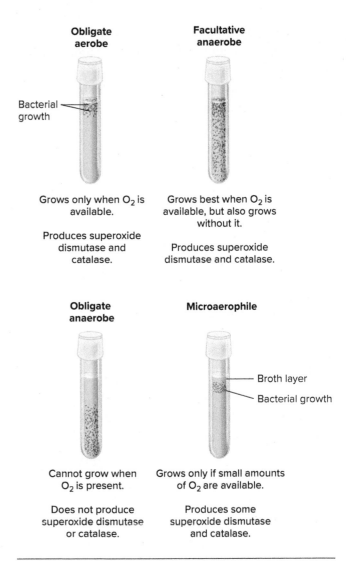

Obligate aerobe

Grows only when O_2 is available.

Produces superoxide dismutase and catalase.

Facultative anaerobe

Grows best when O_2 is available, but also grows without it.

Produces superoxide dismutase and catalase.

Obligate anaerobe

Cannot grow when O_2 is present.

Does not produce superoxide dismutase or catalase.

Microaerophile

Grows only if small amounts of O_2 are available.

Produces some superoxide dismutase and catalase.

Figure 9.3 The appearance of aerobic and anaerobic growth in thioglycollate tubes.

EXERCISE

9

Laboratory Report: Aerobic and Anaerobic Growth

Results

Culture	A	B	C	D
Appearance of growth on slant or Brewer's plates (1+, 2+, etc.)				
Was an anaerobic state achieved?	Yes/No	Yes/No	Yes/No	Yes/No
Appearance and location of growth in deep or thioglycollate tubes				
Oxygen requirements? (Use proper terms)				
Microbe Identity (scientific name) of the culture:				

Questions

1. Would you expect an obligate anaerobe to grow on a slant or plate incubated aerobically? Why?

2. Which kind of organism would you expect to grow in or on both types of media you used?

3. Which kind of organisms grow aerobically on slants or at the top of thioglycollate tubes?

4. Name the genus of the organism(s) that could grow throughout the agar deeps or thioglycollate broth tubes.

5. Why did you place the Brewer's anaerobic plates into the anaerobe system?

6. If air can diffuse into agar and broth, how were the obligate anaerobes probably grown in the broth for the class?

10

The Effect of Incubation Temperature on Generation Time [ASM 8]

Definitions

Absorbance. A measure of turbidity as light rays are absorbed by cells and not transmitted through a solution.

Colorimeter. An instrument used to measure the turbidity of bacterial growth.

Death phase. When more bacteria are dying than replicating in the fourth growth phase.

Doubling time. The time it takes for one cell to divide into two cells or for a population of cells to double the generation time.

Lag phase. The time bacteria adapt to the growth medium and are not yet replicating in the first growth phase.

Log phase. The time bacteria actively replicate in a logarithmic fashion during the second growth phase. Also known as the exponential phase.

Optical density (O.D.). A measure of the amount of turbidity. Frequently called *absorbance* as light rays are absorbed by cells and not transmitted through a solution.

Stationary phase. The time when the number of bacterial cells dying equals the number of bacterial cells replicating during the third growth phase.

Synchronous growth. Bacterial replication occurring at a constant rate, such as every 20 minutes.

Turbidity. Cloudiness in a liquid.

Objectives

1. Describe the phases of a growth curve.
2. Interpret the effect of temperature on generation time.[ASM 8]
3. Describe the calculation of generation time.
4. Explain the use of semi-log paper.

Pre-lab Questions

1. What are the four phases of a bacterial growth curve?
2. What is the generation time of a bacterial cell?
3. Why is generation time only measured in the log phase?

Getting Started

Every bacterial species has an optimal temperature—the particular temperature resulting in the fastest cell replication, or growth. Normally, the optimal temperature for each organism reflects the temperature of its environment, and the proper terms associated with certain temperature requirements are shown in **table 10.1**. Organisms associated with animals usually grow fastest at about 37°C, the average body temperature of most warm-blooded animals. Organisms can divide more slowly at temperatures below their optimum, but there is a minimum temperature below which no growth occurs. Bacterial growth is often inhibited at temperatures slightly higher than their optimum temperature.

The effect of temperature on cell replication can be carefully measured by determining the **doubling time** at different temperatures. Doubling time, or generation time, is the time it takes for one cell to divide into two cells; on a larger scale, it is the time required for the population of cells to double (**figure 10.1**). The shorter the generation time is, the faster the growth rate.

Environmental Factors That Influence Microbial Growth

Table 10.1 Microbial Growth Temperature Classifications

Temperature Terms	Characteristics
Temperature	Thermostability appears to be due to protein structure.
Psychrophile	Optimum temperature between −5°C and 15°C. The soil bacterium *Listeria* is a psychrophile.
Psychrotroph	Optimum temperature between 15°C and 30°C, but grows well at refrigeration temperatures.
Mesophile	Optimum temperature between 25°C and 45°C. The intestinal bacterium *Escherichia* is a mesophile.
Thermophile	Optimum temperature between 45°C and 70°C. The soil bacterium *Bacillus stearothermophilus* is a thermophile.
Hyperthermophile	Optimum temperature of 70°C or greater. The hot springs bacterium *Thermus aquaticus* is a hyperthermophile that functions well at 95°C.

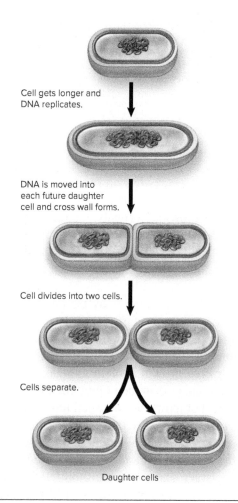

Cell gets longer and DNA replicates.

DNA is moved into each future daughter cell and cross wall forms.

Cell divides into two cells.

Cells separate.

Daughter cells

Figure 10.1 Replication of bacterial cells by binary fission, depicting a generation time.

Generation time can be measured when the cells are dividing at a constant rate, or during **synchronous growth** in a laboratory. To understand when this occurs, one can study the growth curve of organisms inoculated into a fresh broth medium. If plate counts are made from the growing culture, the four growth phases of lag, log, stationary, and death can be seen (**figure 10.2**).

In the **lag phase**, the cells synthesize necessary enzymes and other cellular components needed for growth. The cells then grow as rapidly as conditions permit in the **log phase**; when sufficient nutrients are no longer available or when toxic products accumulate, the cells enter the **stationary phase**. In the stationary phase, the number of viable cells neither increases nor decreases. This is then followed by the **death phase** in which the cells die at a steady rate. The cells are growing at a constant maximum rate for the particular environment only in the log phase.

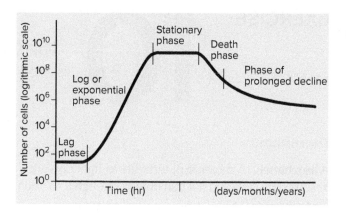

Figure 10.2 Growth curve showing the four phases of growth.

In this exercise, the generation time of *Escherichia coli* will be compared when grown at different temperatures. The cell density can be measured with a spectrophotometer or a Klett **colorimeter**, because the number of cells in the culture is directly proportional to the absorbance (**figure 10.3**). The **absorbance** (also called **optical density**) increases proportionately as **turbidity** (cloudiness) increases

Materials

Per team

Culture of the following:

Escherichia coli (trypticase soy broth cultures in log phase)

Trypticase soy broth, in a tube that can be read in a prewarmed spectrophotometer or Klett colorimeter (blank), 1

4.5 ml sterile trypticase soy broth already in a spectrophotometer tube or cuvette, 4

OR a flask with trypticase soy broth and 4 sterile spectrophotometer tubes or cuvettes

Test tube rack at 25°C or room temperature (used by some of the class)

Test tube rack at 30°C (used by some of the class)

Test tube rack at 37°C (used by some of the class)

Test tube rack at 45°C (used by some of the class)

1 ml pipet with bulb

Kimwipes for wiping off fingerprints

Pipet discard container

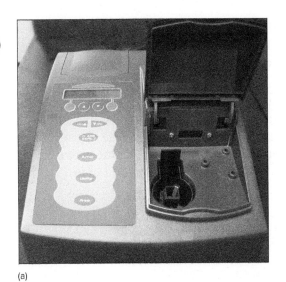
(a)

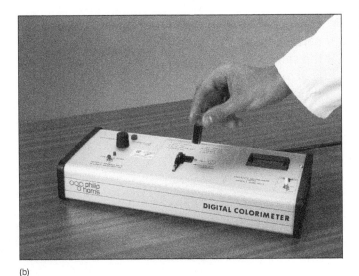
(b)

Figure 10.3 (*a*) Spectrophotometer. (*b*) Klett colorimeter. (*a*) Courtesy of Anna Oller, University of Central Missouri. (*b*) Martyn F. Chillmaid/Science Source.

from the multiplication of the bacteria. Spectrophotometer readings of the cultures incubated at various temperatures will be taken every 20 minutes for 80 minutes. The results are then plotted, and the generation time is determined. As with all experiments, consistency is important in order to draw conclusions.

PROCEDURE

1. Turn on the spectrophotometer or colorimeter to let it warm up, if not already prewarmed.

2. Properly label your broth tube so you know which one(s) is yours.

3. If provided a flask of trypticase soy broth, aseptically pipet 4.5 ml of sterile broth into a spectrophotometer tube. The broth amount can be scaled to fit the tube or cuvette capacity (9 ml, etc.).

4. Using a sterile pipet, aseptically transfer 0.5 ml of a log phase *E. coli* culture to 4.5 ml of trypticase soy broth to make a 1/10 dilution. The broth should be turbid enough to be read at the low end of the scale (0.1 O.D. or a Klett reading of about 50). If you start with an O.D. that is too high, your last readings will reach the part of the scale that is not accurate (an O.D. of about 0.4 or about 200 on the Klett). Discard the pipet into the proper container. Disinfect the tube if any *E. coli* drops ended up on the outside of the tube.

5. With a wavelength of 600 nm, make sure the warmed-up spectrophotometer is set to absorbance.

6. Wipe off any liquid and fingerprints from the uninoculated tube of trypticase soy broth (termed a *blank*), and set the spectrophotometer reading to zero. You will need to measure the blank broth tube and set the spectrophotometer to zero for each reading you take. Place the blank tube into the spectrophotometer in the same orientation each time so glass aberrations do not affect the readings. Your instructor will give specific directions.

7. Wipe off fingerprints from the *E. coli* inoculated tubes before taking a reading. Take a reading of the culture and record it in the Report table as time zero (0).

8. Place the tube in the rack for the assigned temperature as quickly as possible because cooling slows bacterial growth.

9. Read the O.D. of the culture every 20 minutes for 80 minutes. Record the exact time of the reading so the data can be plotted correctly.

10. Record your data—the time and O.D. readings—in your manual and on the blackboard. Be sure to label the *x* and *y* axes. Time is plotted on the linear horizontal *x* axis and the O.D. reading is plotted on the logarithmic vertical *y* axis.

11. Plot the data on semi-log graph paper (page 81). Semi-log paper is designed to convert numbers or data in to $\log_{10}$ as they are plotted on the *y* axis. The same results would be obtained by plotting the $\log_{10}$ of each of the data points on regular graph

paper, but semi-log paper simplifies this by permitting you to plot raw data and obtain the same line. Time is plotted on the linear horizontal x axis. **Draw a straight best-fit line through the data points.** The cells are growing logarithmically; therefore, the data should generate a straight line on semi-log paper (**figure 10.4**).

12. Plot the data using a different color pen from the other temperature settings by compiling the class data.

13. Determine the generation time for *Escherichia coli* at each temperature. This can be done by arbitrarily selecting a point on the line and noting the O.D. Find the point on the line where this number has doubled. The time between these two points is the generation time.

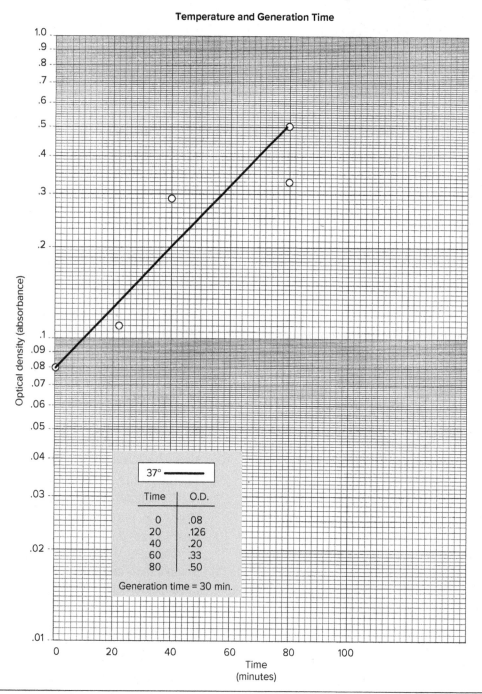

Figure 10.4 Growth curve of cells growing in log phase at 37°C.

EXERCISE

10

Laboratory Report: The Effect of Incubation Temperature on Generation Time ASM 8

Results

Your Assigned Temperature

_____

Temperature Increments: _____ minutes   O.D. = _____ nm

	Time 25°C O.D.	Time 30°C O.D.	Time 37°C O.D.	Time 45°C O.D.
0				
1				
2				
3				
4				
5				

Generation Time = _____   _____   _____   _____

Class Average   _____   _____   _____   _____

Questions

1. Which organism is growing faster: one with a short generation time or one with a longer generation time?

2. Why is it important to keep culture at the correct incubation temperature when measuring generation times?

3. If you didn't have semi-log paper, how could you plot the data on a piece of graph paper that would have resulted in a straight line?

4. If the growth of two cultures, one slower than the other, were plotted on semi-log paper, which would have the steeper slope?

5. What is the relationship between the density of the cells and the O.D.?

6. Where would you place the human population on the growth curve?

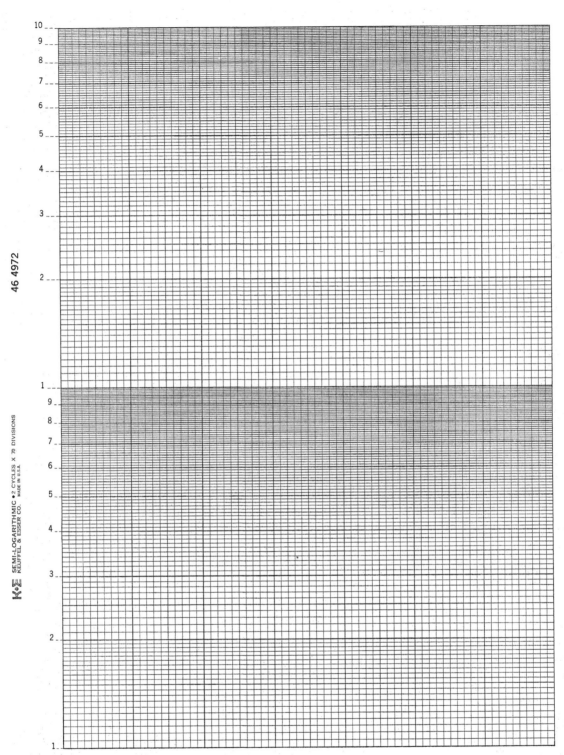

46 4972

SEMI-LOGARITHMIC • 2 CYCLES X 70 DIVISIONS
KEUFFEL & ESSER CO. MADE IN U.S.A.

Source: K & E Semi Logarithmic. 2 Cycles x 70 Divisions, Keuffel & Esser Co. Made in U.S.A.

INTRODUCTION to Control of Microbial Growth

For many microbiologists, control of microbial growth means *maximization* of microbial growth—such as growing baker's yeast for cooking, or growing microorganisms in order to harvest antibiotics. To others, such as physicians and allied members of the medical profession, control means *minimization* of microbial growth—such as using heat and ultraviolet light to destroy microorganisms in growth media, and those on gloves and masks. It can also imply the use of antiseptics and antibiotics to inhibit or destroy microorganisms present on external or internal body parts. Minimizing microbial growth in canned foods is also important in preventing food poisoning. A physicist named Denis Papin invented a steam digester in 1679, which was essentially a pressure cooker.

Historically, Louis Pasteur (1822–1895) contributed to both areas. In his early research, he discovered that beer and wine making required yeast growth for the fermentation process to occur. Later, he demonstrated that a sterile broth infusion in a swan-necked flask showed no turbidity due to microbial growth (**figure I.5.1**) and that upon tilting the flask, the sterile infusion became readily contaminated. The swan-necked flask experiment was both classical and monumental in that it helped resolve a debate of more than 150 years over the possible origin of microorganisms by spontaneous generation (abiogenesis).

The debate was finally squelched by John Tyndall (1820–1893), a physicist, who established an important fact overlooked by Pasteur—some bacteria in hay infusions existed in two forms: a **vegetative form**, readily susceptible to death by boiling of the hay infusion, and a resting form, now known as an endospore, which could survive boiling. With this knowledge, Tyndall developed a physical method of sterilization in 1860 that we now describe as **tyndallization**, whereby both vegetative cells and endospores are destroyed when the infusion is boiled for about 20 minutes on three consecutive days, with cooling between days. In 1879 a microbiologist named Charles Chamberland invented a steam sterilizer, which was the precursor to the autoclave. Tyndallization is a somewhat

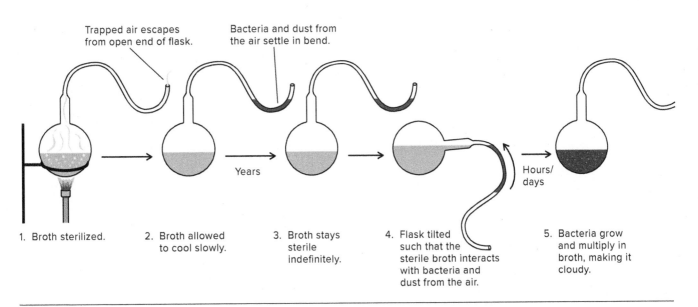

Trapped air escapes from open end of flask.

Bacteria and dust from the air settle in bend.

Years

Hours/days

1. Broth sterilized.

2. Broth allowed to cool slowly.

3. Broth stays sterile indefinitely.

4. Flask tilted such that the sterile broth interacts with bacteria and dust from the air.

5. Bacteria grow and multiply in broth, making it cloudy.

Figure I.5.1 Pasteur's experiment with the swan-necked flask. If the flask remains upright, no microbial growth occurs (1–3). If the microorganisms trapped in the neck reach the sterile liquid, they grow (4 and 5).

lengthy process and can be used to sterilize plant seeds or chemical nutrients degraded by the higher autoclaving temperatures.

At about this same time, chemical disinfectants aiding compound bone fracture healing were introduced by John Lister, an English surgeon, who was also impressed with Pasteur's findings. Lister heard that carbolic acid (phenol) could be used to treat sewage in Carlisle; it prevented odors from farmlands irrigated with sewage, and it destroyed intestinal parasites in cattle fed on such pastures.

The control of microbial growth has many applications. In microbiology, control methods are used to obtain pure cultures, as well as sterilize culture media, antibiotics, bandages, and instruments. In other areas, control methods facilitate plant and mammalian tissue or cell cultures, and molecular applications such as PCR to proceed without contaminating microorganisms. Control methods also ensure fermented beverages and foods can be safely consumed.

Exercises 7, 9, and 10 emphasized maximizing microbial growth. In this section, the exercises focus on minimizing or eliminating microbial growth by heat, ultraviolet light, osmotic pressure, pH, antiseptics, and antibiotics.

. . . we are too much accustomed to attribute to a single cause that which is the product of several, and the majority of our controversies come from that.

Von Liebig

EXERCISE 11

Moist and Dry Heat Sterilization: Thermal Death Point and Thermal Death Time ASM 8

Definitions

Autoclave. A device that uses steam under pressure to sterilize materials, conventionally performed at 121°C for 15 minutes at 15 psi. Destroys endospores.

Decimal reduction time (D value). The time required to kill 90% of a microbial population under specific conditions.

Incineration. A dry heat method used to sterilize inoculating loops and medical wastes at 850°C.

High-temperature-short-time (HTST) pasteurization. A brief heat treatment that reduces spoilage organisms and disease-causing microbes. Does not inactivate endospores. It is commonly used for milk products and is usually performed at 72°C for 15 seconds.

Thermal death point (TDP). The lowest temperature at which all microbes in a culture are killed after a given time.

Thermal death time (TDT). The minimal time necessary to kill all microbes in a culture held at a given temperature.

Thermoduric bacteria. Microbes able to survive conventional high-temperature-short-time pasteurization of 72°C for 15 seconds.

Thermophile. An organism with an optimum growth temperature between 45°C and 70°C.

Tyndallization. The process of using repeated cycles of heating and incubation to kill endospore-forming bacteria.

Ultra-high-temperature (UHT) pasteurization. A moist heat method performed at 140°C for 2 seconds.

Objectives ASM 8 & 10

1. Describe physical sterilization methods requiring either moist heat (autoclaving, boiling, and pasteurization) or dry heat.
2. Explain the susceptibility of different bacteria to the lethal effect of moist heat—thermal death point (TDP) and thermal death time (TDT).
3. Appropriately use laboratory equipment commonly used for physical sterilization of moist

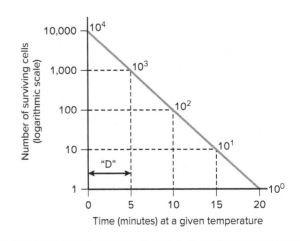

Figure 11.1 The decimal reduction time, or D value, is the time required to kill 90% of a population under specified conditions.

and dry materials: the steam autoclave and the **dry heat** oven.

Pre-lab Questions

1. Name the genus and species of a bacterium that is a thermophile.
2. What is the difference between thermal death point and thermal death time?
3. Name the proper specifications of routine autoclaving.

Getting Started

Physical methods used to kill microorganisms include various heat methods and types of radiation, such as ultraviolet light. Ultraviolet light can be used to kill microbes on biological safety cabinet surfaces. Filtration removes microorganisms from antibiotics, water, air in operating rooms and clean rooms used by pharmaceutical and electronic companies. Requiring employees to wear sterile gloves, masks, and clothing helps control air convection of microorganisms present on an employee's skin and hair.

In this exercise, heat sterilization effects are studied because heat is commonly used to sterilize hospital, laboratory, industrial, and food items. When high heat is applied, most microbes arc killed, whereas cold temperatures inhibit but do not kill bacteria. The sensitivity of a microorganism to heat is affected by its environment and genetics. Environmental factors include incubation temperature, chemicals and nutrients available in the growth medium, and the age and concentration of cells in the growth medium. Genetically, **thermophiles** like *Bacillus stearothermophilus* and *Thermus aquaticus* grow at higher temperatures than others. In addition, *Bacillus* and *Clostridium* species can also produce heat-resistant endospores.

Heat is applied in either a dry or moist form and is shown in **table 11.1**. Keep in mind typical sterilization times and temperatures are 2 hours at 165°C for dry heat compared to 15 minutes at 121°C at 15 psi (pounds per square inch of pressure) for moist heat. The mode of action is the same for both.

Autoclaving is the most commonly used moist heat sterilization method in laboratories (color plate 15). To ensure an autoclave run worked or killed microbes properly, autoclave tape embedded with stripes of *Bacillus stearothermophilus* spores are added to items (color plate 16a). The autoclave tape provides a visual method of identifying when an autoclave run fails and might need maintenance. It is important to leave some space around items placed in an autoclave for proper air circulation. If an autoclave is filled too full, the air cannot circulate properly in the vessel to kill the contaminating microbes. In order to confirm an autoclave is working properly, ampules (color plate 16b) containing *Bacillus stearothermophilus* spores are placed in the autoclave and run with items. The difference from the tape is that the ampule contains broth with a pH indicator. Once autoclaved, the ampule is then placed in a 55°C incubator for 48 hours. If the autoclave worked properly and killed the endospores, the ampule remains a purple color. However, if some spores were not killed,

Table 11.1   Dry and Moist Heat Methods

Method/Item	Mechanism	Temperature/Time	Kill Endospores?	Examples
Dry Heat	Dehydrates microbial cells and denatures essential enzymes			
1. Hot air (oven)		160°C for 2–4 hr	Yes	Glassware like Petri dishes, powdered cosmetics, metal, oils
2. Incineration Bunsen Burners		1800°C for sec.	Yes	Metal inoculation loops/ needles
Microincinerators		800°C for a few sec.	Yes	Metal inoculation loops/ needles
Furnace		6500°C for a few sec.	Yes	Medical waste
Moist heat	Denatures essential enzymes or proteins			
1. Autoclaving		121°C, 15 min., 15 psi	Yes	Culture media, cotton gowns, metals, some plastics
2. Boiling		100°C for 30 min.	No	Water, plastics, clothing, silverware
3. Pasteurization		66°C for 30 min.	No	Juices and dairy products
Flash/High temperature short time (HTST)		72°C for 15 sec.	No	Milk and dairy products
Ultra-high temperature (UHT)		140°C for a few sec.	Yes	Shelf-stable dairy items (creamers)
4. Tyndallization (no pressure)		3 days at 100°C for 30 min.	Yes	Plant seeds

then they become vegetative cells that ferment the sugar supplied and turn the ampule yellow.

Pasteurization, named for Louis Pasteur, is a moist heat process used to reduce the number of **thermoduric** bacteria in beverages like milk, beer, and wine. High-temperature-short-time (HTST) pasteurization heats beverages at 72°C for 15 seconds. **Ultra-high-temperature (UHT) pasteurization** heats dairy products to 140°C for 2 seconds, so items like creamers are stable at room temperature after pasteurization. This allows for a longer shelf-life of the product. However, the lower temperature of HTST helps preserve the flavor of foods that can be degraded by UHT.

Methods for determining the heat sensitivity of a microorganism are the **thermal death point (TDP)** and the **thermal death time (TDT)**. The TDP is defined as the lowest <u>temperature</u> necessary to kill all of the microorganisms present in a culture in 10 minutes. The TDT is defined as the minimal <u>time</u> necessary to kill all of the microorganisms present in a culture held at a given temperature. Used in the food and canning industry, the **decimal reduction time (D value)** is the time required to kill 90% of microbes at a specific temperature (**figure 11.1**). These general principles are commonly used when establishing sterility requirements for various processes, such as for canning.

Materials

Cultures of the following:

24-hours 37°C *Escherichia coli* cultures in 5 ml trypticase soy broth, 1 per team or student

Spore suspension in 5 ml of sterile distilled water of a 4–5 day 37°C nutrient agar slant culture of *Bacillus subtilis*, 1 per team or student

5 ml of trypticase soy broth, 5 tubes per team or student

A community water bath, heat block, hot plate, or incubator, one held at each temperature (37°C, 55°C, 80°C, and 100°C) with a thermometer added

If community items are not used:

Two large beakers, one as a water bath and the other as a reservoir of boiling water (per student)

One thermometer (per student)

A vortex apparatus (if available)

PROCEDURE

First Session: Determination of Thermal Death Point and Thermal Death Time

Note: A procedural culture and heating distribution scheme for eight students is listed in table 11.2. For this scheme, each student receives one broth culture, either *Escherichia coli* or *Bacillus subtilis*, to be heated at one of the assigned temperatures (37°C, 55°C, 80°C, or 100°C). Students will label trypticase soy broth tubes for 0 (C), 10, 20, 30, and 40 minutes.

Note: As an alternative, instead of each student preparing his or her own water bath, the instructor can provide community water baths, heat blocks, or incubators, preset at 37°C, 55°C, 80°C, and 100°C. If community items are used, proceed to step 5 after step 1.

1. Suspend your assigned culture by gently rolling the tube between your hands, followed by aseptically transferring a loopful to a fresh tube of broth (label "0 time control").

2. Fill the beaker to be used as a water bath approximately half full with water, sufficient to immerse the broth culture without dampening the test tube cap.

3. Place your tube of broth culture in the water bath or incubator along with an open tube of uninoculated broth in which a thermometer has been inserted for monitoring the water bath temperature.

4. Place the water bath and contents on the hot plate and heat almost to the assigned temperature. One or two degrees before the assigned temperature is reached, remove the water bath from the heat source and place it on your benchtop. The temperature of the water bath can now be maintained by periodically stirring in small amounts of boiling water obtained from the community water bath.

   Note: Students with the 100°C assignment may wish to keep their water bath on the heat source, providing the water can be controlled at a low boil.

5. After 10 minutes of heating, resuspend the broth culture either by vortexing or by gently

Table 11.2 Culture and Heating Temperature Assignments (Eight Students)

Bacterial Culture Assignment		Temperature Assignment			
		37°C	55°C	80°C	100°C
Escherichia coli		1	2	3	4
Bacillus subtilis		5	6	7	8

tapping the outside of the tube. Aseptically transfer a loopful to a fresh tube of broth (label "10 min.").

6. Repeat step 5 after 20, 30, and 40 minutes.

7. Properly label (your name, microbe, temperature, time, media type, date, and lab section) all five tubes and incubate them in the 37°C incubator for 48 hours.

Second Session

After 48 hours, examine the broth tubes for the presence or absence of turbidity (growth). Write your results in the appropriate place in the table on the blackboard of your classroom. When all the results are entered, transfer them to **table 11.2** of the Laboratory Report.

Demonstration of the Steam-Jacketed Autoclave and Dry Heat Oven

The Steam-Jacketed Autoclave

Note: As the instructor demonstrates the special features of the autoclave, follow the diagram in **figure 11.2**. Note the control valves—their function and method of adjustment (exhaust valve, chamber valve, and safety valve); the steam pressure gauge; the thermometer; and the door to the chamber. Also make note of the following special precautions necessary for proper sterilization. Ensure the gasket and the door are seated and closed tightly. If the door is not tight up against to the gasket, steam can escape causing the autoclave run to fail.

1. Material sensitivity. Certain types of materials, like powders, oils, or petroleum jelly, cannot be steam sterilized because they are water repellent. Instead, dry heat is used. Some materials are destroyed (milk) or changed (medications) by the standard autoclave temperature of 121°C. In such instances, the autoclave may be operated at a lower pressure and temperature for a longer period of time. A heat-sensitive fluid material like antibiotics can be sterilized by filtration. Keep in mind that the smaller the filter pores, the slower the rate of filtration.

2. Proper preparation of materials. Steam must directly contact all materials to be sterilized. Therefore, media container closures such as caps with air passages, loosened screw cap lids, and aluminum foil (heavy grade) are used. A small piece of autoclave tape or an indicator vial should be included in each autoclave run to ensure sterility.

3. Proper loading of supplies and autoclave settings. There must be ample space between packs and containers so that the steam can circulate. The proper exhaust speed needs set. Fast exhaust drops the steam pressure quickly and should be used for dry items like pipet tips or metal instruments. Slow exhaust should be used for liquids.

4. The air must be completely removed from the chamber before it is replaced by steam. This is done automatically by the autoclave.

5. Proper temperature. Autoclaving at a pressure of 15 psi achieves a temperature of 121°C (250°F) at sea level. If the temperature gauge does not register 121°C, either cold air is trapped in the autoclave, the gauge is going out, or there is a problem with the heating mechanism.

6. Adequate sterilization time. After the chamber temperature reaches 121°C, additional time is required for the heat to penetrate the material. The larger the size of individual containers and packs, the more time required. Time must be adjusted to the individual load size.

7. Completion of the autoclaving process. The fast exhaust causes a rapid reduction in steam

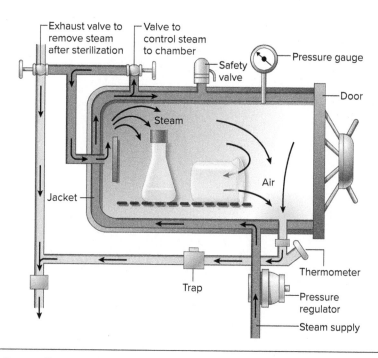

Figure 11.2 Autoclave. Steam first travels in an enclosed layer, or jacket, surrounding the chamber. It then enters the autoclave, displacing the air downward and out through a port in the bottom to the chamber.

pressure, which will cause liquids to boil over, or even glass tubes to explode. Drop the steam pressure gradually as cooling occurs. If the door is opened before the pressure has been released, the steam will rush out of the autoclave and can cause severe burns. After the autoclave has cooled, the water may need to be drained out of the chamber.

8. To ensure items were completely autoclaved, check to see that the autoclave tape turned black. Alternatively, an autoclaved indicator vial may need to be incubated at 55°C for 48 hours.

The Dry Heat Oven

The hot air, or dry heat, oven is used for drying and sterilizing glassware, and for various experiments.

When using the dry heat oven, the following guidelines are important:

1. Materials suitable for sterilization include oil, petroleum jelly (Vaseline), metal containers, and dry, clean glassware.

2. A convection oven (air circulates) takes about half the sterilization time of a static air oven.

3. Proper loading is necessary to ensure air circulates to touch inside surfaces. For example, syringes must be separated from plungers so that all surfaces are exposed to circulating air.

4. Items should be cleaned. The presence of extraneous materials like protein delays the process and may allow bacteria to survive inside the material.

5. In part 2 of the Laboratory Report, prepare a list of materials for your class that are sterilized in the autoclave. For each one, indicate the standard temperature, pressure, and time used for sterilization.

Do the same for materials sterilized in the dry heat oven, indicating the temperature, time, and why it would be sterilized in a hot oven over an autoclave temperature, time, and reasons for sterilizing there.

EXERCISE

11

Laboratory Report: Moist and Dry Heat
Sterilization: Thermal Death Point
and Thermal Death Time ASM 8

Results

1. Determination of thermal death point and thermal death time:

Table 11.2   Bacterial Growth at Assigned Temperatures and Times

Culture	37°C					55°C					80°C					100°C				
	C	10	20	30	40	C	10	20	30	40	C	10	20	30	40	C	10	20	30	40
Escherichia coli																				
Bacillus subtilis																				

Note: C = Control; 10, 20, 30, 40 = minutes of heating the inoculated culture at the assigned temperature; use a + sign for growth and a − sign for no growth.

   a. Determine the thermal death time (TDT) in minutes and the thermal death point (TDP) in °C for each culture.

	TDT	**TDP**
Escherichia coli	_____	_____
Bacillus subtilis	_____	_____

2. Laboratory materials sterilized with moist heat (autoclave) and dry heat (hot air oven) (See Procedure for criteria):

   a. Moist heat   b. Dry heat

3. Plot a graph of the thermal death time.

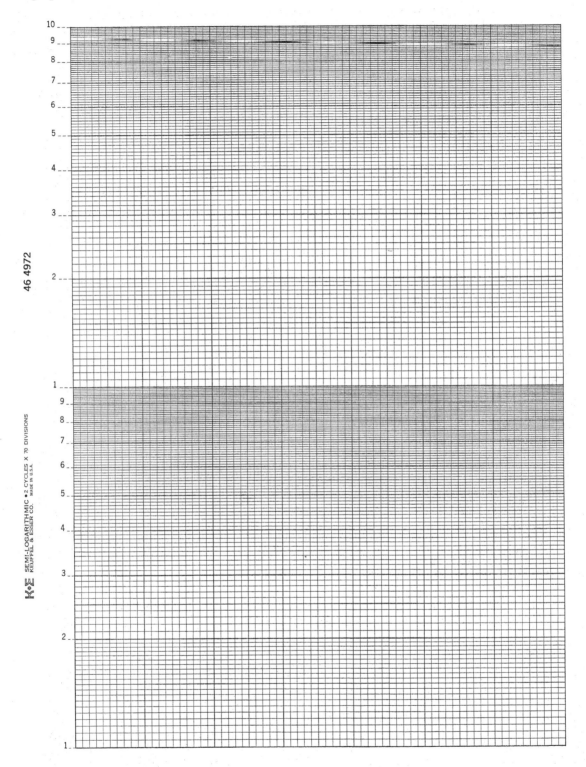

1. Discuss similarities and differences between determining thermal death point and thermal death time.

2. How would you set up an experiment to determine to the minute the TDT of *E. coli*? Begin with the data you have already collected.

3. A practical question related to thermal death time (TDT) relates to a serious outbreak of *E. coli* infection when people ate insufficiently grilled hamburgers. How would you set up an experiment to determine the TDT of a solid such as a hamburger? Assume that the thermal death point (TDP) is 67.2°C (157°F), the temperature required by many states to cook hamburger on an open grill. What factors would you consider in setting up such an experiment?

4. What is the most expedient method for sterilizing a heat-sensitive liquid that contains an endospore-forming bacterium?

5. List one or more items that are best sterilized by the following processes:
   a. Membrane filtration

   b. Ultraviolet light

   c. Dry heat

   d. Moist heat

6. Was the *Bacillus subtilis* culture sterilized after 40 minutes of boiling? If not, what is necessary to **ensure** sterility by boiling?

12

Effects of Ultraviolet Light

Definitions

Endospores. A kind of resting cell characteristic of species in a limited number of bacterial genera like *Bacillus* and *Clostridium*. Endospores are highly resistant to heat, radiation, and disinfectants.

Light repair. The process by which bacteria use the energy of visible light to repair UV damage to their DNA.

Thymine. A pyrimidine; one of the four nucleotide subunits of DNA.

Thymine dimer. A molecule formed when two adjacent thymine molecules on the same strand of DNA are joined together through covalent bonds.

Ultraviolet (UV) light. Electromagnetic radiation with wavelengths between 175 and 350 nm, invisible to humans.

Visible spectrum. Wavelengths of radiation between 400 and 800 nm that humans can see.

Objectives

1. Explain the effects of ultraviolet (UV) light irradiation on bacteria.

2. Explain how to work safely with ultraviolet light.

Pre-lab Questions

1. Why are safety glasses required in this exercise when turning on the UV light?

2. Which component of the cell is affected by UV light?

3. Why was *Bacillus* chosen as one of the bacterial cultures to irradiate?

Getting Started

Ultraviolet (UV) light is a type of non-ionization radiation, and is the component of sunlight that is responsible for sunburns and makes posters and shirts glow in the dark. It can cause mutations in the DNA of microorganisms. It consists of very short wavelengths of radiation (175–350 nm) located just below blue light (450–500 nm) in the **visible spectrum** (**figure 12.1**). UV light is separated into three categories: UV-A (320–400 nm), UV-B (290–320 nm), and UV-C (100–290 nm). UV-A is commonly referred to as black light, and UV-B is what commonly causes sun burns, and ultimately skin cancers. UV-C is very dangerous to cells, and is being used as a disinfectant for food, air, water, and surfaces. Sunscreens protect against UV-A and UV-B skin damage, and sunglasses protect the eyes from developing cataracts.

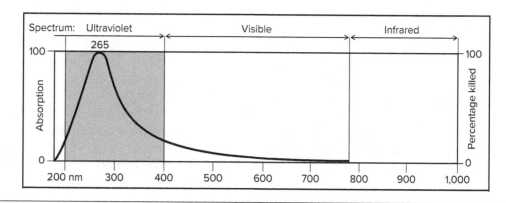

Figure 12.1 The light spectrum and germicidal activity of radiant energy.

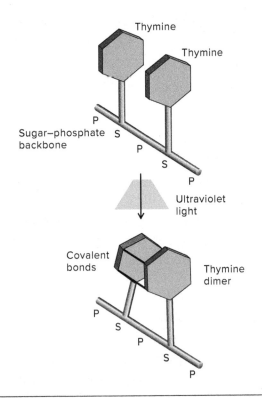

Figure 12.2 Thymine dimer formation. Covalent bonds form between adjacent thymine molecules on the same strand of DNA. This distorts the shape of the DNA and prevents replication of the changed DNA.

The actual mechanism of DNA mutations is the formation of **thymine dimers** (**figure 12.2**). Two adjacent **thymines** on a DNA strand bind to each other. When the DNA is replicated, an incorrect base pair is frequently incorporated into the newly synthesized strand. This may cause mutations and, if there is sufficient radiation, ultimately, the death of the cell.

UV light does not penetrate surfaces and will not go through ordinary plastic or glass. It is only useful for killing organisms *on* surfaces and in the air. Sometimes, UV lights are turned on in operating rooms, biological safety cabinets (hoods), and other places where airborne bacterial contamination is a problem. Because UV light quickly damages eyes, these lights are turned on only when no one is in the irradiated area.

Bacteria vary in their sensitivity to UV light. In this exercise, the sensitivity of *Bacillus* **endospores** will be compared with non-spore-forming cells. You will also irradiate a mixed culture, such as organisms in soil or hamburger, to compare the resistance of different organisms. Some cells severely damaged by UV that would not survive in the dark may undergo **light repair** and recover.

Materials

Cultures of the following:

  Bacillus species on slants producing endospores

  Escherichia coli in trypticase soy broth

  Raw hamburger or soil mixed with sterile water

Trypticase soy agar plates, 4 per team

UV lamp with shielding. A fluorescent bulb with a wavelength of 240–280 nm is ideal.

Sterile swabs, 4

Safety glasses for use with the UV lamp

A dark incubator to store plates after UV exposure

A 3 × 5 card to place over plates

PROCEDURE

First Session

Safety Precautions: (1) The area for UV irradiation should be in an isolated part of the laboratory. (2) Wear safety glasses as a precautionary measure when working in this area. (3) Never look at the UV light after turning it on because it could result in severe eye damage. Skin damage is also a slight possibility.

1. Your instructor will assign one of the cultures to use. For the assigned culture, one plate will be subjected to 0 minutes (Control), a second plate to 1 minute of UV, the third plate to 3 minutes of UV, and the fourth plate to 10 minutes of UV.

2. Label each plate with your name, microbe, media type, temperature, date, UV time, and lab section.

3. Aseptically dip a sterile swab into the assigned bacterial suspension and swab an agar plate in three or four directions to create a lawn of growth, as shown in **figure 12.3** and on the left of **figure 12.4**. Discard the swab in the proper disposal container.

4. Repeat the procedure using a new sterile swab dipped into the assigned bacterial suspension and inoculate a second plate with a lawn of

Figure 12.3 A lawn of growth. Using a swab dipped into a bacterial suspension, rub the swab back and forth across the agar surface in touching lines. Turn the plate in at least three directions to ensure the entire plate surface is covered with bacteria. See exercise 14.

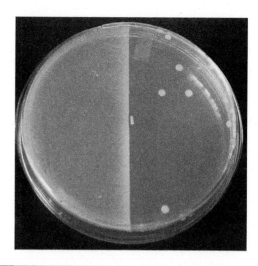

Figure 12.4 A plate demonstrating a lawn of growth, covered on the left and exposed to UV light on the right. Courtesy of Anna Oller, University of Central Missouri.

growth. Discard the swab into the proper disposal container.

5. Repeat the procedure with the remainder of the plates. Discard the swabs into the proper disposal container.

6. Place the plates under a UV lamp propped up about 20 cm from the bench surface. Open the Petri dish(es) and partially cover each plate with the lid or the 3' × 5' card. The part of the plate protected by the lid will be the control because UV radiation does not penetrate most plastic or paper.

7. Put on your safety glasses and turn on the UV light. Expose the plates to UV radiation for 1, 3, and 10 minutes. (Your instructor may assign different exposure times to some teams.)

8. Turn off the UV light after 1, 3, and 10 minutes. Cover the plates, invert them, and place them in a covered container or incubator. Incubate at 37°C for 48 hours.

9. The control plate incubated at 0 minutes ensures the bacterium grew and provides a growth reference if the UV was not performed properly or the lids were all left off.

Second Session

Observe the plates and record your findings in the Laboratory Report. Compare the growth of the covered side of each plate to the times you were assigned. You are looking for changes in colony number, colony color, colony size and shape, etc.

EXERCISE

12

Laboratory Report: Effects of Ultraviolet Light

Results

1. Record your observations for controls and treated sides of Petri dishes exposed to UV light. Record your results on the blackboard.

	Control	**1 min.**	**3 min.**	**10 min.**	**Class Exposures**	**1 min.**	**3 min.**	**10 min.**
Bacillus endospores								
Escherichia coli								
Soil or meat suspension								

Score irradiated growth:

++++ no different than non-irradiated control

+++ slightly different than non-irradiated control

++ significantly less growth than non-irradiated control

+ less than 10 colonies

2. My assigned bacterial culture was _____

3. Which organisms were most resistant to UV light? _____

   Least resistant?_____

Questions

1. Why can't you use UV light to sterilize microbiological media, e.g., agar or broth prepared in glassware?

2. How does UV light cause mutations?

3. What is the result of increasing the time of exposure to UV light?

4. Give a possible reason some organisms in the soil (or meat) were able to grow after exposure to UV light but others were not.

5. Frequently, organisms isolated from the environment are pigmented, whereas organisms isolated from the intestine or other protected places are not. Can you provide an explanation for this?

6. Mutations can lead to cancer in animals. Explain why people who tan using tanning beds have a higher incidence of skin cancer than those who do not use tanning beds.

13

Effects of Osmotic Pressure and pH
on Microbial Growth ASM 8

Definitions

Acidophile. An organism that prefers to grow at an acidic pH, usually below 5.5.

Alkalophile. An organism that prefers to grow at a basic pH, or above 8.5.

Halophile. An organism that prefers or requires a high salt medium to grow.

Hypertonic solution. A solution containing a greater osmotic pressure, causing a cell to shrink away from its cell wall.

Hypotonic solution. A solution containing a lower osmotic pressure so a cell will absorb water and potentially rupture.

Isotonic solution. A solution containing the same osmotic pressure.

Neutrophile. An organism that prefers to grow at a neutral pH, usually between 6 and 8.

Osmotic pressure. The pressure exerted by water on a membrane as a result of a difference in the concentration of solute molecules on each side of the membrane.

Plasmolysis. Dehydration and shrinkage of the cytoplasm from the cell wall as a result of water diffusing out of a cell.

Plasmoptysis. The rupturing of the cell wall from absorbing excess water from the external environment.

Saccharophile. An organism capable of growing in an environment containing high sugar concentrations.

Semipermeable membrane. A membrane, such as the cytoplasmic membrane of the cell, that permits passage of some materials but not others. Passage usually depends on the size of the molecule.

Solute. A dissolved substance in a solution.

Objectives ASM 8

1. Describe osmotic pressure and show how it can be used to inhibit growth of less osmotolerant microbes while allowing more osmotolerant microbes to grow, although often at a considerably slower growth rate.

2. Interpret if microorganisms either require or grow more in an environment containing high concentrations of salt (halophilic) or sugar (saccharophilic).

Pre-lab Questions

1. If a cell is placed in a hypotonic solution, what will happen to the cell?

2. What is the proper term given to a microbe that is able to grow at a high pH?

3. What growth differences between the lowest and highest salt concentrations do you hypothesize you will see for *Escherichia coli*?ASM 7

Getting Started

Osmosis, which is derived from the Greek word "alter," refers to the process of flow or diffusion that takes place through a **semipermeable membrane**. In a living cell, the cytoplasmic membrane, located adjacent to the inside of the cell wall, represents such a membrane (**figure 13.1a**).

Both the cytoplasmic membrane and the cell wall help prevent the cell from either bursting (**plasmoptysis**) (figure 13.1a) or collapsing (**plasmolysis**) (figure 13.1b) due to either entry or removal of water from the cell, respectively. The **solute** (dissolved substances) concentrations both inside and outside the cell determine which, if any, process happens. A cell will remain intact if the solute concentration inside of the cell is the same as outside of the cell, termed **isotonic**. If the solute concentration outside the cell is less than the

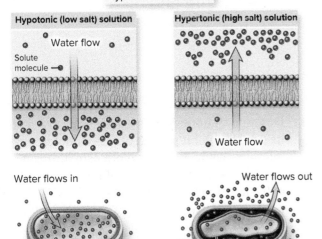

Water flows across a membrane toward the hypertonic solution.

Hypotonic (low salt) solution

Water flow
Solute molecule —

Hypertonic (high salt) solution

Water flow

Water flows in

Cytoplasmic membrane is forced against cell wall.

Water flows out

Cytoplasmic membrane pulls away from cell wall.

(a) In low salt water, water molecules move into the cell to equalize the number of salt particles on both sides of the membrane. The cell membrane is pushed against the cell wall (plasmoptysis).

(b) In high salt water, the water molecules leave the cell so the cytoplasmic membrane shrinks away from the cell wall (plasmolysis).

Figure 13.1 Movement of water into and out of cells. (*a*) Low and (*b*) high salt-containing solutions. The cytoplasmic membrane is semipermeable and only allows water molecules to pass through freely.

solute concentration inside the cell, an inward **osmotic pressure** occurs, and water enters the cell in an attempt to equalize the solute concentration on either side of the cytoplasmic membrane. A cell will absorb water and sometimes burst (plasmoptysis) if the solute concentration outside of the cell is low, termed **hypotonic**. However, this rarely occurs, due to the rigidity of the cell wall. In the reverse phenomenon, cell shrinkage is followed by cell lysis (plasmolysis), and occurs when the cell is placed in a more concentrated, or **hypertonic**, solution. The cell can die if too much water is removed (figure 13.1*b*).

When placed in an isotonic solution, some cells recover, but many genera die once the external osmotic pressure exceeds their limitations. This concept is the basis for food preservation methods—the use of high salt concentrations (for cheese and pickle brine) and high sugar concentrations (in honey and jams).

In general, fungi (yeasts and molds) are much more resistant to high external solute concentrations than bacteria, which is one reason fungi can grow in or on jelly, cheese, and fruit. The genus *Halobacterium* in *Archaea* is a bacterial exception found in nature that grows in water containing a high salt content, like Great Salt Lake in Utah; and the microbe *Nesterenkonia halobia* in the family *Micrococcaceae* is sometimes found in nature growing on highly salted (25%–32%) protein products, like fish and animal hides. These bacteria also produce a red pigment.

Some salt-tolerant *Staphylococcus* strains are capable of growing in salt concentrations greater than 10% (w/v), enabling them to grow on skin surfaces. Salt-loving bacteria are termed **halophiles** and are unique in that, like all Archaea, they lack muramic acid as a bonding agent in their cell walls. Instead, their cell walls are believed to contain sodium and potassium ions. These ions help confer cell wall rigidity, perhaps helping explain the reason they require such high salt concentrations for good growth.

Yeasts and molds are able to grow in high sugar (50%–75%) concentrations, and are termed **saccharophiles**. Sometimes fungi can also grow in salt concentrations between 25% and 30%. Some of these yeast and fungus genera are *Saccharomyces*, *Aspergillus*, and *Penicillium*.

Besides water, the pH of food and other substances becomes an important factor in bacterial, yeast, and mold growth. The pH is a measure of acidity or alkalinity of a substance or solution. Most bacteria grow at a neutral pH of 7, and are termed **neutrophiles**. The pH terms are shown in **table 13.1**. However, a few bacteria, like *Lactobacillus acidophilus,* possess the ability to lower the pH of its environment, such as in yogurt, and are termed **acidophiles**. *Helicobacter pylori* in particular grows well in the acidity of the stomach. Finally, the

Table 13.1 Microbiology Terms Associated with pH

pH	Prokaryotes that live in pH extremes maintain a near-neutral internal pH by pumping protons out of or into a cell.
Neutrophile	Multiples in the range of pH 5 to 8. Ex. *Escherichia coli, Staphylococcus aureus*
Acidophile	Grows optimally at a pH below 5.5. Ex. *Acetobacter aceti, Lactobacillus acidophilus*
Alkalophile	Grows optimally at a pH above 8.5. Ex. *Saccharomyces cerevisiae, Penicillium chrysogenum*

term **alkalophile** refers to organisms that prefer a basic environment, such as yeasts and molds.

In this exercise, you will examine the ability of some of the previously mentioned halophiles and saccharophiles to grow on the surface of TYEG (trypticase yeast extract glucose) agar plates containing increasing concentrations of salt and sugar. *Escherichia coli* serves as a salt-sensitive Gram-negative rod control. You will examine any cell morphology changes with increasing salt and sugar concentrations. You will also have an opportunity to examine the pH effects on microbial growth.

Keep in mind that the halophiles and saccharophiles require an increased lab time and a decreased growth rate. In some ways, their growth curve (exercise 10) parallels being grown at a temperature below their optimal growth temperature. For example, halobacteria have a doubling time of 7 hours and halococci double at 15 hours.

Materials

Per team

Cultures of the following:

Trypticase yeast extract glucose (TYEG) agar slants of *Escherichia coli, Micrococcus luteus* (or *Staphylococcus aureus*), and *Saccharomyces cerevisiae*

American type culture collection (ATCC) medium 213 of *Halobacterium salinarium*

TYEG agar plates containing 0, 0.5%, 5%, 10%, and 20% salt (NaCl), 5

TYEG agar plates containing 0, 0.5%, 10%, and 25% sucrose, 4

pH tubes of 2, 5, 7, and 9, one tube of each

PROCEDURE

First Session

Note: One student of the team can inoculate the five trypticase yeast extract glucose agar plates containing 0, 0.5%, 5%, 10%, and 20% NaCl, whereas the other student inoculates the four plates

containing 0, 0.5%, 5%, and 10% sucrose. Your instructor may assign you one of the microbes for the pH experiment.

1. First, use a marking pen to divide the bottom of the nine plates into quadrants and label each quadrant. Label the bottom of each plate with the salt or sugar concentrations, your name, date, temperature, and lab section.

2. Label the pH tubes with the pH, your name, date, temperature, and lab section.

3. Using aseptic technique, remove a loopful from a culture and streak the appropriate quadrant of your plate in a straight line approximately 1 inch long. Then reflame your loop, cool it for a few seconds, and make a series of cross streaks approximately 1/2 inch long in order to initiate single colonies for use in studying colonial morphology (**figure 13.2**). Repeat the inoculation procedure used for the first culture in the appropriate quadrant of the remaining agar plates.

4. Repeat the inoculation procedure for the remaining three test organisms.

5. Invert the plates and incubate for 48 hours at 30°C,

6. Using aseptic technique, remove a loopful of organism and inoculate each labeled pH tube. Incubate the tubes of *E. coli, M. luteus* (or *S. aureus*) for 24 hours at 37°C, *S. cerevisiae* for 24–48 hours at 25°C, and *H. salinarum* at 35°C for a week.

7. Observe the cultures periodically (up to 1 week or more if necessary) for growth.

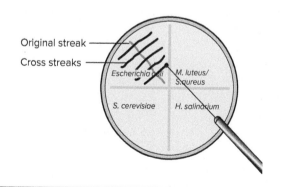

Figure 13.2 Streaking procedure for isolation of single colonies.

Second Session

1. Examine your plates for the presence (+) or absence (−) of growth. For growth, use one to three + signs (+ = little, but isolated colonies, + + = some isolated colonies, and + + + = lawn of growth). Enter results in table 13.2 (various salt concentrations) and **table 13.3** (various sugar concentrations) of the Laboratory Report. Use semiquantitative growth of 1+, 2+, 3+, and 4+ to record the pH results in **table 13.4**.

2. Compare the colonial growth characteristics of cultures grown on agar media containing increasing salt and sugar concentrations. Make notes of any marked changes in colony color, colony size (in mm), and colony texture: dull or glistening, rough or smooth, and flat or raised. Record your findings in the Laboratory Report.

3. Prepare wet mounts of bacteria and yeast colonies showing marked changes in visual appearance with increasing salt and sugar concentrations. Examine bacteria and yeasts with the high–dry objective lens. Look for plasmolyzed cells and other changes, such as cell form and size. Prepare drawings of any such changes in the Laboratory Report.

EXERCISE

13

Laboratory Report: Effects of Osmotic Pressure and pH on Microbial Growth ASM 8

Results

1. Examine Petri dish cultures for the presence (+) or absence (−) of growth in the presence of increasing salt (table 13.2) and sugar concentrations (table 13.3). Use a 1+ for little growth with isolated colonies, 2+ for some isolated colonies, and 3+ for lawn of growth to describe the amount of growth.

Table 13.2   Presence or Absence of Growth on TYEG Agar Plates Containing NaCl and Incubated for 48 Hours–1 Week

Culture	NaCl Concentration (%)									
	0		0.5		5		10		20	
	48 hr	1 wk	48 hr	1 wk	48 hr	1 wk	48 hr	1 wk	48 hr	1 wk
Escherichia coli										
Halobacterium salinarium										
Micrococcus luteus										
Saccharomyces cerevisiae										
Staphylococcus aureus										

Table 13.3   Presence or Absence of Growth on TYEG Agar Plates Containing Sucrose and Incubated for 48 Hours–1 Week

Culture	Sucrose Concentration (%)							
	0		0.5		5		10	
	48 hr	1 wk	48 hr	1 wk	48 hr	1 wk	48 hr	1 wk
Escherichia coli								
Halobacterium salinarium								
Micrococcus luteus								
Saccharomyces cerevisiae								
Staphylococcus aureus								

Table 13.4   Microbial Growth in pH Tubes

Culture	pH											
	2			5			7			9		
	24 hr	48 hr	1 wk	24 hr	48 hr	1 wk	24 hr	48 hr	1 wk	24 hr	48 hr	1 wk
Escherichia coli												
Halobacterium salinarium												
Micrococcus luteus												
Saccharomyces cerevisiae												
Staphylococcus aureus												

Record growth as 1+, 2+, 3+, 4+

2. Comparison of the colonial growth characteristics of cultures inoculated on agar media containing increasing amounts of salt or sugar.

Microbe	NaCl %	Growth	Colony Color	Colony Size	Colony Texture
	0.5				
	5				
	10				
	20				

Microbe	Sucrose %	Growth	Colony Color	Colony Size	Colony Texture
	0				
	0.5				
	5				
	10				

Wet Mounts	Cell Appearance	Plasmolysis or Plasmoptysis?
Microbe:		
Microbe:		

Microbe	NaCl %	Growth	Colony Color	Colony Size	Colony Texture
	0.5				
	5				
	10				
	20				

Microbe	Sucrose %	Growth	Colony Color	Colony Size	Colony Texture
	0				
	0.5				
	5				
	10				

Wet Mounts	Cell Appearance	Plasmolysis or Plasmoptysis?
Microbe:		
Microbe:		

Questions

1. From your studies, which organism(s) tolerate salt best? _____ Least? _____
2. Which organism(s) tolerate sugar best? _____ Least? _____
3. Which organism(s) grew best at pH 7? _____ Least? _____
4. Compare bacteria and yeast with respect to salt tolerance. Bear in mind both colonial and cellular appearance in formulating your answers.

5. Compare bacteria and yeast with respect to sugar tolerance. Bear in mind both colonial and cellular appearance in formulating your answer.

6. What evidence did you find of a *nutritional requirement* for salt or sugar in the growth medium?

7. Matching
   Each answer may be used one or more times or not at all.
   a. *Halobacterium*   _____ osmosensitive   _____ Neutrophile
   b. *Saccharomyces*   _____ long generation time   _____ Acidophile
   c. *Escherichia coli*   _____ saccharophilic   _____ Alkalophile
   d. *Micrococcus/Staphylococcus* _____ osmotolerant

8. Matching
   Choose the best answer. Each answer may be used one or more times or not at all.
   a. Plasmolysis   _____ isotonic solution
   b. Plasmoptysis   _____ hypotonic solution
   c. Normal cell growth   _____ hypertonic solution
                            _____ swelling of cells

EXERCISE

14

Effects of Antibiotics

Definitions

Antibiotic. A chemical produced by certain bacteria and fungi that can inhibit or kill other organisms.

Antimetabolite. A substance that inhibits the utilization of a metabolite necessary for growth.

Coenzyme. A heat-stable, non-protein organic compound that assists some enzymes, acting as a loosely bound carrier of small molecules or electrons. One example is flavin adenine dinucleotide (FAD).

Cofactor. A non-protein compound required for the activity of some enzymes. Examples include magnesium and zinc.

Competitive inhibition. Enzymatic inhibition that occurs when the inhibitor competes with the normal substrate for binding to the active (catalytic) site on the enzyme.

Essential metabolic pathway. A metabolic pathway necessary for growth; if inhibited, the organism usually dies. The Krebs cycle and Embden–Meyerhof pathway are examples.

Essential metabolite. A chemical necessary for proper growth.

Pathogen. A disease-causing organism or agent.

Resistant. When referring to antibiotics, a bacterium not killed by the chemical compound.

Susceptible. When referring to antibiotics, a bacterium killed by the chemical compound.

Synergism. When two drugs work better together than when used individually.

Zone of inhibition. The region around an antibiotic disc where bacteria cannot grow. Measured in millimeters (mm).

Objectives

1. Describe the origin and use of antibiotics.
2. Describe the mechanism of action of the first synthetic antibiotics, the sulfa drugs.

3. Determine the inhibitory activity of antibiotics on bacteria using a modified Kirby–Bauer test.

Pre-lab Questions

1. What is the definition of *antibiotic*?
2. What was the first synthetic antibiotic? What is the mechanism or mode of action?
3. How can you determine if an antibiotic is an effective therapeutic agent for a patient?

Getting Started

Chemicals such as antibiotics have been in existence a long time. However, the therapeutic properties of antibiotics simply were not recognized until Alexander Fleming's discovery of penicillin in the 1930s.

By definition, **antibiotics** are chemicals produced and secreted by bacteria and molds, which inhibit growth or kill other microorganisms. While some antibiotics used today are naturally produced, like some forms of penicillin, the majority of antibiotics prescribed are either semisynthetic or completely synthetic. Semisynthetic antibiotics are made by chemically modifying a naturally made antibiotic, such as amoxicillin, a penicillin derivative. Ciprofloxacin and sulfa drugs represent synthetic antibiotics; they are completely synthesized in a laboratory.

The first chemicals used as antibiotics were the synthetic sulfa drugs, such as sulfanilamide, which originated from the azo group of dyes (**figure 14.1**). The inhibitory action of sulfa drugs is termed **competitive inhibition**, in which the sulfanilamide acts as an **antimetabolite**. Antimetabolites are molecules that resemble essential cellular compounds but block cellular function when incorporated into the cell's metabolism. Sometimes a **coenzyme** like flavin adenine dinucleotide (FAD) is required. Some enzymes require a **cofactor** like magnesium or zinc in order to function properly. The sulfanilamide replaces para-aminobenzoic acid (PABA) (figure 14.1), an

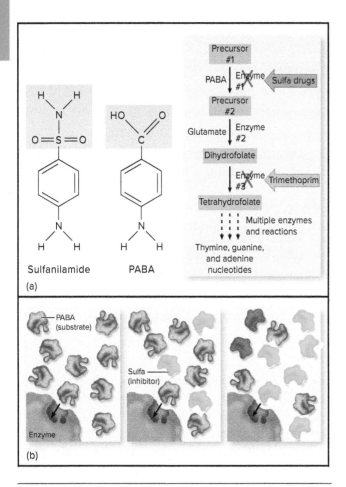

Figure 14.1 (*a*) Structures of sulfanilamide (sulfa drug) and of para-aminobenzoic acid (PABA). The portions of the molecules that differ from each other are shaded. The pathway on the right shows where the drugs sulfa and trimethoprim inhibit the metabolic pathway and formation of nucleotides. (*b*) Reversible competitive inhibition of folic acid synthesis by sulfa drugs. The higher the concentration of sulfa drug molecules relative to PABA, the more likely that the enzyme will bind to the sulfa drug, and the greater the inhibition of folic acid synthesis.

essential metabolite required to generate folic acid. In the presence of sulfanilamide, bacterial cells cannot synthesize folic acid and cannot grow. Thus, folic acid is an **essential metabolite**, and an **essential metabolic pathway** is inhibited. This drug is selectively toxic to bacterial cells and not human cells because it inhibits a bacterial pathway necessary to generate folic acid. Humans obtain folic acid from dietary sources such as dark green leafy vegetables, beans, and meat.

Antibiotic resistance has become a public health issue and threatens people's lives. We use the term **resistant** if the bacterium is not killed and **susceptible** if the bacterium is killed. The microbes that fall in the middle are termed intermediate. Antibiotics, generally, only kill bacteria and a few select parasites. Thus, prescribing an antibiotic for a viral infection will not effectively inhibit the virus from replicating, and it makes the bacteria on and in the body more resistant to antibiotics over time. **Figure 14.2** shows how a few microbes are resistant to antibiotics by random genetic selection. Then, when an antibiotic is taken, many of the bacterial cells are susceptible and killed. Usually the infection symptoms of pain, redness, heat, etc. subside, but a few drug-resistant cells remain. The next time an infection presents, more of the bacterial population will be antibiotic resistant.

Antibiotics usually kill or inhibit bacterial cells by working on a specific bacterial cell structure and usually inhibit one of five main cell structures: the cell wall, cell membrane, DNA or RNA, protein synthesis, and metabolic pathways. The antibiotic, penicillin, inhibits the bacterial cell wall from being formed and repaired, whereas polymyxin antibiotics cause the bacterium to lose cell membrane permeability. Ciprofloxacin inhibits the enzyme gyrase, thus preventing DNA formation. Tetracycline and streptomycin both inhibit the 30S ribosomal subunit used in protein synthesis, whereas erythromycin inhibits the 50S ribosomal subunit. Sulfonamides and trimethoprim inhibit metabolism by competing with PABA for folic acid production.

The bacterial cell mechanisms responsible for antibiotic resistance are shown in **figure 14.3**. These mechanisms include bacterial efflux pumps that can recognize and pump the antibiotic back out of the

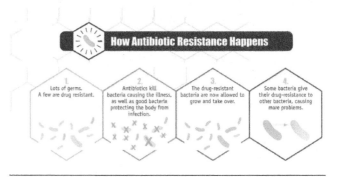

Figure 14.2 How bacteria become resistant to antibiotics. **Source:** CDC/Melissa Brower

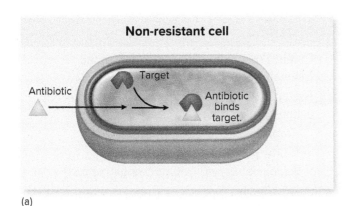

Non-resistant cell

Target

Antibiotic

Antibiotic binds target.

(a)

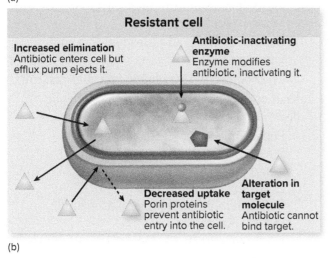

Resistant cell

Increased elimination
Antibiotic enters cell but efflux pump ejects it.

Antibiotic-inactivating enzyme
Enzyme modifies antibiotic, inactivating it.

Decreased uptake
Porin proteins prevent antibiotic entry into the cell.

Alteration in target molecule
Antibiotic cannot bind target.

(b)

Figure 14.3 Antibiotic resistant mechanisms.

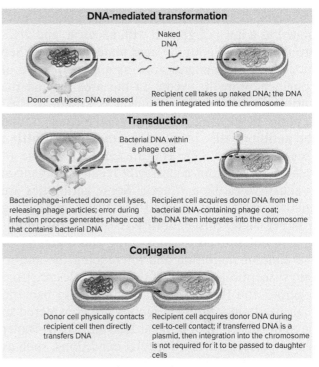

DNA-mediated transformation

Naked DNA

Donor cell lyses; DNA released

Recipient cell takes up naked DNA; the DNA is then integrated into the chromosome

Transduction

Bacterial DNA within a phage coat

Bacteriophage-infected donor cell lyses, releasing phage particles; error during infection process generates phage coat that contains bacterial DNA

Recipient cell acquires donor DNA from the bacterial DNA-containing phage coat; the DNA then integrates into the chromosome

Conjugation

Donor cell physically contacts recipient cell then directly transfers DNA

Recipient cell acquires donor DNA during cell-to-cell contact; if transferred DNA is a plasmid, then integration into the chromosome is not required for it to be passed to daughter cells

Figure 14.4 Horizontal gene transfer mechanisms.

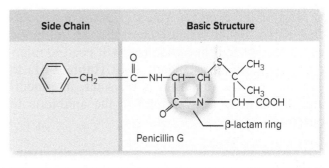

Side Chain   Basic Structure

β-lactam ring

Penicillin G

Figure 14.5 The structure of the antibiotic penicillin, showing the β-lactam ring that the enzyme β-lactamase degrades.

bacterial cell, thus rendering the antibiotic ineffective. A bacterium can also contain many efflux pumps, thus making the bacterium multidrug resistant. The bacteria can also produce enzymes like beta-lactamase, which can inactivate the antibiotic penicillin (**figures 14.3** and **14.5**). Some bacteria have porins in the cell wall that change the cell permeability and block the antibiotic entry into the cell. Some bacterial cells alter their metabolic pathways so the drug no longer inhibits the cell. Finally, some bacteria alter the target so antibiotics will not bind and be able to interrupt cell activities. There are actually many other drug-resistant mechanisms being continually discovered.

Scientists now know some of the resistance mechanisms that can be transferred between bacteria (**figure 14.4**). When one cell degrades, we now know that living cells can uptake the dead cell's DNA and transfer resistance that way. Plasmids, which are extra pieces of genetic material that are not required for replication, can also contain resistance genes to many

drugs. We also know that viruses can transfer material via transduction, and the sex pilus used in conjugation can transfer the DNA to another cell. Further, bacterial cells mutate quickly and pass the mutation onto daughter or progeny cells via vertical transmission.

The Kirby–Bauer disc diffusion method is still used in many laboratories to test the efficacy of antibiotics against **pathogens**. In this assay, filter paper discs are impregnated with standardized antibiotic concentrations similar to concentrations dissolved in human tissues. The agar plate is inoculated as a lawn of growth with bacteria, and discs are then placed onto the agar plate surface. When incubated, the bacteria grow as a smooth lawn of confluent growth, except in the area around the antibiotic disc, if the organism is killed, or susceptible to the drug. The clear zone formed around the antibiotic disc is called the **zone of inhibition** (see color plate 17). The presence of a zone of inhibition does not necessarily mean an organism is sensitive to the antibiotic. Some antibiotics are smaller molecules and diffuse faster, producing a larger zone. Susceptibility to a specific antibiotic is determined by measuring the diameter of the zone of inhibition and referencing the antibiotic susceptibility chart. If two antibiotics are used together and they are more inhibitory to bacteria than used alone, the antibiotics are considered to be **synergistic**. In a clinical lab, the Kirby–Bauer test is performed using special conditions, such as 18–24 hours cultures, a controlled inoculum size, and short incubation periods. Because these conditions are difficult to achieve within the time frame of most classrooms, a modified Kirby–Bauer test will be performed.

In this lab exercise, you will perform a modified Kirby–Bauer test to determine the antibiotic sensitivity of four bacteria. You will need to reference the antibiotic standard chart (**table 14.1**) to determine if the organism is resistant, intermediate, or sensitive (susceptible) to the antibiotic in question.

Materials

Cultures (per team of 2–4 students) of the following:

Bacteria (24-hour 37°C trypticase soy broth cultures)

   Staphylococcus epidermidis (a Gram-positive coccus)

   Escherichia coli (a Gram-negative rod)

   Alcaligenes faecalis or *Pseudomonas aeruginosa* (a non-fermenting Gram-negative rod)

   Bacillus subtilis (an endospore-forming Gram-positive rod)

   Mycobacterium smegmatis (an acid-fast rod)

Vials or dispensers of the following antibiotic discs:

   Penicillin (P-10), 10 U;
   Streptomycin (S-10), 10 μg;
   Tetracycline (Te-30), 30 μg;
   Ciprofloxacin (CIP-5), 5 μg;
   Sulfanilamide (or another sulfonamide) (SXT), 300 μg;
   Erythromycin (E-15), 15 μg

Mueller–Hinton agar, 5 large plates *or* use 8 regular size plates

Sterile cotton swabs, 5

Small forceps, 1 per student

Ruler divided in mm

Table 14.1 Chart Containing Zone Diameter Interpretive Standards for Determining the Sensitivity of Bacteria to Antimicrobial Agents (Zone Diameters in mm)

Antimicrobial Agent	Disc Content	Susceptible	Intermediate	Resistant
Penicillin G **(P-10)** when testing	10 U			
Staphylococci and				
Streptococci		≥29	None	≤28
when testing *Enterococci*		≥15	None	≤14
Streptomycin **(S-10)**	10 μg	≥15	12–14	≤11
Tetracycline **(Te-30)**	30 μg	≥19	15–18	≤14
Ciprofloxacin **(CIP-5)**	5 μg	≥21	16–20	≤15
Trimethoprim & sulfamethoxazole* **(SXT)**	1.25 + 23.75 μg	≥16	11–15	≤10
Erythromycin **(E-15)**	15 μg	≥23	14–22	≤13

***Note:** Sulfamethoxazole (a sulfa drug) is usually given in combination with trimethoprim.

PROCEDURE

First Session: Filter Paper Disc Technique for Antibiotics

Note: *Pseudomonas aeruginosa* is a pathogen. Handle with extra care, if used.

1. Divide the five broth cultures among team members so that each student sets up at least one susceptibility test.

2. With a permanent pen, divide the bottom of four large Mueller–Hinton plates into six pie-shaped sections (**figure 14.6a**). If using standard size Petri plates, divide the plates into three sections, placing three of the six antibiotics on each plate.

3. Record the codes of the six antibiotic discs on the bottom side of the four plates, one code for each section (**table 14.2** shows code designations).

4. Label the bottom of each plate with the name of the respective bacterium, your name, media type (MH), 25°C, date, and lab section.

5. Using aseptic technique, streak the first broth culture with a sterile swab in horizontal and vertical directions and around the edge of the agar plate (as shown in **figure 14.6b**) to create a lawn of growth. The remaining three cultures should be streaked on separate plates in a similar manner.

6. Heat sterilize forceps by briefly touching them to the flame of the Bunsen burner. Be careful to avoid burns. Prolonged flaming can cause heat to rapidly travel down the forceps to your fingers and can cause burns. Air cool the forceps and remove an antibiotic disc from the container. Antibiotic discs are held in the spring-loaded cartridge by a plastic cap. You may need to gently push on one side of the disc and then pull from the opposite side to grasp the disc from the cartridge. Place the disc gently, in the center of one of the pie-shaped sections of the agar plate (figure 14.6a). Tap gently with forceps to fix in position.

7. Continue placing the remaining five antibiotic discs in the same manner.

   Note: Be sure to flame the forceps after placing each disc because it is possible to contaminate stock antibiotics with resistant organisms.

   Note: If a disc dispenser is used, follow the manufacturer's instructions.

8. Repeat steps 5–7 with the remaining three cultures.

9. Invert and incubate the plates at 25°C for 24–48 hours.

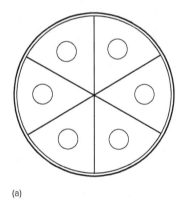

(a)

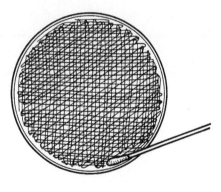

(b)

Figure 14.6 Antibiotic susceptibility test. (*a*) The underside of a Mueller–Hinton agar plate showing the marking of sections and the arrangement for placement of antibiotic discs on the agar surface. (*b*) Procedure for streaking an agar plate in three or more directions with a swab inoculum in order to achieve a uniform lawn of growth (see figures 6.4 and 12.3). The swab is moved (*1*) up and down over the entire agar surface, (*2*) left to right over the surface, and (*3*) diagonally over the surface. Make sure no gaps can be seen.

Second Session: Filter Paper Disc Technique for Antibiotics

1. On the bottom of the *Staphylococcus epidermidis* plate, and with a ruler calibrated in millimeter, determine the diameter of the clear zone (zone of inhibition) surrounding each disc. Measure from one edge of the circle to the other (diameter) as shown in **figure 14.7**. Repeat this process with the other bacterial plates. In addition, make note of any small colonies present in the clear zone of inhibition surrounding each antibiotic disc as they may be resistant mutants. If small colonies are within the clear zone, you should measure the area truly void of any growth (from colony to colony directly across, or diagonally, from the disc).

Note: It may be necessary to illuminate plates in order to define the boundary of the clear zone.

2. Record your findings in table 14.2 of the Laboratory Report.

3. Compare your results where possible with table 14.1 and indicate in table 14.2 the susceptibility of your test cultures (*when possible*) to the antibiotics as resistant (R), intermediate (I), or susceptible (S).

Note: Your answers may not agree exactly with those in table 14.1 because this is a modified Kirby–Bauer test.

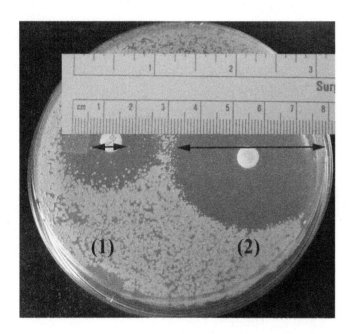

Figure 14.7 How to measure a zone of inhibition in millimeter. 1. Place a ruler at one end of the zone of inhibition. 2. Determine the entire diameter from one end of a clear zone until the other, and record the results in mm. The zone on the left (*1*) shows some resistant mutants so only the area from mutant to mutant can be measured. On the right, (*2*) the zone of inhibition distance is 45 mm. Then the zone can be looked up on the chart to determine resistant, intermediate, or susceptible. Courtesy of Anna Oller, University of Central Missouri.

EXERCISE

14

Laboratory Report: Effects of Antibiotics

Results

Table 14.2 Antibiotic Susceptibility (Modified Kirby–Bauer Test)

Antibiotic	Pen (P-10)		Str (S-10)		Tet (Te-30)		Cip (CIP-5)		Sul (SXT)		Ery (E-15)	
	Zone (mm)	S/I/R	Zone (mm)	S/I/R	Zone (mm)	S/I/R	Zone (mm)	S/I/R	Zone (mm)	S/I/R	Zone (mm)	S/I/R
S. epidermidis												
E. coli												
M. smegmatis												
A. faecalis or P. aeruginosa												
Bacillus subtilis												

R = Resistant, I = Intermediate, and S = Susceptible.

Questions

1. What relationship did you find, if any, between the Gram-staining reaction of a microorganism and its susceptibility to an antibiotic?

2. Did you observe any small colonies growing in the zone of inhibition around the antibiotic disc? Explain how this might occur.

3. When performing the Kirby–Bauer procedure to evaluate a clinical specimen, why is it important to have a standardized inoculum size, 18- to 24-hour cultures, and a short incubation period?

4. Compare your antibiotic sensitivity results for the antibiotic penicillin with *E. coli* and *M. smegmatis*. Why is there a difference in sensitivity?

5. You test a clinical specimen of *Streptococcus* with the antibiotics penicillin and tetracycline. The zones of inhibition for penicillin and tetracycline are both 20 mm. What does this tell you about comparing zone sizes and the use of the antibiotic zone diameter chart? Is this organism sensitive to both of these antibiotics?

EXERCISE

15

Effects of Antiseptics and Disinfectants

Definitions

Antiseptic. A non-toxic disinfectant used on the skin.

Bacteriocidal. A substance that kills bacteria.

Bacteriostatic. A substance that prevents the growth of bacteria for a time but does not kill them.

Contact time. The length of time a chemical needs to be in contact with a microorganism in order to kill the microbe.

Disinfectant. A chemical that destroys many, but not all microbes. It kills the vegetative cells of bacteria, but usually will not inactivate endospores.

Endospore. A kind of resting cell characteristic of species in a limited number of bacterial genera like *Bacillus* and *Clostridium*. Endospores are highly resistant to heat, radiation, and disinfectants.

Enveloped virus. Viruses that have a lipid bilayer surrounding their nucleocapsid. HIV and Herpes viruses are enveloped viruses.

High-level disinfectant. A chemical that destroys viruses and vegetative microorganisms, but does not inactivate endospores.

Intermediate-level disinfectant. A chemical that destroys vegetative microorganisms, fungi, and most viruses. They can kill *Mycobacteria* species but do not inactivate endospores.

Low-level disinfectants. A chemical that kills vegetative bacteria, fungi, and enveloped viruses. They do not kill *Mycobacterium* species, endospores, or non-enveloped viruses.

Minimum inhibitory concentration. The lowest concentration of a chemical or drug that inhibits bacteria.

Non-enveloped virus. A virus that does not have a lipid envelope and is often called a naked virus. Hepatitis A is a non-enveloped virus.

Sterilization. The destruction or removal of all microbes, including endospores, through physical or chemical means.

Objectives

1. Describe the differences between an antiseptic and disinfectant.

2. Describe the factors that influence the effectiveness of chemicals and choose the appropriate chemical for a particular situation.

3. Determine the effectiveness of a chemical to inhibit microbial growth.

Pre-lab Questions

1. In what situation would you use an antiseptic? In what situation would you use a disinfectant?

2. List one example of how chemicals inhibit bacterial growth.

3. In this lab, how will you determine if an organism is sensitive to a chemical?

Getting Started

Joseph Lister, the pioneer of aseptic surgery, recognized early in the 1860s that the use of chemicals, such as carbolic acid (phenol), dramatically reduced the incidence of post-surgical infections. Since this time, the use of antiseptics and disinfectants has been essential for controlling infections in humans and animals in order to limit the spread of disease.

Antiseptics (**table 15.1**) are chemicals considered safe for use on the skin and mucous membranes. Examples of antiseptics include wiping an alcohol swab on the skin before a fingerstick to help prevent an infection, and gargling with Listerine®.

The same chemicals that are antiseptics can also serve as disinfectants. **Disinfectants** kill the vegetative cells of many microbes, but usually do not inactivate endospores or the mycolic acid of acid-fast microbes. Disinfectants have been divided into three levels depending on if the chemical is in contact with the skin, as well as the **contact time** needed to kill microorganisms. **Low-level disinfectants** such as alcohols have a contact time of less than 1 minute. **Intermediate-level disinfectants** include phenolics and iodophors and have a contact time greater than 1 minute. Finally, **high-level disinfectants** such as glutaraldehyde, hydrogen peroxide, and sodium hypochlorite require a contact time between 12 and 30 minutes. Hydrogen peroxide and sodium hypochlorite can corrode metals so consideration should be given to the item being disinfected.

Many different chemicals are used as antiseptics and disinfectants in the medical, veterinary, and industrial fields. Dimethyl benzyl ammonium saccharinate is the active ingredient in Lysol, which is effective at killing many microorganisms including *Mycobacterium* species and some viruses. Zinc and silver are compounds found in household products and medical items.

Zinc oxide is the key ingredient in sunscreens and diaper rash creams. Silver in the form of silver nanoparticles is used in topical antiseptic gels, bandages (Curad® brand), and surgical mesh dressings. Samsung has created a line of products called Silver Nano®, which includes silver nanoparticles integrated in the surface of household appliances. Triclosan is a bisphenol historically used in toothpastes, toys, and soaps. The FDA banned the use of triclosan in antiseptic washes like soaps, effective in 2017.

Choosing the proper chemical for a proper situation is very complex. Factors that influence a compound's effectiveness are the environmental pH, temperature, the concentration of the chemical, the presence of organic matter (such as blood), and the type of organism targeted. For example, some chemicals work optimally at a neutral pH and the product's effectiveness is reduced in an acidic or alkaline environment. A temperature increase can increase the rate at which microbes are destroyed. Organic material will inhibit the activity of

Table 15.1   Chemicals Used to Control Microbial Growth in the Medical, Veterinary, and Industrial Fields

Chemical	Action	Use
Sodium hypochlorite (5%)	Disinfectant	External surfaces, such as tables, bedpans, and floors
Iodine (1% in 70% alcohol)	Disinfectant	External surfaces, such as tables
Iodophors (70 ppm avail. I_2)	Disinfectant	External surfaces, such as tables
Quaternary ammonium compounds	Disinfectant	External surfaces, such as tables and floors
Phenol (5%), (also known as carbolic acid)	Disinfectant	External surfaces, such as tables
Hexachlorophene (pHisoHex®)	Disinfectant	Presurgical hand washing
Formaldehyde (4%)	Disinfectant	Embalming and tissue preservation
Iodophors (12.5–25 ppm avail. I_2)	Disinfectant and Antiseptic	Food equipment
Alcohol, ethanol (70%)	Antiseptic	Skin
Iodine (tincture in alcohol with KI) (12.5–25 ppm avail. I_2)	Antiseptic	Skin
Organic mercury compounds (merthiolate, mercurochrome)	Antiseptic	Skin
Hydrogen peroxide (3%)	Antiseptic	Superficial skin infections
Potassium permanganate	Antiseptic	Skin fungal infections
Phenol (1.4%)	Antiseptic	Throat sprays
Zinc oxide paste	Antiseptic	Diaper rash
Zinc salts of fatty acids (Desenex®)	Antiseptic	Athlete's foot
Ethylene oxide gas (12%)	Sterilant	Linens, syringes, medical instruments
Formaldehyde (20% in 70% alcohol)	Sterilant	Tissue fixation
Glutaraldehyde (pH 7.5 or more)	Sterilant	Metal instruments and tissue fixation

hypochlorite, whereas phenolics remain effective in their presence. **Endospores, non-enveloped viruses**, and *Mycobacteria* are the most resistant to general-purpose disinfectants and often require high-level disinfectants or sterilants, such as ethylene oxide or glutaraldehyde, to destroy them. High-level sterilants, such as glutaraldehyde, are also very toxic.

Most chemicals inhibit microbial growth by altering the cell wall structure, damaging cytoplasmic membranes (quaternary ammonium compounds), or denaturing essential proteins (alcohols and heavy metals) such as enzymes. If the proteins or lipids that are part of the cell wall structure are compromised, the cell cannot retain its shape and cell lysis occurs when the microbe is placed in a hypotonic solution. Cytoplasmic membrane damage prevents nutrients and wastes from being transported across the membrane. Membranes that are part of the envelope of viruses are also sensitive to chemicals and, if altered, hinder the ability of the virus to bind to target cells. This is why **enveloped viruses** are more sensitive to chemicals than non-enveloped viruses. The protein ceases to function when it has been denatured. Silver and mercury disrupt disulfide bridges. Some chemicals also lower the surface tension, as seen in antibacterial soaps.

The concentration of various chemicals, including antibiotics and disinfectants, will have a minimum concentration that inhibits bacterial cells. A known concentration of a bacterium is added to broth tubes and the chemical of interest is added to the broth tubes in a serial dilution (color plate 18). The first tube in which no growth is seen is considered to be the **minimum inhibitory concentration**, or MIC.

In this lab, you will test the effectiveness of antiseptics and disinfectants on bacteria. Filter paper discs are soaked with test chemicals, then placed on the surface of TSA plates inoculated with a lawn of growth of one organism. The plates are incubated at 37°C for 24–48 hours and then evaluated for the presence of zones of inhibition, or clear zones, around the discs containing the test chemicals. To compare the effectiveness of the test chemicals toward different organisms, you will measure the diameter of the zones of inhibition in millimeters Students may be able to use their own disinfectants or cleaners to test.

Per team cultures of the following:

Bacteria (24-hour 37°C trypticase soy broth cultures)

Staphylococcus epidermidis (a Gram-positive coccus)

Escherichia coli (a Gram-negative rod)

Alcaligenes faecalis (non-fermenting Gram-negative rod)

Bacillus cereus endospores in saline (**Note:** best from 5-day-old culture)

Mycobacterium smegmatis (an acid-fast rod)

Beakers or large test tubes containing 5–10 ml aliquots of:

70% ethanol

3% hydrogen peroxide

Antiseptic mouthwash or throat spray containing phenol®

Disinfectant like Lysol®

Filter paper discs (Whatman #1) autoclaved on a dry cycle in a sterile container

Trypticase soy agar plates, 5

Sterile cotton swabs, 5

Small forceps, 1 per student

Ruler divided in millimeter

PROCEDURE

First Session: Filter Paper Disc Technique for Antiseptics and Disinfectants

1. With a marking pen, divide the underside of five trypticase soy agar plates into quadrants.

2. Record codes for the four antiseptics and disinfectants on the bottom sides of the five agar plates, one code for each quadrant: 70% ethanol—E; 3% hydrogen peroxide—HP; Oral Agent—O; and Lysol®—Ly.

3. Label the bottom of each plate with the bacterium, your name, TSA, date, 25°C, and lab section.

4. Suspend the *Staphylococcus epidermidis* culture, by gently swirling. Insert and moisten a sterile

swab into the broth tube, and press against the side of the tube to remove excess liquid. Next, roll the swab in three or four directions on the surface of the agar plate to create a lawn of growth as shown in figure 14.2. Discard swab in the appropriate waste container.

5. Repeat step 4 with the remaining cultures, adding one bacterium per plate.

6. Sterilize forceps by briefly flaming in the Bunsen burner. Air cool.

7. Using forceps, remove one of the sterile filter paper discs from the container and dip the disc about half way into the 70% ethanol. By dipping the disc half way into the solution, capillary action should wick the chemical all the way up the disc without having excess liquid run across the agar surface.

8. Place the disc into in the center of quadrant 1 of the dish labeled *Staphylococcus epidermidis* (**figure 15.1** and color plate 19). Tap the disc gently onto the agar surface. Flame the forceps.

9. Repeat steps 6–8 and add discs containing 70% ethanol to the center of quadrant 1 of the other three plates.

10. Repeat steps 6–8 for the remaining three compounds, using 3% hydrogen peroxide first, then the oral mouthwash or phenol, and lastly the Lysol®.

11. Invert the Petri dishes and incubate at 25°C for 24–48 hours.

Second Session: Filter Paper Disc Technique for Antiseptics and Disinfectants

1. Turn over the *Staphylococcus epidermidis* plate, and with a ruler calibrated in millimeter, determine the diameter of the clear zone (zone of inhibition) surrounding each disc. Repeat this process with the three other plates.

2. The presence of a zone of inhibition reflects the organism's susceptibility to that chemical. Record your results in **table 15.2** of the Laboratory Report.

Note: It may be necessary to illuminate the plate in order to define the clear zone boundary.

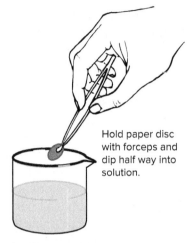

Hold paper disc with forceps and dip half way into solution.

(a) Antiseptic solution

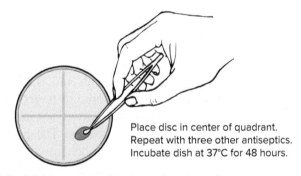

Place disc in center of quadrant. Repeat with three other antiseptics. Incubate dish at 37°C for 48 hours.

(b) Petri dish inoculated with a lawn of one bacterium

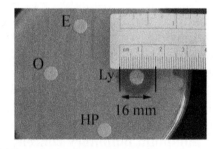

(c) Measure the clear zone of inhibition surrounding each disc

Figure 15.1 Filter paper disc technique for antiseptics and disinfectants. (c) Courtesy of Anna Oller, University of Central Missouri.

EXERCISE

15

Laboratory Report: Effects of Antiseptics and Disinfectants

Results

Zones of inhibition are measured in millimeters.

Table 15.2   Inhibitory Activity of Various Antiseptics and Disinfectants

Antiseptic or Disinfectant	Zone of Inhibition (mm)				
	S. epidermidis	B. cereus endospores	E. coli	A. faecalis	M. smegmatis
70% ethanol (E)					
3% hydrogen peroxide (HP)					
Oral agent or throat spray (O)					
Lysol® (Ly)					
Others:					

Questions

1. What general conclusions can you make from this study? What differences, if any, did you observe on your plates between antiseptics and disinfectant preparations?

2. What relationship did you find, if any, between the Gram-staining reaction of a microorganism and its susceptibility to antiseptics and disinfectants?

3. Were any of the organisms especially resistant to the chemicals tested? If so, which one(s)? Propose a reason for these results.

4. Based on your test results, which of the test chemicals is the most effective antiseptic overall? Which of the test chemicals is the most effective disinfectant overall?

5. When choosing a chemical for a particular application, what should be considered?

6. What type of chemical (list an example of a specific chemical) would you use to
   a. prepare the skin before an injection? _____
   b. disinfect lab benchtops before/after class? _____
   c. prepare the skin before surgery? _____
   d. disinfect drinking water? _____
   e. destroy *Bacillus anthracis* endospores contaminating a room? _____

INTRODUCTION to Microbial Culturing and Identification

It is easy to think of pathogens as deadly, vicious forces—especially when the diseases they cause kill or debilitate many young people or wipe out entire populations. The organisms, however, are simply growing in a favorable environment, which may actually be a human body.

If a microbe becomes too efficient at taking advantage of its host, the host dies and the organism dies with it. Thus, the most successful pathogens are those that live in balance with their hosts. When a new pathogen enters a population, the microorganism is often very virulent, but over time the pathogen often becomes less virulent. Often, the host's immune system adapts to the microorganism, so fewer signs and symptoms are present or mortality rates become lower. There is a selection toward less virulent pathogens and also a selection in the hosts for increased resistance.

Medical microbiology continues to offer challenges in medicine and in pathogenic microorganisms. These next exercises provide an introduction to culturing microbes, many of which are types of organisms frequently encountered in a clinical laboratory. Not only you will study the characteristics of the organisms, but you will also learn some strategies for isolating and identifying them. One strategy for growing microbes from a sample includes using basic types of agar, some of which are shown in **table I.16.1**. A strategy for growing microbes from samples containing many types of microorganisms uses selective and differential media. **Figure I.16.1** shows how the different media types influence the presence or absence of bacterial growth. In addition, these exercises will help you learn to differentiate between normal biota or flora found in various locations on and within the body and other microorganisms responsible for causing certain diseases or conditions. Thus, these exercises focus on the culturing of microorganisms and identification methods utilized therein.

Table I.16.1 Media Used to Cultivate Bacteria

Medium	Characteristic
Blood agar	Complex medium used routinely in clinical labs. Differential because colonies of hemolytic organisms are surrounded by a zone of red blood cell clearing. Not selective.
Chocolate agar	Complex medium used to culture fastidious bacteria, particularly those found in clinical specimens. Not selective or differential.
Glucose-salt agar	Chemically defined medium. Used in laboratory experiments to study nutritional requirements of bacteria. Not selective or differential.
MacConkey agar	Complex medium used to isolate Gram-negative rods that typically reside in the intestine. Selective because bile salts and dyes inhibit Gram-positive organisms and Gram-negative cocci. Differential because the pH indicator turns pink-red when the sugar in the medium, lactose, is fermented.
Nutrient agar	Complex medium used for routine laboratory work. Supports the growth of a variety of non-fastidious bacteria. Not selective or differential.

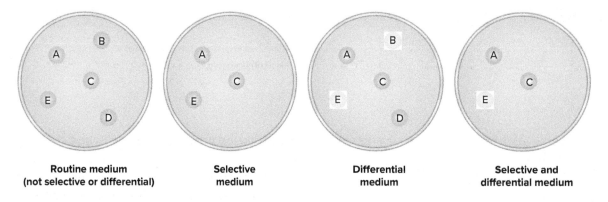

Figure I.16.1 How selective and differential media works. Routine medium will grow most of what is in a given sample. The selective medium will grow one particular type of microorganism like Gram-positive or Gram-negative bacteria due to an ingredient like salt or the dye eosin. The differential medium contains an ingredient like the pH indicator phenol red to facilitate bacteria types looking different from one another. A selective and differential medium contains a combination of ingredients so only certain microorganisms grow, but then different types will also appear differently.

16

Definitions

Alpha-hemolytic (α-hemolytic). A type of hemolysis characterized by a greenish clearing around colonies grown on blood agar. This hemolysis indicates partial lysis of red blood cells.

Beta-hemolytic (β-hemolytic). A type of hemolysis characterized by a clear zone around colonies grown on blood agar. This hemolysis indicates complete lysis of red blood cells.

Beta-lactamase. An enzyme that degrades the beta-lactam ring of a beta-lactam antibiotic, like penicillin and methicillin, rendering the antibiotic inactive. Bacteria that produce this enzyme will test positive for beta-lactamase.

Catalase. An enzyme that breaks down hydrogen peroxide (H_2O_2) to produce water (H_2O) and oxygen gas (O_2).

Coagulase test. A plasma-clotting virulence factor produced by *Staphylococcus aureus* that serves as an identifying characteristic. Organisms are added to plasma and if the cells clump together, or clot, the microorganism is coagulase positive.

Commensals. A relationship between two organisms in which one benefits and the other is not affected.

Diphtheroid. A Gram-positive, club-shaped organism resembling *Corynebacterium diphtheriae*. *Propionibacterium* species are another example.

Gamma-hemolytic (γ-hemolytic). A type of hemolysis characterized by no clearing around colonies grown on blood agar. This hemolysis indicates no lysis of red blood cells, so the agar remains red.

Lyse. To burst. Lyse is normally used in reference to a cell that ruptures.

Nosocomial infection. A secondary infection acquired while at a health care facility, unrelated to the initial reason a patient was admitted. Catheter-associated urinary tract infections are an example.

Also called health care-associated infections (HAIs).

Oxidase. A rapid biochemical test used to detect activity of cytochrome C oxidase. This reagent reacts with air, so reading the test within a minute will avoid a false-positive test. A positive test is purple.

Transients. Organisms that live in an area for a short period of time.

Zone of inhibition. Lack of bacterial growth around an antibiotic disc. The clear zone diameter is measured in millimeters and indicates bacteria were killed.

Objectives

1. Differentiate organisms making up the normal skin biota.
2. Describe the importance of skin biota.
3. Classify Staphylococci using the coagulase, mannitol salt, beta (β)-lactamase, and antibiotic sensitivity tests.

Pre-lab Questions

1. Resistance to which antibiotic helps distinguish Staphylococci from one other?
2. Why is the beta-lactamase test important to conduct?
3. Which *Staphylococcus* is coagulase positive and why is this test important?

Getting Started

Normal biota are organisms that grow on body surfaces and in various body locations, like the nose, sinuses, intestines, vagina, etc. They are usually considered **commensals** because they do not harm their hosts and are beneficial. Pathogens, however, have difficulty colonizing the skin because the normal flora are already established there (they out compete the pathogens) and they can utilize available nutrients. Some bacteria produce enzymes or other substances that inhibit non-resident organisms. **Transient** micro-organisms are present for short periods of time, but they cannot grow at the location, and soon disappear.

Normal skin bacterial commensals include *Staphylococcus, Micrococcus, Sarcina, Sporosarcina, Propionibacterium,* and *Bacillus* species and are listed in **figure 16.1**. Yeasts like *Candida albicans* and fungi like *Aspergillus* are also often present. *Staphylococcus aureus* is an opportunistic skin and nasal bacterium that causes food poisoning infections, as well as many wound and **nosocomial infections**, often referred to as health care-associated infections (HAIs). Nosocomial infections are a secondary infection unrelated to the initial problem a patient was being seen for. They occur in hospitals, clinics, surgical centers, and long-term care centers. *Staphylococcus aureus* can be resistant to the antibiotic methicillin, which is called methicillin resistant *Staphylococcus aureus*, or MRSA. Since methicillin is no longer produced, oxacillin is used in clinical laboratories instead. However, the abbreviation MRSA has continued to be used. This drug resistance makes infections caused by *S. aureus* more difficult to treat effectively than infections caused by *Staphylococcus epidermidis*. In addition, many people are colonized by MRSA and do not know it. Wounds infected by MRSA can be extremely painful and can swell to the size of a golf ball overnight.

Skin continually flakes off the body, and the bacteria on those skin cells floating in the air can land in Petri dishes and grow, especially when lids are left off the dish for a prolonged period of time. You want to minimize moving your hand or arm repeatedly across an open Petri dish. Thus, recognizing potential contaminants like *Staphylococcus* and *Micrococcus* is important.

The microbiological tests performed in this lab will help distinguish similar organisms from one another, and will include the coagulase, mannitol salt, blood hemolysis, beta-lactamase, and bacitracin antibiotic tests. *Staphylococcus aureus* needs to be distinguished from *S. epidermidis* because *S. aureus* is often more antibiotic resistant and can cause life-threatening infections.

Although the Gram stain was previously performed in exercise 4, another method utilizing

3% potassium hydroxide (KOH) can also be utilized. A drop of 3% potassium hydroxide is added to a slide. Bacteria are added to a slide, and if the bacteria lyse, the cellular material causes the solution to form a "string." Potassium hydroxide will lyse Gram-negative cells, which then produce the stringy appearance. Gram-positive cells, however, are not lysed due to the thick peptidoglycan.

The **coagulase test** determines if a bacterium produces the enzyme coagulase, which causes plasma to clump together (color plate 20). A latex agglutination test is also available and detects clumping factor and protein A (color plate 21). Only a few bacteria produce coagulase, one of which is *S. aureus* (or *S. hyicus* and *S. intermedius*), so other bacteria are often referred to as coagulase-negative Staphylococci (CoNS). Another latex agglutination test is available that detects the presence of the Penicillin-binding protein 2a (PbP 2a), which is found in methicillin resistant (MRSA) strains.

The mannitol salt plate is considered both selective and differential as it uses 7.5% salt to inhibit most Gram-negative bacteria, and a phenol red pH indicator turns the tomato red agar to yellow if the sugar mannitol is fermented (color plate 22). *Staphylococcus aureus* turns the surrounding mannitol salt agar yellow and can be distinguished from *S. epidermidis,* as the agar around the colonies remains red. Ignore the color of the colonies as they can turn a yellow coloration, but it is not indicative of a positive result.

Staphylococcus aureus is resistant to the antibiotic bacitracin (antibiotic effectiveness is discussed in exercise 14), whereas *S. epidermidis* is susceptible. The antibiotic disk can be placed in the middle of the first isolation streak area.

The **beta-lactamase** test (color plate 23) detects if a bacterium produces the enzyme beta-lactamase, which breaks the bonds of the beta-lactam ring of penicillins, rendering them ineffective. This test can be used for both Gram-positive and Gram-negative bacteria. *Micrococcus* species generally do not produce beta-lactamases.

The beta-lactamase test is performed on a yellowish nitrocefin dry slide card (a chromogenic cephalosporin) or a cefinase disc (resembles an antibiotic disc) to which a drop of sterile water has been added. The bacterium of interest is stamped onto a sterile applicator stick, and the microbe is added/stamped onto the card or disc. Results are read at 15 minutes for the dry slide card or up to an hour for the disc. No

Bacterium	Location	Colony Color on Blood Agar	Appearance of cells	Gram Stain & Morphology	Oxygen Requirements	Hemolysis	Catalase	Coagulase	Antibiotic Susceptibility
Staphylococcus epidermidis	Skin	White or light gray		Gram +, coccus, clusters	Facultative	α or γ	+	−	Bacitracin S
Staphylococcus aureus	Skin	Light yellow		Gram +, coccus, clusters	Facultative	β	+	+	Bacitracin R
Staphylococcus saprophiticus	Skin	White		Gram +, coccus, clusters	Facultative	α or γ	+	−	Bacitracin S
Kocuria rosea (formerly Micrococcus roseus)	Skin, environment	Pink	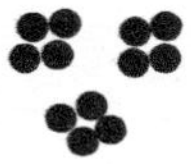	Gram +, coccus, tetrads, pink	Obligate aerobe	α or γ	+	−	
Micrococcus luteus	Skin, environment	Medium yellow		Gram +, coccus, tetrads, yellow	Obligate aerobe	γ	+		
Propionibacterium acnes	Skin	White		Gram +, bacillus, palisades	Anaerobe/aerotolerant	α, β, or γ	+	−	
Propionibacterium granulosum	Skin	Buff, gray		Gram +, bacillus, palisades	Anaerobe/aerotolerant	α, β, or γ	+	−	
Bacillus spp.	Skin, environment	Buff, gray, white, light yellow		Gram +, bacillus, endospores	Facultative	α, β, or γ	+	−	
Sarcina aurantiaca	Skin, environment	Buff, opaque, light yellow		Gram +, coccus, cuboid packet, endospores	Anaerobe/aerotolerant	γ	+	−	
Sarcina ventriculi	Skin, environment	Buff, opaque, light yellow		Gram +, coccus, cuboid packet, endospores	Anaerobe/aerotolerant	γ	−	−	
Sporosarcina ureae	Skin, environment	Buff, light yellow to orange	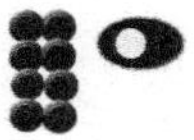	Gram +, coccus to short bacillus, tet rad, endospores	Obligate aerobe	γ	+	−	
Yeasts, like Candida albicans	Skin, environment	Off-white, matte, may see hyphae		Gram +, budding coccus	Facultative	β or γ	+	−	

Figure 16.1 Gram stain appearance and usual test results for Gram-positive flora.

color change (yellow to yellow-orange) is considered a negative result, whereas a color change to pink, red, or maroon is a positive result and indicates drug resistance to penicillin. Many *S. epidermidis* strains will test negative, whereas *S. aureus* usually tests positive for beta-lactamase. These tests help classify Staphylococci once a Gram stain and **catalase** test (see exercise 17) have been performed, and catalase positive, Gram-positive cocci in clusters are seen microscopically. The catalase test detects if a bacterium produces the enzyme catalase that breaks down hydrogen peroxide into non-toxic water and oxygen. This test helps differentiate Staphylococci from Streptococci.

The **oxidase** test detects if a microbe produces cytochrome oxidase and can utilize the electron transport chain/system on its own. One Gram-positive microbe that produces oxidase is the *Bacillus* species and Gram-negative microbes include *Neisseria* and *Pseudomonas* species.

Some microorganisms produce hemolysins, which **lyse** red blood cells (erythrocytes), particularly when grown on a blood agar plate. Exercise 17 contains a comprehensive description of blood agar. Blood agar contains a base agar rich in vitamins and nutrients. Sheep's blood is added to give additional nutrients, and this also allows microorganisms to be distinguished from one other.

Microorganisms that produce hemolysins that completely lyse red blood cells are termed **beta-hemolytic** (β-hemolytic) (color plate 24), and a clear zone around the colony is seen. This clear zone is truly clear, and not just a yellow coloration. You can actually read through the agar. Microorganisms that partially (incomplete) lyse red blood cells are termed **alpha-hemolytic** (α-hemolytic) (color plate 25), and a greenish coloration is seen. Microorganisms that do not lyse red blood cells are termed **gamma-hemolytic** (γ-hemolytic), so the agar remains the original red coloration. Holding the blood agar plate up to the light allows hemolysis to be seen more easily. *Staphylococcus aureus* grows as large, light yellow β-hemolytic colonies, whereas *S. epidermidis* colonies are white or gray with γ- or α-hemolysis.

Some bacteria you may encounter include:

Bacillus subtilis and ***Bacillus cereus*** colonies are usually large, are light yellow or gray, appear rough (similar to colony (1) on color plate 8), can be transients or commensals on skin, and normally do not cause skin infections.

Micrococcus colonies are usually medium sized and appear light yellow or pink, and normally do not cause infections.

Propionibacterium acnes colonies are a few millimeter in diameter, usually a buff or gray color, and are known to cause acne on the human body.

Sarcina and ***Sporosarcina*** colonies usually appear matte yellow or orange, and can be commensals or transients on skin.

In this lab exercise, you will inoculate media with skin flora, but since you may culture some pathogens, we recommend instructor-made demonstration plates and instructor performed coagulase tests. Alternatively, the instructor may provide an "unknown" tube to utilize for performing the tests.

Materials

Per team of two students

First Session

Trypticase soy cultures (24 hours, 25°C) labeled A, B, C, and D of the following control organisms:

Bacillus subtilis or *B. cereus*

Micrococcus luteus or *Kocuria (Micrococcus) roseus*

Staphylococcus epidermidis

Sporosarcina ureae

Blood agar plates, 2

Mannitol salt plate, 2

Bacitracin antibiotic discs, 1

Forceps

Second Session

Demonstration plates (24 hours, 25°C) of skin flora grown on blood agar plates:

Bacillus subtilis or *B. cereus*

Micrococcus luteus or *Kocuria (Micrococcus) roseus*

Staphylococcus aureus (β-hemolytic strain), with bacitracin disc added

Staphylococcus epidermidis

Ruler divided into mm

Sterile deionized distilled water in test tubes

Sterile droppers or pipettes to transfer water

Wooden applicator sticks

Demonstration nitrocefin cards or cefinase discs (+: *Staphylococcus aureus*) and (–: *Micrococcus luteus*)

Demonstration coagulase cards or tubes of plasma

Precleaned glass slides

Dropper bottles of fresh H_2O_2

Dropper bottles of fresh 3% KOH

Sterile swabs or filter paper

Oxidase reagent

Gram stain reagents

PROCEDURE

Safety Precautions: There may be colonies of *Staphylococcus aureus* on the agar plates of normal biota. Handle plates with special care. Seal plates with parafilm prior to student's viewing. Any plates should be promptly autoclaved. If any culture material is spilled, notify your instructor.

Skin Swab

First Session

1. Divide one blood agar and one mannitol salt plate in half, label them as skin flora, and place your initials on one side and your partner's initials on the other side. Label the blood agar and the mannitol salt plate. Your instructor may combine exercises 16 and 17 to conserve blood agar plates.

2. Gently press your dry fingertips onto half of 1 blood agar and 1 mannitol salt agar plate. Label your plates completely, and place them to be incubated.

3. Incubate the blood agar plate and the mannitol salt plate at 37°C for 24 hours. You or the instructor should then refrigerate or parafilm the plates to prevent them from drying out. *Staphylococcus*, *Bacillus*, and *Micrococcus* species can be observed after 24 hours.

4. Your instructor may assign you one of the cultures labeled A, B, C, and D. Label the other blood agar plate and mannitol salt plates with skin flora, the culture letter you were assigned, your name, date, medium, and temperature.

5. Perform an isolation streak on the two plates.

6. Gently flame the forceps and obtain the bacitracin antibiotic disc cartridge.

7. Add a bacitracin disc to the blood agar plate by gently pressing the disc to the first quadrant.

8. Invert the plates and incubate at 25°C for 24–48 hours.

Second Session

1. Examine the mannitol salt plate (color plate 22) of the assigned control microbe (A, B, C, or D) and record the results in the Laboratory Report. Although you do not need to perform a Gram stain as organisms that grow on this plate should be Gram-positive, you will need to determine the shape and arrangement either by a simple stain, negative stain, or Gram stain. Demonstration plate results should be placed in **table 16.1**.

2. Examine your skin flora mannitol salt plates. Count the number of colonies that tested positive and negative for mannitol fermentation and record the results. If the plates are parafilmed, do not remove the seal unless instructed to do so. Discard the plates in the appropriate container.

3. Examine the blood agar plate of the assigned control microbe (A, B, C, or D) and record results of colony appearance and hemolysis.

4. Measure the zone of inhibition in millimeters for each colony on the blood agar plates at 24 hours. If the zone size is >14 mm, the bacterium is considered to be susceptible. Were any of the colonies resistant to bacitracin?

5. Choose one large isolated colony to perform additional tests on. Using a needle is recommended, as you will perform several tests on one colony (Gram stain, catalase test, etc.).

6. Circle the chosen colony on the plate bottom and place a #1 beside it. You may need to choose a second large colony to label as #2 that has a similar appearance to complete the beta-lactamase and oxidase tests.

7. H_2O_2: Obtain a clean slide and add a drop of hydrogen peroxide (H_2O_2) to the slide. Be careful handling the slide so you do not add your hand flora to where you will add the reagent, as it can give a false positive reaction. Touch the chosen colony using a cooled, flamed needle

and hold the needle in the hydrogen peroxide reagent. Bubbling is a positive test and discussed more in exercise 17 (color plate 27). Discard the slide in the proper container.

8. **Gram Reaction/String Test:** Obtain a clean slide and add a drop of 3% potassium hydroxide (KOH) to the slide. Touch the chosen colony using a cooled, flamed loop and transfer the microbes to the drop of KOH. Swirl the loop around in the solution and if a string forms, the microorganism is Gram-negative. Lack of a string indicates a Gram-positive microorganism. In order to determine the bacterial arrangement, the String test can be followed by a simple stain using crystal violet or a Gram stain to confirm your test results.

9. **Coagulase:** If a colony is a Gram-positive coccus:

   a. Perform a coagulase test (color plates 20 and 21) by placing one full-sized drop of reagent into the center of the packaged card, then touch a sterile applicator stick to the colony. Press straight up and down in a "stamping" motion. Transfer the applicator stick with microbes to the reagent and gently make circles in the liquid until the liquid is spread around the entire circle to the defined border. Gently rock the card in a circular motion for approximately 3 minutes. Examine for clumps (color plate 21). Discard according to your instructor's directions.

   b. Optionally: Add microbes to a coagulase tube and incubate at 37°C for up to 24 hours. Look for a solid clump (color plate 20).

10. **Beta-lactamase:** Add one drop of sterile distilled water to one square of a nitrocefin dry slide card or cefinase disc if it has not already been done. Using a sterile applicator stick, gently acquire the bacteria by stamping the colony first, and then stamping the stick onto the moistened area of the nitrocefin dry slide card or cefinase disc. Place the stick in the proper discard. A positive test is a pink to maroon color that forms (a) within 15 minutes for the nitrocefin dry slide card, or (b) within 1 hour for the cefinase disc. No color change (yellowish-orange) is a negative result (color plate 23).

11. **Oxidase:** Obtain a sterile swab for each of your colonies and obtain a small amount of colony from your blood agar plate. Add a drop of the oxidase reagent to each swab and look for a purple color to develop within about a minute (color plate 26). Alternatively, add a drop of reagent to filter paper and use a sterile applicator stick to stamp the microbe onto the paper. Properly discard the swab into disinfectant.

Coagulase Test

Method 1

1. Place a drop of water on a slide and make a *very* thick suspension of cells from a yellowish colony (not bright yellow).
2. Place a drop of plasma next to it, and mix the two drops together. Look for clumping; clumped cells indicate a coagulase-positive result.
3. Drop the slide in the proper discard container.

Method 2

1. Mix a loopful of yellowish cells with 0.5 ml undiluted rabbit plasma in a small tube. The plasma can also be diluted 1:4 with saline.
2. Label the tube. Incubate at 37°C, and examine after 4–24 hours. Tip the tube sideways. A solid clot is a positive test (color plate 20).

Method 3 (Purchased Coagulase Kit)

1. Obtain a small card with a circle (provided with the kit).
2. Place a drop of the reagent into the middle of the circle.
3. Using an applicator stick, gently touch three colonies that appear the same color and size. You will want to gently press straight up and down to "stamp" them onto the end of the stick.
4. Mix the colonies from the stick with the reagent in a circular motion, being sure to spread the reagent and colonies across the entire surface of the circle.
5. Holding the card by the edge, gently move the card in a circular manner for 3 minutes.
6. Look for clumping (color plate 21), which is a positive test. Dispose of the cards in the proper container.

Table 16.1 Demonstration Test Results

Test	Organism: _____	_____
Mannitol salt		
Appearance		
Test result		
Blood agar plate		
Appearance		
Test result		
Coagulase		
Appearance		
Test result		
Beta-lactamase		
Appearance		
Test result		

EXERCISE 16

Laboratory Report: The Skin Flora:
Staphylococci and Micrococci ASM33

Results

Table 16.2   Mannitol Salt and Blood Agar Plates

Mannitol plates Sample:	A	B	C	D	Your flora Partner's flora Number of colonies	
Mannitol positive						
Mannitol negative						
Colony appearance (color, size)						
Bacterial shape and arrangement						
Possible identity						
Blood agar plates Sample:	**A**	**B**	**C**	**D**	**Your flora Partner's flora Number of colonies**	
Hemolysis color					α _____	
Type of hemolysis					β _____	
					γ	
Colony appearance (color, size)						
Bacitracin zone (nun) Resistant or susceptible						
Catalase Gram reaction Bacterial shape and arrangement Coagulase	+ or – + or – + or –	+ or – + or – + or –	+ or – + or – + or –	+ or – + or – + or –		
Beta-lactamase Oxidase	+ or – + or –	+ or – + or –	+ or – + or –	+ or – + or –		
Possible identity						

1. Although this was not a quantitative procedure, what organism seemed to be the most numerous on your fingertips?

2. How many estimated *Staphylococcus aureus* did you isolate from your fingertips? _____

Optional Class Results

3. Number of students in the class. _____

4. Number of students most likely isolating *Staphylococcus aureus*. _____

5. Conclusion:

Questions

1. How could normal skin biota be helpful to the host?

2. How can you immediately distinguish *Staphylococcus* from *Micrococcus* in a Gram stain?

3. Prepare a flowchart that would help you identify your skin biota isolates.

17

The Respiratory Flora: Streptococci and Enterococci

Definitions

Alpha-hemolytic (α-hemolytic). A type of partial lysis of red blood cells characterized by a greenish clearing around colonies grown on blood agar.

Beta-hemolytic (β-hemolytic). A type of complete lysis of red blood cells characterized by a clear zone around colonies grown on blood agar.

Catalase. An enzyme that breaks down H_2O_2 to produce water (H_2O) and oxygen gas (O_2).

Diphtheroid. A Gram-positive, club-shaped (palisades) organism resembling *Corynebacterium diphtheriae*.

Enterococci. Gram-positive cocci found in the intestines.

Gamma-hemolytic (γ-hemolytic). A type of hemolysis in which red blood cells were not lysed, so no clearing is seen around colonies grown on blood agar, and the medium remains red.

Glomerulonephritis. Inflammation of the renal glomeruli in the kidneys (inflammation of the capillaries caused by toxins produced elsewhere in the body).

Lyse. To burst. Lyse is normally used in reference to a cell that ruptures.

Non-enterococci. Cocci found in other locations besides the intestines.

Oxidase. A rapid biochemical test used to detect activity of cytochrome C oxidase. This reagent reacts with air, so reading the test within a minute will avoid a false-positive test. A positive test is purple.

Oxygen-labile hemolysin. A hemolysin destroyed by oxygen. The agar is stabbed in order to check for oxygen-labile hemolysins that would not be seen otherwise.

Viridans streptococci. Streptococci that are α-hemolytic, considered normal flora, and not usually associated with causing disease. Viridans in Latin means green.

Objectives

1. Describe the importance of Group A β-hemolytic streptococci and distinguish them from normal throat biota.

2. Interpret hemolysis results from blood agar plates.

3. Classify streptococci using the bile esculin, NaCl, and antibiotic sensitivity tests.

Pre-lab Questions

1. Which main test helps distinguish Staphylococci genera from Streptococci?

2. Name another test besides hemolysis that helps distinguish streptococci from one another.

3. Name an important *Streptococcus* that is α-hemolytic.

Getting Started

A diverse group of microorganisms inhabit the human respiratory tract, which includes species of *Candida, Corynebacterium, Haemophilus, Moraxella, Neisseria, Prevotella, Staphylococci,* and *Streptococci* (**figure 17.1**). Pathogens commonly found in the respiratory tract include *Streptococcus pneumoniae, Neisseria meningitidis, Haemophilus influenzae, Bordetella pertussis,* and *Klebsiella pneumoniae,* and are usually present in large numbers if a patient is symptomatic.

Rebecca Lancefield (1933) grouped streptococci serologically into 18 groups (A–R) based on group-specific carbohydrate antigens present in their cell wall. Some Lancefield groups are shown in **figure 17.2**. Group D streptococci are one exception because a non-carbohydrate antigen, called teichoic acid, is found in the cytoplasmic membrane. In 1984, molecular biology advances divided Streptococci into *Streptococcus, Enterococcus,* and *Lactococcus.*

Bacterium	Appearance of cells	Gram Stain and Morphology	Oxygen Requirements	Hemolysis on Blood Agar	Catalase	Antibiotic Susceptibility	Oxidase	Bile Esculin	6.5% Salt
Corynebacterium and diphtheroids		Gram +, bacillus, palisades	Facultative	α or γ	+		–	–	–
Streptococcus pyogenes		Gram +, coccus, chains	Facultative	β	–	Bacitracin S	–	+	–
Streptococcus equi/equisimilis		Gram +, coccus, chains	Facultative	β	–		–	+	–
Enterococcus (Streptococcus) bovis		Gram +, coccus, chains	Facultative	α or γ	–		–	+	–
Enterococcus faecalis/ E. faecium		Gram +, coccus, chains	Facultative	β or γ	–		–	+	+
Viridans Streptococci		Gram +, coccus, chains	Facultative	α (usually)	–		–	–	–
Streptococcus pneumoniae		Gram +, diplococci, capsule	Facultative	α	–	Optochin S	–	–	–
Staphylococcus aureus/ S. epidermidis		Gram +, coccus, clusters	Facultative	β / γ	+		–	+/–	+/–
Neisseria sicca		Gram –, diplococci	Obligate aerobe	α or γ on chocolate agar	+	Colistin S	+	–	–
Neisseria subflava ser. perflava		Gram –, diplococci	Obligate aerobe	γ	+	Colistin S	+	–	–
Moraxella catarrhalis		Gram –, diplococci	Obligate aerobe	γ	–	Colistin R	+	–	–
Lactobacillus sp.	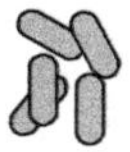	Gram +, bacillus, chains	Facultative	α	–		–	–	+/–
Yeast, Candida albicans		Gram +, budding coccus	Facultative	α, β, or γ	+		+	–	–
						S = susceptible, most other species are resistant			

Figure 17.1 Gram stain appearance and usual test results for Gram-positive respiratory flora.

Lancefield Group	Subcategory	Example Bacterium	Recovered From:	Distinguishing Characteristics
A		*Streptococcus pyogenes*	Strep throat, Scarlet and Rheumatic fever	Bile+, Bacitracin S
B		*Streptococcus agalactiae*	Neonatal meningitis or sepsis	Bile+, Carrot broth +
C		*Streptococcus equi*	Pharyngitis	Bile+, Bacitracin R
D:	Enterococcus	*Enterococcus faecalis*	Endocarditis, wounds, UTIs, systemic	Bile+, NaCl+
	Non-enterococcus	*Enterococcus (Streptococcus) bovis*	Endocarditis	Bile+, NaCl-
None		*Streptococcus pneumoniae*	Upper respiratory tract infections	α-hemolysis, Optochin S
Viridans:	M, O	*Streptococcus mitis group*	Oral caries/cavities	
	K	*Streptococcus salivarius group*	Oral caries/cavities	
	None	*Streptococcus mutans group*	Oral caries/cavities	
	None	*Streptococcus anginosus group*	Oral caries/cavities	
	None	*Streptococcus sanguinus group*	Oral caries/cavities	
	R, S, T	*Streptococcus suis*	Endocarditis, arthritis	

Figure 17.2 Lancefield groups of Streptococci.

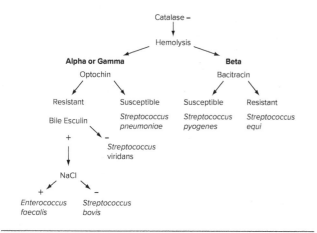

Figure 17.3 A flowchart for identification of streptococci.

In 2002, sequencing allowed some **viridans** streptococci to be re-classified.

Streptococci are commonly found in the respiratory, gastrointestinal, and reproductive tracts and are a common cause of strep throat and opportunistic infections. *Streptococcus* infections need properly identified as they can progress to rheumatic or scarlet fever, causing heart valve damage or acute **glomerulonephritis**, a disease of the kidney. The intestinal bacterium *Enterococcus faecalis* can cause urinary tract and wound infections. With one exception— *S. pneumoniae*—all streptococci appear as chains when grown in a broth medium. *Streptococcus pneumoniae* appear as diplococci.

Many sore throats are caused by viruses. Because viruses do not have a cell wall or any metabolic machinery, antibiotics are not effective on viruses. Therefore, an accurate diagnosis is important so antibiotics can be properly used. Fortunately, streptococci are usually sensitive, or susceptible, to penicillins, and antibiotic susceptibility can help identify some Streptococci. *Streptococcus pyogenes* is susceptible to the antibiotic bacitracin (color plate 28), but other β-hemolytic species are resistant. In addition, *Streptococcus pneumoniae* is susceptible to the antibiotic optochin, whereas other α-hemolytic streptococci are resistant. *Moraxella* and *Neisseria* can be distinguished using the antibiotic colistin, as *Moraxella* is resistant, but *Neisseria* is susceptible.

Strep throat is identified by taking a throat swab from a person with a sore throat and strawberry-appearing tongue (color plate 29). A rapid test (color plate 30) is performed in a doctor's office. A commercial kit contains specific antibodies that bind to *Streptococcus pyogenes* bacterial surface antigens if the person's throat contains the bacterium. If the rapid test is negative, the results should be verified by culturing the swab on a blood agar plate to determine hemolysis.

Blood agar contains a base agar rich in vitamins and nutrients. Before melted, trypticase soy agar is poured into Petri dishes, 5% sheep's blood is added. (Companies keep sheep for this purpose.) The blood adds nutrients, and helps visually distinguish various Streptococci and Enterococci. Even with additional nutrients, colonies can take 48–72 hours to be visible as pinpoint colonies.

Some streptococci produce streptolysins, also called hemolysins, which **lyse** red blood cells (erythrocytes). The streptolysins are designated as O (**oxygen labile**) and S (oxygen stable). On a throat culture, stabbing the agar with the loop allows visualization of Streptolysin O, produced by *Streptococcus pyogenes*. If red blood cells are completely lysed, a clear zone surrounding the colony is created, which is called **beta-hemolysis (β-hemolytic)** (color plate 24). If red blood cells are incompletely lysed, a green to light-brown color is seen (color plate 25), termed **alpha-hemolysis (α-hemolytic).** Many α-**hemolytic** colonies are made by throat bacteria like viridans streptococci (like *S. mutans*) or **diphtheroids**, and rarely cause disease. However, *Streptococcus pneumoniae* is α-**hemolytic** (color plate 25), but is not classified into a Lancefield group. If no blood cells are lysed, the agar remains red and is called **Gamma-hemolysis (γ-hemolytic).** Since blood agar is not selective, a Gram stain needs to be performed to ensure the presence of streptococci on blood agar. Beta-hemolysis is not always associated with pathogenicity. For example, non-pathogenic strains of *Escherichia coli* can be β-hemolytic and not cause disease.

All staphylococci and streptococci test **oxidase** negative (color plate 26), meaning they cannot make cytochrome C oxidase, or directly use the electron transport system. However, *Moraxella, Neisseria,* and *Haemophilus* are all oxidase positive.

The **catalase** test distinguishes catalase-positive *Moraxella, Neisseria, Staphylococcus,* and Enterobacteria (intestinal bacteria) from catalase-negative Streptococci and Enterococci. If a microbe produces the enzyme catalase, hydrogen peroxide is broken down into water and non-toxic oxygen, producing bubbles (color plate 27).

The bile esculin and sodium chloride (6.5% NaCl) tests can be utilized in conjunction with one another to differentiate Lancefield Group D into **enterococci** and **non-enterococci**. The bile esculin test differentiates Lancefield Group D enterococci (*Enterococcus faecalis* or *E. faecium*) from other Lancefield groups. Bile salts make this media selective, and the hydrolysis of the esculin changes the media color from light tan to black, making it differential (color plate 31). Some Group B streps (*Streptococcus agalactiae*) and other bacteria will also test positive for bile esculin.

A tryptic soy broth tube containing 6.5% NaCl further differentiates Group D enterococci from non-enterococci such as *E. bovis* (color plate 32). *Enterococcus faecalis* or *E. faecium* can grow in high salt concentrations, but *S. bovis* cannot as shown in **figure 17.3**.

In this exercise, you will have an opportunity to observe a throat culture and use the testing methods discussed to separate respiratory tract microorganisms. Some expected organisms that you may grow and observe in this lab include the following (figure 17.1):

Yeasts In a Gram stain, yeast cells appear purple, are larger than bacteria at 3-6 μm, and can have buds. *Candida* species are common oral biota forming large, white, matte colonies that may have short hyphae penetrating the agar.

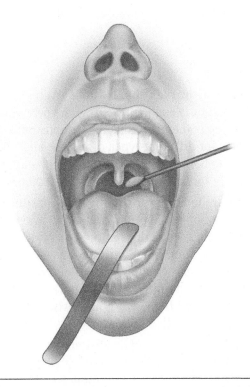

Figure 17.4 Diagram of open mouth. Shaded areas indicate places to swab.

Enterococcus bovis or *S. durans*

Streptococcus pyogenes (must be β-hemolytic strain)

Enterococccus faecalis

Streptococcus pneumoniae (must be α-hemolytic strain)

Clean slides, 6–8

H_2O_2 in dropper bottles

Gram stain reagents

Forceps

Ruler divided into mm

Sterile swabs or filter paper

Oxidase reagent

PROCEDURE

Safety Precautions: Colonies of β-hemolytic streptococci (*Streptococcus pyogenes*) or *Staphylococcus aureus* may grow on agar plates of normal biota. Handle plates with special care. Seal plates with parafilm prior to student's viewing. Any cultures should be promptly autoclaved. If any culture material is spilled, notify your instructor.

Throat Swab

First Session

1. Obtain two blood agar plates and label as throat swab, BAP (blood agar plate), and properly label the rest of the plate for a 37°C incubation temperature.

2. Swab your own throat using a mirror. Carefully remove a sterile swab from the wrapper, being careful not to touch any object other than the throat. If needed, depress the tongue with a sterile tongue depressor, say "ah," and swab the tonsillar area on the side of the throat (**figure 17.4**). Do not swab the hard palate directly in the back behind the uvula and do not touch the tongue or lips. Do this quickly to avoid the gag response.

3. Swab the first third of a streak plate on the blood agar plate, rolling the swab over the agar

to be sure to inoculate all sides. Properly discard the swab.

4. Continue streaking the rest of the plate with your loop as usual for isolated colonies. Stab the first third of the agar two or three times straight up and down with your loop (puncture the agar). This will check for **oxygen-labile hemolysins.**

5. Using sterile forceps, add a bacitracin antibiotic disc to the first streak of each throat swab plate.

6. Incubate the plates aerobically at 37°C for 24–48 hours.

7. <u>For Control Cultures</u>: Obtain one blood agar plate, bile esculin slant, and 6.5% NaCl tube.

8. Obtain one blood agar plate and label with the assigned control culture letter (A, B, C, or D), 25°C, and properly label the rest of the plate.

9. Inoculate the control culture you were assigned as a streak plate.

10. Using sterile forceps, add a colistin antibiotic disc to the first streak of the control plate.

11. Using a sterile loop, inoculate the bile esculin and 6.5% NaCl tubes and properly label them.

12. Incubate the plates aerobically at 25°C for 24–48 hours.

Second Session

1. Examine the parafilmed throat swab plates at 24–48 hours and compare it to the pure cultures of α- and β-hemolysis on the demonstration plates for different types of hemolysis. You will need to hold the plates up to the light and look closely for hemolysis around small colonies.

2. You are looking for small yellow or green pinpoint colonies. Medium and large colonies that are bright white are usually not streptococci. They are often Staphylococci or yeast. If you observe small colonies with β-hemolysis, are they the predominant colony type or only type on the plate? Is there β-hemolysis in the stab lines? In a clinical laboratory:

    a. If the α- (mucoid or glistening) or β-hemolysis (must also have clear zones around where the loop punctured the media) were the most numerous or only colony type, a culture and sensitivity would be determined, the physician notified, and proper antibiotic therapy would be given.

b. If only a few colonies are present, then the results would be reported as "ruled out β strep."

3. Measure the inhibition zone of each bacitracin disc with a ruler in mm on the blood agar plate at 24 hours. If the zone of inhibition size is >14 mm, the bacterium is considered susceptible. Record the results as susceptible (S) or resistant (R).

4. Examine the control culture (A, B, C, or D) blood agar plate. Record the hemolysis results in the Laboratory Report.

a. Measure the inhibition zone of the colistin disc with a ruler in mm on the blood agar plate at 24 hours. If the zone of inhibition size is >11 mm, the bacterium is considered susceptible, and <8 mm is resistant. Record the results as susceptible (S) or resistant (R).

b. On an isolated colony, perform a catalase test. Using a loop or needle, obtain a small amount of cells and place the needle in the hydrogen peroxide and look for bubbling. If bubbling is present, the colony is not a streptococcus. Flame your needle. Discard the slides in the proper discard container.

c. Perform an oxidase test on the colonies on the blood agar plate.

d. Prepare a smear on a glass slide at 24 hours using a colony from the control blood agar plate. Gram stain the slide. Examine the cell morphology with the oil immersion objective lens and record the results in the Laboratory Report.

5. Record the bile esculin and salt tube test results.

6. See Getting Started for a description of some of the organisms you might see and utilize the flowchart in figure 17.3 to help identify the microbes.

Control Microorganism Results

Your instructor may have known cultures for you to record results from (recommended).

1. Measure the inhibition zone of each bacitracin and optochin disc with a ruler in mm on a blood agar plate at 24 hours. If the zone of inhibition size is >14 mm, the bacterium is considered susceptible. Record the results as susceptible (S) or resistant (R).

2. Record the rest of the results in the Laboratory Results table.

Catalase Test

1. Briefly clean slides with alcohol and allow to dry. This prevents the microbes from your fingers creating a false positive result. Your instructor may have you write +, −, or T (1, 2, 3, etc.) on the slide so you know what each drop contains.

2. Place 2–3 single drops of hydrogen peroxide across a clean slide. Using a sterile loop or needle, obtain cells from a *Streptococcus durans* or *Enterococcus bovis* control colony and place it into the drop of hydrogen peroxide. It should not bubble.

3. Flame a loop well and repeat with a *Staphylococcus, Moraxella,* or *Neisseria* positive control colony, which should bubble.

4. Repeat with the colony you are testing. If bubbles form, the culture is catalase positive (**figure 17.5a** and color plate 27). Place slides in the designated discard container.

Oxidase Test

1. Place a small piece of filter paper on a glass slide or in a plastic Petri dish, and moisten with freshly prepared oxidase reagent.

2. Remove cells from a colony to be tested with a sterile wooden stick and rub the stick on the moistened filter paper.

a. *Enterococcus bovis* serves as a negative control.

b. *Neisseria subflava/N. sicca* serves as a positive control.

3. If a blue to purple color appears within about a minute, the cells are oxidase positive (**figure 17.5b** and color plate 26). Note: The longer the reagent is in contact with air, the more it

oxidizes, and the longer the reaction takes for a positive test to develop.

4. Place the paper in the proper discard container.

5. Alternative 1 uses sterile cotton tipped applicators:

   a. Using a swab, obtain desired colony on a sterile swab.

   b. Aseptically add a drop of oxidase reagent to the area where bacteria were placed (color plate 26).

   c. Repeat with new swab for each colony. Properly discard swabs.

6. Alternative 2 uses Pathotec strips:

   a. Place a Pathotec strip on a glass slide.

   b. Add sterile water to the strip to activate the reagent.

c. Using a wooden stick, obtain desired colony on a sterile wooden applicator stick.

d. Rub the colony from the wooden applicator stick onto the moistened strip.

e. A dark pink color should develop within a minute if the test is positive for oxidase.

7. Alternative 3 uses a nutrient agar plate:

   a. If the bacterium was grown on a nutrient agar plate, a drop of the oxidase reagent can be directly dropped onto the agar.

   b. A dark blue to purple color is positive.

   c. Other agar types contain ingredients that can cause false positive tests.

   d. Place all sticks or swabs in the proper discard container.

Catalase and Oxidase Tests

(a) Make sure slides are briefly cleaned with alcohol and allowed to dry. This prevents the microbes from your fingers creating a false positive result. Place 2–3 drops of hydrogen peroxide on a slide. Using a sterile loop or needle, place some cells from a *Staphylococcus aureus, Neisseria, or Moraxella* control colony into the drop of hydrogen peroxide. It should bubble. Now flame your loop or needle well and repeat with the colony you are testing. (b) Obtain a sterile applicator stick and touch 4 or 5 colonies and stamp the cells onto filter paper. Alternatively, touch a sterile swab to 4 or 5 colonies. Add a drop of the oxidase reagent to the filter paper or the swab. If it turns purple, the test is considered to be positive.

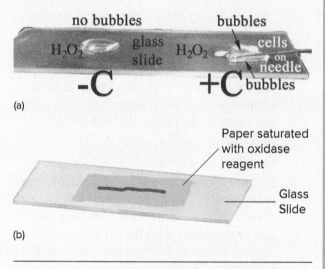

Figure 17.5 (a) Catalase and (b) oxidase tests.
(a) Courtesy of Anna Oller, University of Central Missouri.

EXERCISE 17

Laboratory Report: The Respiratory
Flora: Streptococci and Enterococci

Results

Indicate the number of colonies. Since you performed an isolation streak, you can use 1+, 2+, 3+ to semiquantitate. If few colonies are present you can count the numbers too. Some instructors prefer +++ for many colonies and + for few colonies present.

Blood agar plates Samples:	A	B	C	D	Your flora	Partner's flora
					Number of colonies	
Hemolysis color					α _____	_____
Type of hemolysis					β _____	_____
					γ _____	_____
Colony appearance (color, size)						
Bacitracin zone (mm)						
Colistin zone (mm)						
Resistant or Susceptible						
Catalase	+ or -	+ or -	+ or -	+ or -		
Gram reaction	+ or -	+ or -	+ or -	+ or -		
Bacterial shape and arrangement						
Oxidase	+ or -	+ or -	+ or -	+ or -		
Bile esculin	+ or -	+ or -	+ or -	+ or -		
6.5% NaCl	+ or -	+ or -	+ or -	+ or -		
Possible identity						

Demonstration Results

Organism	Cell Morphology	Hemolysis Appearance	Bacitracin Measurement	Bacitracin Result	Optochin Measurement	Optochin Result	Bile Esculin Color	Bile Esculin Result	6.5% NaCl (Quantitate 1+, 2+, etc.)
Enterococcus bovis or S. durans				R or S		R or S			
Streptococcus pyogenes				R or S		R or S			
Enterococcus faecalis				R or S		R or S			
Streptococcus pneumoniae				R or S		R or S			

Questions

1. Give two reasons why it is very important to correctly diagnose and treat strep throat.

2. Name one genus of Gram-negative cocci. _____

3. If a student had a cold and sore throat caused by a virus, how would the virus appear on the blood agar plate?

4. Why are Group D streptococci classified into the enterococcus and non-enterococcus groups?

5. Why should antibiotic sensitivity testing be performed on β-hemolytic streptococci? Which antibiotic(s) should be used?

18

Enzymes and Hydrolysis

Definitions

Activation energy. The initial energy required to break a chemical bond.

Amylase. An enzyme that degrades starches.

Caseinase. An enzyme that degrades casein.

Disaccharides. A carbohydrate molecule consisting of two monosaccharide molecules. Lactose and sugar are examples.

DNase. An enzyme that degrades deoxyribonucleic acids, or DNA.

Enzyme. A molecule that functions as a catalyst, speeding up a biological reaction.

Exoenzymes. Enzymes that are secreted outside of the bacterial cell into the cell's surrounding environment.

Gelatinase. An enzyme that degrades gelatin.

Hydrolysis. A chemical reaction in which a molecule is broken down as H_2O is added.

Kovac's reagent. A reagent that is added to a medium of tryptone to detect indole production.

Lipase. An enzyme that degrades fats or lipids.

Monosaccharide. A simple sugar, the basic unit of a carbohydrate. Glucose and fructose are examples.

Product. Resulting components of a chemical reaction.

Polysaccharide. A long chain of monosaccharide sub-units. Cellulose, glycogen, and dextran are examples.

Protease. An enzyme that degrades proteins.

Substrate. A substance on which an enzyme acts to form products.

Tryptophanase. An enzyme that breaks down tryptophan into pyruvic acid and indole.

Urease. An enzyme that converts urea into ammonia.

Objectives

1. Differentiate organisms that are positive for caseinase and amylase from those that are gelatin positive.

2. Differentiate organisms that are indole positive from those that are lipase and DNase positive.

3. Classify bacteria that are urease positive in conjunction with bacteria that are also indole positive.

Pre-lab Questions

1. Which main test helps distinguish *Proteus* and *Providencia* genera from most other bacteria, both Gram positive and Gram negative? What is one exception?

2. Which bacteria are indole positive?

3. Name a bacterium that tests positive for the caseinase, lipase, amylase, and DNase tests.

Getting Started

Due to structural limitations, bacteria must find ways to obtain molecules out in the environment to serve as a food, and thus energy, source. Since many different **substrates** can serve as a food source, bacteria can make different **enzymes** to break down carbohydrates, proteins, and fats. Enzymes, which are usually proteins, function to speed up a reaction compared to the time needed to contact them on their own, and for the reaction to occur. Thus, enzymes are catalysts for each part, or step, of each metabolic pathway. Substrates are converted to **products**, and those products become the new substrate for the next step of the pathway (**figure 18.1*b***). For example, in glycolysis, glucose serves as the substrate for the first reaction and glucose-6-phosphate is the product. Then, glucose-6-phosphate becomes the new substrate and fructose-6-phosphate is the product. Enzymes are also specific to the substrate they break down.

Enzymes lower the **activation energy**, which is how much energy is needed for the reaction to occur (**figure 18.1*a***). Think of driving to your favorite location. How long would it take for you to drive there without pushing down on the gas pedal? It would probably take a long time regardless of miles. Your car

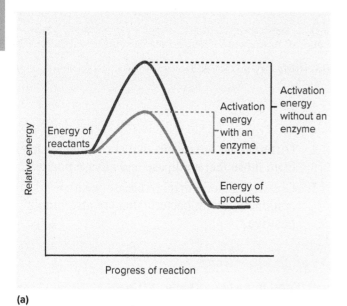

(a)

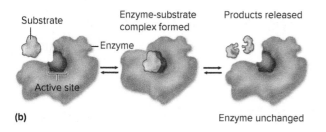

(b)

Figure 18.1 (*a*) An enzyme catalyzes a chemical reaction by lowering the activation energy. (*b*) Substrate binds to active site, complexing with the enzyme. Products are released, leaving the enzyme unchanged. © McGraw-Hill Education/Lisa Burgess, photographer

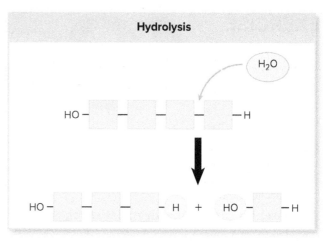

Figure 18.2 The process of hydrolysis, showing a water molecule added to the reaction and the substrate separated into two molecules.

would move very slowly, especially going uphill. The gas pedal serves to lower your activation energy. You get to where you are going *much* faster. Another example is trying to roll a stone up a mountain. Lowering the activation energy makes the mountain into a small hill that can be climbed more easily. The activation energy increases how quickly a substrate is converted to a new product.

Bacteria have the capability to secrete various **exoenzymes** into their surroundings when they sense certain types of molecules. They can then break large molecules like **polysaccharides** into smaller, simple sugars (also called **monosaccharides**) like glucose, which is then brought inside of the cell and

used in various metabolic pathways like glycolysis. **Disaccharides** include lactose and sucrose, of which lactose is made up of glucose and galactose joined together. Sucrose is made of glucose and a fructose joined together. Some biochemical tests utilizing disaccharides are in exercise 19.

The term **hydrolysis** is commonly used together with the molecule the enzyme is breaking down, such as *starch hydrolysis,* because a water molecule was added during the reaction, and the substrate was separated into two molecules (**figure 18.2**). The different hydrolysis reactions you will have an opportunity to perform in this lab are discussed in the next few paragraphs. The paragraphs also include how positive and negative results should appear.

The enzyme **amylase** breaks down starches. Humans also produce amylase in their saliva, which breaks down polysaccharides in our food so we can use glucose for making energy in glycolysis as well. Once bacteria have grown on a starch plate, iodine needs to be added. Be careful so you do not turn any plates upside down and spill contaminated cultures on yourself! Iodine binds to any starch left in the agar and turns dark purple—a negative test result. Where starch was used up around the colony or streak, no iodine can bind, and it remains yellow (color plate 33). Yellow indicates a positive result for amylase presence. You are looking for a clear zone in the

Hydrolysis Reactions	Starch	Casein/Skim milk	Spirit Blue	DNase	Urea	Tryptophan	Gelatin
Bacterium							
Bacillus cereus or *B. subtilis*	+	+	+	+	−	−	variable
Staphylococcus epidermidis	−	−	+	−	−	−	−
Serratia marcescens or *S. liquefaciens*	−	−	−	−	−	−	+
Proteus vulgaris or *Providencia Stuartii*	−	−	−	−	+	+	−
Escherichia coli	−	−	−	−	−	+	−
Alcaligenes faecalis	−	−	−	−	−	−	−

Figure 18.3 Expected hydrolysis reactions.

medium around the bacteria, and not the bacteria themselves appearing yellow as iodine stains them.

Proteases break down proteins like casein and gelatin into amino acids. Casein makes up the white coloration in milk, so plates look like solidified milk. A clear zone or clearing around colonies is where casein was degraded (color plate 34), and is a positive result for **caseinase**. No clearing is a negative result.

Gelatinase is an enzyme that degrades gelatin. Gelatin is derived from collagen and is soluble in hot water, but will solidify once cooled. If a bacterium produces gelatinase, the gelatin tube is liquid, even after refrigeration. Thus, once the bacterium has been incubated and allowed to grow, this medium requires refrigeration for 20–30 minutes. A solid tube when tipped is a negative result, and a liquefied tube after refrigeration is a positive result (color plate 35).

Lipases break down lipids (fats) into fatty acids and glycerol. The fatty acids are used in the tricarboxylic acid (TCA) cycle to generate indirect energy in the electron transport chain (ETC). The medium often appears as if oil droplets are on the agar surface. If a microbe can degrade fats, the colonies are dark blue with a clear zone around them. The plate will appear either lightened (oil droplets remain) or the light blue of the uninoculated plate (color plate 36), giving a negative result.

The **DNase** test hydrolyzes the reaction with Deoxyribonucleic acid, or DNA. Methyl green is often used as an indicator, giving a light green coloration to the plate. A clear zone around colonies is a positive result, and no clear zone is a negative result (color plate 37).

Urease is an enzyme that breaks urea into two ammonia molecules and one carbon dioxide molecule. This causes the tube medium to turn pink, which indicates a positive test. If the medium remains tan (color plate 38), then it is a negative test result.

Tryptophanase breaks tryptophan into indole and pyruvate. Although pyruvate is used in downstream pathways like the TCA cycle, indole is not. Thus, indole production can be detected as an end product. **Kovac's reagent** is added to the tube, and a cherry red color is positive for indole production. The original light yellow color of the reagent will remain for a negative test. Tryptone broth and Sulfur-Indole-Motility (SIM) tubes can both detect indole (color plate 39).

For tests performed in this lab (**figure 18.3**), *Bacillus cereus* serves as a positive control for amylase, lipase, caseinase, and DNase. *Staphylococcus epidermidis* is lipase positive. *Serratia* and *Pseudomonas* test positive for gelatinase, and *Proteus* and *Providencia* species serve as urease positive controls. *Sporosarcina ureae* gives a weak urease positive result. *Escherichia coli*, *Proteus,* and *Providencia* are indole positive controls, and *Escherichia coli* serves as a negative control for all other tests in this exercise. *Alcaligenes faecalis* is a negative control for all the tests in this exercise.

Materials

First Session
Per team of two

Demonstration (24–48 hours, 37°C) control cultures of:

Bacillus cereus or *B. subtilis* (starch, casein, spirit blue, DNA)

Staphylococcus epidermidis (spirit blue)

Serratia marcescens or *S. liquefaciens* (gelatin)

Proteus vulgaris, Providencia stuartii, or *Sporosarcina ureae* (urea)

Escherichia coli (indole)

Alcaligenes faecalis (negative control)

An unknown culture

Plates of the following (2 plates per team):

_____Starch

_____Casein/Skim milk

_____Spirit blue

_____DNA

Tubes of the following (3–4 tubes per team):

_____Gelatin

_____Urea

_____Tryptophan (may be combined with a Sulfur-Indole-Motility or SIM tube in exercise 19)

Second Session

An uninoculated plate or tube of each for controls:

_____Starch

_____Casein/Skim milk

_____Spirit blue

_____DNA

_____Gelatin

_____Urea

_____Tryptophan

Kovac's reagent

Iodine

PROCEDURE

Safety Precautions: If your instructor has you use plates from exercises 16 or 17, colonies of β-hemolytic streptococci (*Streptococcus pyogenes*) or *Staphylococcus aureus* may be present. Handle these parafilmed plates and the demonstration plates containing β-hemolysis with special care. Any slants and plates should be promptly autoclaved. If any culture material is spilled, notify your instructor.

For the unknown cultures, your instructor may have you do one per person or one per team.

Hydrolysis Reactions

First Session

1. Label your starch, casein, spirit blue, and DNA plates completely. Divide plates into half, using a positive control like *Bacillus subtilis* on one half, and a negative control like *E. coli* or *Alcaligenes faecalis* on the other.

2. Before inoculating unknowns, be sure to label the extra plate and tubes appropriately.

3. Using a sterilized loop, transfer a negative control like *Alcaligenes faecalis* to the plates using a straight line or S streak.

4. Using a sterilized needle, transfer a negative control, like *Alcaligenes faecalis*, to a labeled gelatin tube.

5. Repeat with the urea and tryptophan tubes.

6. Using a sterilized loop, transfer a positive control like *Bacillus cereus* to the plates using a straight line or S streak.

7. Using a sterilized needle, transfer the positive control to labeled gelatin (*Serratia*), urea (*Proteus* or *Providencia*), and tryptophan (*E. coli, Providencia*, or *Proteus*) tubes.

8. Incubate plates and tubes at 37°C for 24–48 hours.

9. Your instructor may have you incubate the *Bacillus* at 25 °C.

10. You or the instructor should then refrigerate the plates to prevent them from drying out.

Second Session

1. Obtain your gelatin tubes and place them in a rack to be placed into the refrigerator for 20–30 minutes.

2. Record all of your results in the Laboratory Report.

3. Examine the starch plate for growth. Flood the plate with iodine and determine your results. The color change is almost instantaneous. The plates will fade in color within 10–15 minutes. Keep your plates upright to avoid spilling contaminated culture on you. Discard the plates into the appropriate container.

4. Examine the spirit blue, casein, and DNA plates for clear zones.

5. Examine the urea tubes for a color change to pink.

6. Examine the tryptophan tubes for growth and add 10–15 drops of Kovac's reagent to the tube. The cherry red color should develop within a minute.

7. Examine the refrigerated gelatin tubes by tipping them sideways and look to see if the tube is still liquid.

Name _____ Date _____ Section _____

EXERCISE

18

Results: Starch	Uninoculated media	Reagent to add (if applicable)	Media color (after reagent added, if applicable)
+ Control used:			
– Control used:			
Unknown #:			
Casein/Skim milk			
+ Control used:			
– Control used:			
Unknown #:			
Spirit Blue			
+ Control used:			
– Control used:			
Unknown #:			
DNA			
+ Control used:			
– Control used:			
Unknown #:			
Gelatin			
+ Control used:			
– Control used:			
Unknown #:			
Urea			
+ Control used:			
– Control used:			
Unknown #:			
Tryptophan			
+ Control used:			
– Control used:			
Unknown #:			

Questions

1. Why might hydrolysis reactions be important to identifying microbes from (1) clinical samples and (2) environmental samples?

2. Why is Kovac's reagent added to the tryptophan tubes? What does a positive test indicate?

Definitions

Coliform. Gram-negative rod found in the intestine that ferments lactose—for example, *E. coli, Enterobacter,* and *Klebsiella.*

Enteric. Associated with the intestine.

Fermentation. A metabolic process that stops short of oxidizing glucose or other organic compounds completely, using an organic intermediate such as pyruvate or a derivative as a final electron acceptor.

Opportunist. An organism that normally is non-pathogenic but can cause disease in hosts with impaired defense mechanisms or when introduced into an unusual location.

Objectives

1. Classify bacteria using biochemical tests.
2. Describe the physiological basis for the tests.
3. Describe the organisms commonly seen in a clinical laboratory, especially enteric organisms causing urinary tract infections.

Pre-lab Questions

1. Name a bacterium that will test positive for urea hydrolysis.
2. How can you tell if a microbe is motile?
3. Why is acetoin detected in the Voges–Proskauer test instead of another end product?

Getting Started

In this exercise, you will learn to identify bacteria using biochemical tests (**figure 19.1**). Gram-negative rods are excellent organisms to use because they are frequently seen in the clinical laboratory, either as pathogens or contaminants from the normal biota of the large intestine.

The organisms in the colon are present in enormous numbers (about 10^{11}/gram of fecal material), but in healthy individuals these bacteria are part of the normal biota. Their presence prevents other organisms from becoming established and causing disease. An example of this situation occurs when the normal biota is eliminated after the therapeutic use of antibiotics. This permits *Clostridium difficile*, an anaerobic Gram-positive endospore-former bacterium, to colonize the colon and become life-threatening.

If the normal biota enters the blood or infect the bladder or wounds, they can cause serious disease, especially in immunocompromised patients. These organisms, which are normally non-pathogenic but can cause disease under special circumstances, are called **opportunists**. The increasing antibiotic resistance seen by these microbes is also of concern.

Other **enteric** organisms are pathogens, such as *Salmonella,* the cause of diarrhea and typhoid fever, and *Shigella,* the cause of dysentery. *Salmonella* and *Shigella* are often found in food poisoning cases from improperly cooked poultry products, or from handling pet chickens and then not hand-washing. It is important to identify these microorganisms in a clinical laboratory.

The following list describes the enteric organisms used in this exercise:

1. *Alcaligenes faecalis* is found in the colon and in soil in the environment. It can cause urinary tract infections.

2. *Citrobacter freundii* is found in the colon but also isolated from plant material and the environment. It can cause urinary tract infections.

3. *Escherichia coli* is a constant part of the intestinal biota of almost all humans. It is usually non-pathogenic, but as an opportunist, it is the major source of urinary tract infections from the patient's own biota. Many strains can cause diarrhea, especially in children and travelers. The notorious strain O157:H7 causes bloody diarrhea and kidney failure. However, most strains are harmless and easily grown, making *E. coli* one of the major microorganisms used in research.

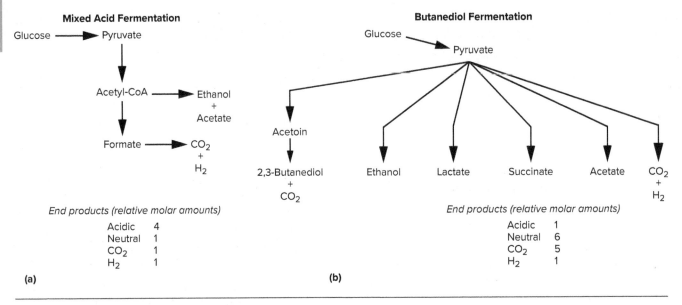

Figure 19.1 Fermentation pathways.

4. *Enterobacter aerogenes* and *E. cloacae* are found in the colon but also isolated from plant material and other sources in the environment. They can cause urinary tract infections.

5. *Klebsiella pneumoniae* is found in the environment and the colon, but some strains cause pneumonia, especially in immunocompromised individuals. It can cause urinary tract infections.

6. *Morganella morganii* is found in the colon, but can cause urinary tract infections.

7. *Proteus mirabilis* and *P. vulgaris* are found in the colon, but can cause urinary tract infections, especially from catheters.

8. *Providencia stuartii* is found in the colon and soil, but it can cause urinary tract infections.

9. *Pseudomonas* is a Gram-negative rod found in the colon, and in soil. It can cause serious wound infections, especially in burn patients. It is especially resistant to antibiotics and can produce pigments (color plate 40).

10. *Serratia liquefaciens* and *S. marcescens* are found in the colon, but are also isolated from plant material and the environment. They can cause urinary tract infections.

Your instructor will determine if you will also be using *Salmonella* or *Shigella* species in your inoculations for this lab. Demonstration test results are recommended.

All of these organisms, except *Alcaligenes* and *Pseudomonas,* belong to the family Enterobacteriaceae, defined as Gram-negative rods that are facultative anaerobes and able to ferment sugars. *Alcaligenes* and *Pseudomonas* are obligate aerobes and cannot ferment sugars.

The major component of the colon biota is *Bacteroides,* an obligate anaerobic Gram-negative rod. *Bacteroides* are rarely seen because most laboratory procedures involving enteric bacteria do not include incubating plates anaerobically, unless requested.

Identification Procedures

Biochemical tests determine the differences in an organisms' ability to ferment various sugars or produce different end products, the presence of enzymes, and physical characteristics such as motility. An organism then can be identified by comparing the results of the tests to the results of known bacteria in reference books like *Bergey's Manual of Systematic Bacteriology.*

The following describes the tests you will be using and how they work. Record the color, as well as the interpretation of + or − to help you remember how the tubes or plates appeared.

Nitrate Reduction This test contains potassium nitrate as its nitrogen source and requires an anaerobic state. Certain bacteria reduce nitrate to nitrite (**figure 19.2**), and some to ammonia or molecular nitrogen.

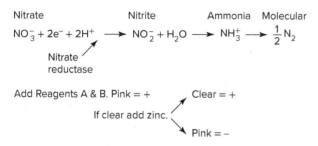

Nitrate Reduction

Nitrate   Nitrite   Ammonia   Molecular

$$NO_3^- + 2e^- + 2H^+ \longrightarrow NO_2^- + H_2O \longrightarrow NH_3^+ \longrightarrow \frac{1}{2}N_2$$

Nitrate
reductase

Add Reagents A & B. Pink = +   Clear = +

If clear add zinc.

Pink = −

Figure 19.2 A simplified nitrate reductase pathway.

Add 2–3 drops of sulfanilic acid (Reagent A) and then 2–3 drops of dimethyl-α-naphthylamine (Reagent B). If the clear broth turns pink, then the test is positive for nitrate reductase and nothing else needs to be done (color plate 41). This means that the nitrites present reacted with Reagent A to form nitrous acid, which then reacted with Reagent B to turn pink. If no color change develops, wait 5 minutes to be sure no reaction occurred.

After 5 minutes, if no color develops, use a scapula to add a small amount of zinc dust to the broth tube and wait 15 minutes. Once zinc dust is added to a tube, the reaction colors become opposite from before. If a color change to pink occurs, the test is now **negative** because the color is due to the leftover nitrate reacting with zinc. If no color change is seen and it remains clear, it means nitrite was converted to ammonia or molecular nitrogen. *Escherichia coli* is only one bacterium that tests positive for nitrates. Did you know a nitrate test is found on the quick urine dipsticks for urinalysis?

Methyl Red and *Voges–Proskauer* These two tests identify different fermentation pathways by detecting different end products produced by bacteria grown in buffered peptone glucose broth (MR-VP broth). Some pathways are simple, such as the conversion of pyruvate to lactic acid (also termed *lactate*). Other pathways are more complex, yielding a variety of products and perhaps additional energy (figure 19.1).

The methyl red test measures pH. *Escherichia coli* and other organisms ferment sugars by the mixed-acid pathway. Acetic and lactic acids, some organic compounds, and CO_2 and H_2 are the primary end products, which reduce the pH to less than 4.3. When methyl red is added, the broth turns pink or red, a positive result (color plate 42).

Other bacteria, such as *Enterobacter*, use the butanediol pathway when fermenting sugars. A small amount of acid, as well as CO_2 and H_2, is produced, so the pH is not low enough to change the broth red; these bacteria are methyl red negative.

Acetoin is the intermediate in the butanediol pathway that the Voges–Proskauer VP reagents detect. If the test is positive, acetoin is present, and the organism used the butanediol pathway. The reagents VP A (α-naphthol) and B (potassium hydroxide, KOH) are added to the broth culture after incubation. If a red layer or brick-red to dark brown precipitate forms, it is VP positive (color plate 43).

Phenylalanine Deaminase (PAD) This is a differential medium. If a bacterium can produce phenylalanine deaminase, it produces phenylpyruvic acid, which reacts with 12% ferric chloride when it is added to the medium, turning the medium a dark green color (color plate 44). *Proteus* species test positive for phenylalanine deaminase.

Motility Organisms are stabbed into a semisolid agar deep containing tetrazolium, an electron acceptor that turns red wherever there is growth. After incubation, a diffuse red color throughout the medium indicates motility. If there is a red streak only where the agar was stabbed, the organism is non-motile.

Sulfur Indole Motility Alternatively, SIM agar deeps may be used. Organisms are stabbed into a semisolid agar deep and motility is visualized as cloudiness away from the stab line. Hydrogen sulfide forms a black precipitate in this agar where organisms are located. Kovac's reagent is added to the top of the agar, and a red ring is indole positive, whereas yellow is negative (color plate 45).

Triple Sugar Iron (TSI) Slant This differential medium determines sugar fermentation. Glucose, sucrose, and lactose are the three sugars the tube contains. Phenol red is the pH indicator, so slants are initially red-orange and turn yellow if acid is produced from fermentation. A needle is used to puncture the butt to inoculate it, and then inoculations are made up the slant. Gas is detected if the medium is pushed up away from the bottom of the tube (designated with a circled bottom letter) and hydrogen sulfide turns the medium black (designated with a positive sign) (color plate 46). Examples: red slant, yellow butt, sulfur produced = K/A+ Ex: yellow slant, yellow butt lifted off the bottom =A/Ⓐ

Fermentation Tubes Carbohydrate fermentation tubes consist of a nutrient (complete) broth, a carbohydrate (like glucose), and a pH indicator. **Fermentation** is the energy-yielding pathway a facultative organism can use if oxygen is unavailable. A small Durham tube is added upside down inside the large tube. The small tube fills with broth under autoclaving pressure. If the organism produces a gas during fermentation, such as hydrogen, some of it will accumulate in the Durham tube as a bubble.

Your instructor will tell you which pH indicator was used.

- Andrade's indicator: If the bacterium ferments the sugar, acidic products are made and the pH falls, turning the medium from yellow to pink (color plate 47).
- Bromocresol purple indicator: If the bacterium ferments the sugar, acidic products are made, turning the medium from purple to yellow (color plate 48).
- Phenol red indicator: When the sugar is fermented and the pH falls, this medium turns from red to yellow (color plate 49).

After incubation, ensure the tubes have growth (sometimes you have to swirl the tubes slightly because the cells might have settled). If for some reason there is no growth, the test is considered invalid and must be repeated. If there is growth, record the results as follows:

A = Acid production

A/G or Ⓐ = Acid and gas—the indicator turned color and a bubble is seen in the Durham tube

N/C = No change—if neither gas nor acid has formed (gas production without acid formation is not of significance)

The sugars, glucose, sucrose, and lactose, are especially useful in the identification of the enteric Gram-negative rods. Lactose fermentation aids in the preliminary differentiation of enteric pathogens like *Salmonella* and *Shigella*, which do not ferment lactose, from the normal **coliforms** *E. coli, Enterobacter aerogenes,* and *Klebsiella pneumoniae,* which do ferment lactose. Lactose fermentation is not related to pathogenicity; it is simply a convenient characteristic for identifying organisms.

Citrate Utilization The Simmons citrate medium is a mineral medium with citrate as the sole carbon source. The agar also contains the pH indicator bromothymol blue. If the organism can utilize the citrate, the pH rises and the indicator turns a deep blue (color plate 50). *Citrobacter* is positive for this test.

Urea Hydrolysis Organisms are grown on agar or in broth containing urea and a pH indicator. If the organism produces the enzyme urease, urea is split, forming ammonia and CO_2. This raises the pH of the medium, turning it bright pink. *Proteus* and *Providencia* species always produce urease (color plate 38).

Indole Some organisms can cleave the amino acid tryptophan, producing indole. After incubation, the broth is tested for the presence of indole by adding Kovac's reagent. A red ring forms on the top of the medium if the organism is indole positive (color plate 39). *Escherichia coli* and *Proteus* test positive for this test.

MacConkey Agar This is both a selective and differential medium. Bile salts and crystal violet make it selective, allowing only Gram-negative organisms to grow on it, and it is differential because colonies that ferment lactose are dark pink colored, and non-lactose-fermenting colonies are cream colored (color plate 51).

Eosin Methylene Blue (EMB) This is both a selective and differential medium. It is selective in that Eosin inhibits Gram-positive bacteria so only Gram-negative organisms can grow on it, and the methylene blue makes it differential in that colonies that ferment lactose are maroon colored, certain strains of *E. coli* have a green metallic sheen, and non-lactose-fermenting colonies are colorless (color plate 52).

Xylose Lysine Deoxycholate (XLD) This medium is both selective and differential. Sodium deoxycholate makes it selective because only Gram-negative organisms can grow on it, and it is differential due to the xylose and lysine. Colonies that ferment lactose, sucrose, and xylose are yellow (they cannot decarboxylate lysine), and non-fermenting colonies that also produce lysine decarboxylase are red (color plate 53). Hydrogen-sulfide-producing colonies are differentiated by a black precipitate in the middle of the colony.

Lysine Decarboxylase (LDC) and *Ornithine Decarboxylase (ODC)* These two tests are identical except for the amino acid added to the broth tube. This test determines whether a bacterium can make either lysine or ornithine decarboxylase, an enzyme that removes CO_2 as it breaks down the specific amino acid. This medium begins as a green-purple color mix. Once incubated with a bacterium, the medium should turn either purple for a positive result or yellow for a negative result (color plate 54). These tests help differentiate *Proteus* (negative for both), *Citrobacter* (positive for both), and *Morganella* (negative for LDC, but positive for ODC) from other Gram-negative enterics when used in conjunction with other tests from this lab.

Hektoen Enteric (HE) Agar This is both a selective and differential medium. Bile salts, bromthymol blue, and acid fuchsin select for Gram-negative organisms. It is differential because colonies that ferment lactose and/or sucrose are orange and non-fermenting colonies are blue-green. Hydrogen-sulfide-producing colonies are differentiated by forming a black precipitate in the middle of the colony (color plate 55).

Salmonella Shigella (SS) Agar This is both a selective and differential medium. Bile salts and brilliant green inhibit many intestinal bacteria, making it selective for Gram-negative organisms. It is differential because colonies that ferment lactose are pink and non-lactose-fermenting colonies are cream colored. Hydrogen-sulfide-producing colonies are differentiated by forming a black precipitate in the middle of the colony (color plate 56).

In this exercise, each team will inoculate a series of biochemical tests with a labeled organism and an unlabeled "unknown" organism. (The "unknown" is one of the organisms listed.) There may seem to be a large number of tubes to inoculate, but if they are labeled and lined up in a test tube rack, inoculation can be done fairly quickly. Good organization is very helpful.

Note: Commercial test systems, such as *Enteropluri System,* are available and are discussed in exercise 21. A whole series of tests are inoculated at the same time with such systems (color plate 57). After incubation, the results are read and the organism can be identified.

Materials

First Session

Cultures on trypticase soy agar slants/plates or in broths of the following:

Note: Cultures provided in broth, the TSI, citrate, and SIM slants may need to be incubated longer for reactions to occur. The oxidase test may need to be performed in the second lab session.

Alcaligenes faecalis

Citrobacter freundii

Escherichia coli

Enterobacter aerogenes or *E. cloacae*

Klebsiella pneumoniae

Morganella morganii

Proteus mirabilis or *P. vulgaris*

Providencia stuartii

Serratia liquefaciens or *Serratia marcescens*

Pseudomonas aeruginosa

Salmonella species

Shigella species

Per team of students testing 1 known control and 1 unknown

Unknown broth, slant or plate, 1

Nitrate broth tubes, 2

Methyl red (MR) tubes, 2

Voges–Proskauer (VP) tubes, 2

Phenylalanine deaminase (PAD) agar plates or tubes, 2

Motility agar deeps (or Sulfur-Indole-Motility (SIM) deep tubes), 2

Tryptone slants (for indole test if not performing SIM tubes), 2

Triple sugar iron (TSI) slants), 2 OR

Glucose fermentation tubes, 2
Lactose fermentation tubes 2
Sucrose fermentation tubes, 2

Simmons citrate slants, 2

Urea slants, 2

MacConkey or EMB agar plates, 2

Xylose lysine deoxycholate (XLD) agar plates, 2

ODC tubes, 2

LDC tubes, 2

Hektoen enteric (HE) or *Salmonella Shigella* (SS) agar plates, 2

Oxidase reagent (freshly prepared) and small squares of white filter paper

Second Session

Demonstration parafilmed plates and tubes already grown in/on tubes/plates of *Pseudomonas, Salmonella*, and *Shigella*

Reagents in dropper bottles:

Nitrate reagents A (sulfanilic acid) and B (dimethyl-α-naphthylamine)

Methyl red

Voges–Proskauer (Barritt's) reagents A (α-naphthol) and B (KOH)

Zinc dust with scupula

Ferric chloride (12%)

Kovac's reagent

PROCEDURE

First Session

1. Choose one labeled culture and one "unknown" culture (or the cultures you are assigned).

2. Label a set of tubes for each organism with the organism name, the medium, temperature, date, and your name. Note that you will need 1 MR-VP broth for the methyl red test and another MR-VP broth for the Voges–Proskauer test. You should have a total of _____ tubes for each organism to be identified.

3. Flame an inoculating needle well all the way to the handle. Use cells from a slant or plate to ensure enough cells and inoculate the motility deep (or SIM) by stabbing the agar almost to the bottom of the tube with your inoculating needle. Inoculate the tube as straight as possible; if your stab is crooked, you may need to

mark the stab line with a wax pencil so you can accurately read your results next time.

4. Flame an inoculating needle well and inoculate a TSI slant with a needle by stabbing the butt of the slant with cells and in one movement making an "S" or zigzag up the slant. You should only enter the tube once.

5. Inoculate the remaining tubes and plates with a loopful of organisms. Inoculate agar slants by gliding the loop over the surface of the agar, starting at the bottom of the slant. Streak the plates to obtain isolated colonies. Use good aseptic technique to avoid contamination with unwanted organisms.

6. Perform an oxidase test using cells from the slant or plate (color plate 26, figure 17.4*b*, exercise 17)

7. Incubate the tubes at 37°C for 48 hours. Incubate the motility deeps at room temperature. Some organisms are not motile at 37°C.

Second Session

1. Record the results of your unknown organism in the chart on page 160 and on a similar chart drawn on the board or Googledoc. These results will establish the reactions.

2. Add 2–3 drops of nitrate reagents A (sulfanilic acid) and B (dimethyl-α-napthylamine) to the nitrate tubes. If a pink color forms it is positive (color plate 41).

   a. If there is no color change, wait 5 minutes to ensure no pink color develops.

   b. Add a small amount of zinc dust to the tube using a scupula and wait another 5–15 minutes to see if a pink color develops. A pink color is now a negative result, but a clear color is now positive.

3. Add about 5 drops of methyl red to the MR broths. A red color indicates a positive test (color plate 42).

4. Add 10–15 drops of VP (or Barritt's) reagent A (α-naphthol) and 10–15 drops of VP reagent B (40% KOH) to the VP broths. Gently swirl and let it stand for 15 minutes—1 hour, maximum. The appearance of a red layer, or a brick-red to dark brown color, indicates a positive test (color plate 43).

   Safety Precautions: Alpha-naphthol is toxic.

5. Add a dropper of 12% ferric chloride to the colonies on a PAD plate/tube. A green color is positive for producing phenylalanine deaminase (color plate 44).

6. Add a dropper full of Kovac's reagent to the tryptone broth or SIM tube. The tryptone broth tube needs swirling. A red layer on the top of the broth indicates a positive test for indole (color plates 39 and 45).

7. Examine the motility deep. If the tube appears pink throughout the agar, it is positive for motility. If only the original stab line appears pink, the test is negative for motility.

   a. SIM: Cloudiness away from the stab line is positive for motility. If you see black throughout the tube, it indicates that sulfur was produced and the organism is motile.

8. TSI slants: Record an "A" if acid (yellow) was produced or a "K" if the area is alkaline (red) (color plate 46). You will record a letter for the slant first, then a letter for the butt. Circle the bottom letter if gas was produced (A / Ⓐ = acid and gas) and place a plus sign (+) if hydrogen sulfide was produced.

9. After incubation for 48 hours, examine the fermentation tubes and record as "A" if the pH indicator has turned colors (see above description, "G or O" for gas production, or N/C" for no change if neither acid nor gas has formed. Be sure there is growth in the tube before recording) (color plates 47–49).

10. Observe the citrate slant. A deep blue color is positive for citrate utilization (color plate 50).

11. Observe the urea tube. A bright pink color is positive for urea hydrolysis (color plate 38).

12. Observe the colonies on the MacConkey plate. Dark pink colonies indicate lactose fermentation (color plate 51).

    a. Observe the colonies on an EMB plate. Maroon colonies indicate lactose fermentation (color plate 52). A green metallic sheen indicates *E. coli* (assuming no contamination occurred).

13. Observe the colonies on an XLD plate. Yellow colonies are positive for sugar fermentation and decarboxylate lysine. Black colonies indicate sulfur production (color plate 53).

14. Observe the LDC and ODC tubes. A purple color is positive for producing decarboxylase and a yellow color is a negative result (color plate 54).

15. Observe the colonies on an HE plate. Orange colonies indicate lactose and/or sucrose fermentation. Green colonies do not ferment sugars (color plate 55).

    a. Observe the colonies on an SS plate. Pink colonies ferment lactose, whereas cream colonies do not. Black colonies indicate sulfur production (color plate 56).

16. Record the class results on the board or on a Googledoc. Also record the results of your unknown organism.

KNOWN RESULTS

	Nitrate	MR	VP	PAD	Sulfur	Indole	Motility	TSI	Fermentation Glu	Lac	Suc	Cit	Urea	Mac/EMB	XLD	ODC	LDC	Ox	HE/SS
Alcaligenes faecalis																			
Citrobacter freundii																			
Escherichia coli																			
Enterobacter _____																			
Klebsiella pneumoniae																			
Morganella morganii																			
Proteus _____																			
Providencia stuartii																			
Pseudomonas aeruginosa																			
Salmonella _____																			
Shigella _____																			
Unknown # _____																			

EXERCISE

19

Laboratory Report: Identification of Enteric Gram-Negative Rods

RESULTS

	Nitrate	MR	VP	PAD	Sulfur	Indole	Motility	TSI	Glu	Lac	Suc	Cit	Urea	Mac/ EMB	XLD	ODC	LDC	Ox	HE/ SS
Alcaligenes faecalis																			
Citrobacter freundii																			
Escherichia coli																			
Enterobacter _____																			
Klebsiella pneumoniae																			
Morganella morganii																			
Proteus _____																			
Providencia stuartii																			
Pseudomonas aeruginosa																			
Salmonella _____																			
Shigella _____																			
Unknown # _____																			

Fermentation (Glu Lac Suc)

Questions

1. What is the identity of your unknown organism? _____

2. Can you determine whether or not an organism can ferment a sugar if it does not grow in the broth? Explain.

3. How can an organism have a positive test for acid from glucose in a fermentation tube but have a negative methyl red test, which is also a glucose fermentation test?

4. Were there any organisms that did not ferment any sugars? If yes, which organisms?

5. When comparing a lactose fermentation tube with a MacConkey plate,
   a. what additional information does a fermentation tube give?

   b. what additional information does a MacConkey agar plate give?

EXERCISE

20

Definitions

Differential media. Media that permit the identification of organisms based on colony appearance or agar color.

Selective media. Media that permit only certain organisms to grow and that aid in isolating one type of organism from a mixture of organisms.

Note: See exercises 17–19 for more information on these organisms and tests.

Objectives

1. Apply your knowledge to solve a microbiological problem.
2. List the procedures used to isolate and identify specimens.
3. Differentiate the presence of contaminants or non-pathogens in clinical specimens from the bacteria that should be present.

Pre-lab Questions

1. Why is it important to do a Gram stain first on your unknown?
2. Why is it important to perform an isolation streak on your unknown?
3. Name a bacterium from this lab that will be catalase negative.

Getting Started

In this exercise, you will have an opportunity to utilize the knowledge and techniques you have learned in the previous four exercises to identify unknown organisms. You are given a simulated (imitation) specimen containing two organisms, and your goal is to separate them into two pure cultures and identify them using different media and tests. The organisms are bacteria that could be associated with disease, normal biota, or are common environmental contaminants of food, water, or even pharmaceutical samples.

Your unknown specimen represents a urine or wound infection, food or food poisoning, or a respiratory water or soil environmental sample. In actual laboratory cases, standardized isolation and culturing procedures exist for each kind of specimen. However, you will be identifying only a limited number of organisms. With some careful thought, you can plan logical steps to use in identifying your organisms.

The following are characteristics useful in identifying your unknown organism.

Bacterial Cell Morphology The size, shape, arrangement, and Gram-staining characteristics of the bacteria as determined by a Gram stain. It may include the presence of special structures, such as endospores.

Colonial Morphology The size, shape, and consistency of isolated colonies on nutrient media, such as TS agar or blood agar.

Growth on Selective Media The ability of organisms to grow on selective media, such as on mannitol salt, which selects for organisms tolerating 7.5% salt, or EMB and MacConkey, which both select for Gram-negative bacilli.

Reactions on Differential Media The color of colonies grown on EMB agar or MacConkey agar is based on lactose fermentation (lactose fermenters are maroon or dark pink, respectively). Lactose fermentation allows different types of bacterial colonies to be visualized on EMB or MacConkey plates (lactose fermenters are maroon or dark pink, respectively), whereas mannitol fermentation turns the media yellow. Blood agar hemolysis shows the enzymes some bacteria can produce, also allowing bacteria to be differentiated.

Biochemical Capabilities These include the ability to ferment different carbohydrates and the production of various end products, as well as the formation of indole from tryptophan. Tests include methyl red, Voges–Proskauer, citrate utilization, urease, catalase, oxidase, and coagulase.

Approach the identification of your "unknown" clinical specimen with the following steps. Your instructor will tell you if you were given pure cultures or a mixture of two unknowns and you have to do the following:

1. Make a Gram stain of the specimen.
2. Streak the cultures on the appropriate complete and selective mediums.
3. After incubation, identify two different colony types and correlate with their Gram reaction and shape. Also correlate the growth and appearance of the colonies on selective media with each of the two organisms.
4. Restreak for isolation. It is useless to do any identification tests until you have pure cultures of the organisms.
5. After incubation, choose a well-isolated colony and inoculate a TS agar slant to be used as your stock culture. Prepare a stock culture for each organism.
6. Inoculate or perform various tests that seem appropriate. Keep careful records. Record your results on the worksheets as you observe them.
7. Identify your organisms from the test results.

Materials

First Session

24-hours unknown cultures (mixtures or pure culture(s)) labeled with hypothetical source (for each student or a team of two students)—plates, slants, or broth tubes

For a **mixed** culture:

Blood agar plate or trypticase soy agar plate, 1 per student

MacConkey agar plate (or EMB agar plate)

Mannitol salt agar plate

Gram-staining reagents

For a **pure** culture:

All tubes and plates listed, and oxidase, H_2O_2, and staining reagents

Second and Third Sessions

Trypticase soy agar plates

Bile esculin agar slants

Trypticase soy broths with 6.5% NaCl

Nutrient agar slants

Citrate agar slants

Urea slants

Glucose + bromocresol purple agar slants

Fermentation broths of glucose, lactose, and sucrose (or TSI slants)

MR-VP broths for the methyl red and Voges–Proskauer test

Tryptone broths for the indole test (or SIM tubes)

Thioglycollate broth tubes and Brewer's anaerobe plates (oxygen requirements)

XLD plates

PAD plates or slants

HE (or SS) plates (optional)

ODC tubes

LDC tubes

Nitrate broth tubes

Reagents

Oxidase reagent

Kovac's reagent

Voges–Proskauer reagents A and B

Methyl red

Coagulase test plasma or reagents

H_2O_2

12% ferric chloride

Nitrate reagents A & B, and zinc dust

Optional: Staining material for the endospore, acid-fast, and capsule stains

PROCEDURE

First Session

Your instructor will tell you if you were given a mixed or a pure culture.

If you are given a **mixed** culture:

1. Make a Gram stain of the culture. Observe it carefully to see if you can see both organisms. Record your result in the Laboratory Report. You can save the slide and observe it again

later if you have any doubts about it. You can also save the broth, but one organism may overgrow the other.

2. Inoculate a complete medium agar plate, such as trypticase soy agar or blood agar, and appropriate selective and differential agar plates. Use MacConkey agar (if you suspect the possibility of a Gram-negative rod in a urine specimen) or a mannitol salt agar plate (if you suspect *Staphylococcus* in a wound infection). Streak the plates for isolated colonies.

3. Incubate at 37°C for 24–48 hours.

If you are given a **pure** culture:

1. Make a Gram stain of the culture. Record your result in the Laboratory Report. You can save the slide and observe it again later if you have any doubts about it.

2. Inoculate a complete medium agar plate, such as trypticase soy agar or blood agar, oxygen shake tubes, and appropriate selective and differential tests (based on your Gram stain, cell shape, and cell arrangement). Streak plates for isolated colonies.

3. Incubate at 37°C for 24–48 hours.

Second Session

If you were given a **mixed** culture:

1. Examine the streak plates after incubation and identify the two different colony types of your unknown organisms, either on the complete medium or the selective media, wherever you have well-isolated colonies. Gram stain each colony type (organisms usually stain better on non-selective media). Also identify each colony type on the selective and differential media so that you know which organisms grow on the various media. Record their appearance on the differential media as well. It is helpful

to circle colonies that you Gram stain on the bottom of the Petri plate with a marking pen.

2. Inoculate and perform other appropriate tests on your isolated microbes.

3. Restreak each organism on a complete medium (instead of selective media) for isolation. This technique ensures that all organisms will grow and that you will be able to see if you have a mixed culture. Do not discard the original streak plates of your isolates—store at room temperature. If at some point your isolate does not grow, you will be able to go back to the old plates and repeat the test.

If you were given a **pure** culture:

1. Examine your tubes and plates and record your test results in the Laboratory Report.

2. Finish any stains or tests that you need to perform on your unknown.

Third Session

If you were given a **mixed** culture:

1. Observe the plates after incubation. If your organisms seem well isolated, inoculate each one on a TS agar slant to use as your stock culture. If you do not have well-isolated colonies, restreak them. It is essential that you have a pure culture. Possible steps in identifying Gram-positive cocci follow.

2. Using the information you currently have, use **figure 20.1** to identify potential bacteria that could be your unknown. Plan work carefully and do not waste media using tests that are not helpful. For example, a urea slant would not be useful for distinguishing between *Staphylococcus epidermidis* and *Staphylococcus aureus*.

A partial listing of unknown organisms' possible identities is given in **figures 20.2** and **20.3**. Others can be added.

Simulated Wound Samples	Simulated Urine Samples	Simulated Food Poisoning	Simulated Respiratory Samples	Simulated Environmental Samples
Micrococcus luteus	Alcaligenes faecalis	Bacillus cereus	Corynebacterium glutamicum	Bacillus subtilis Corynebacterium
Pseudomonas aeruginosa*	Citrobacter freundii	Clostridium sporogenes	Enterococcus bovis Moraxella catarrhalis Mycobacterium phlei	Enterococcus faecalis
Staphylococcus aureus	Edwardsiella tarda	Escherichia coli	Neisseria sicca	Escherichia coli
Staphylococcus epidermidis	Enterobacter aerogenes	Lactobacillus acidophilus	Staphylococcus aureus	Listeria innocua Micrococcus luteus Kocuria (Micrococcus) roseus Klebsiella pneumoniae
	Enterococcus faecalis	Lactococcus lactis	Streptococcus mutans Streptococcus durans	Pseudomonas aeruginosa*
	Enterococcus bovis			
	Escherichia coli	Listeria innocua		Sarcina aurantiaca
	Klebsiella pneumonia	Salmonella enteritidis*		Sporosarcina ureae
	Morganella morganii			
	Proteus vulgaris	Shigella flexneri*		Enterococcus bovis
	Providencia stuartii	Staphylococcus aureus		
	Serratia marcescens			
	Enterococcus bovis (plus wound organisms)			

Figure 20.1 Some possible organisms that may serve as unknowns. The * denotes pathogens.

Gram-Positive Cocci (Non-motile, Non-sulfur producing)	
Enterococcus bovis	Cocci in chains, BAP: α- or γ-hemolysis, facultative anaerobe; **Positive tests**: bile esculin, glucose, VP
Often a contaminant	**Negative tests**: catalase, coagulase, gelatin, indole, oxidase, salt, sulfur
Enterococcus faecalis	Cocci in chains, BAP: β- or γ-hemolysis, facultative anaerobe; **Positive tests**: bile esculin, glucose, lactose, mannitol, nitrate, salt, starch, sucrose, VP
Found in urine	**Negative tests**: catalase, citrate, coagulase, gelatin, indole, oxidase, sulfur
Kocuria (Micrococcus) roseus	Cocci in tetrads, BAP: α- or γ-hemolysis, microaerophile/aerobe, TSA: pink colonies; **Positive tests**: casein, catalase, glucose, mannitol, starch, sucrose
Found on skin/in the environment	**Negative tests**: bile esculin, citrate, coagulase, gelatin, indole, ODC, sulfur; Variable: oxidase
Micrococcus luteus	Cocci in tetrads, BAP: α- or γ-hemolysis, aerobe, TSA: yellow colonies; **Positive tests**: casein, catalase, citrate, starch, oxidase
Often a contaminant	**Negative tests**: bile esculin, coagulase, gelatin, glucose, indole, lactose, mannitol, sucrose, ODC, sulfur; Variable: salt
Sarcina aurantica	Cocci in tetrads or sarcina, BAP: γ-hemolysis, aerotolerant; **Positive tests:** catalase

Found on skin/in the environment	**Negative tests**: casein, coagulase, gelatin, glucose, indole, lactose, LDC/ODC, nitrate, sucrose, sulfur
Sporosarcina ureae	Cocci in tetrads or sarcina, BAP: γ-hemolysis, obligate aerobe; **Positive tests:** catalase, endospores, nitrate, oxidase, urease
Found in the environment	**Negative tests:** casein, citrate, coagulase, indole, gelatin, glucose, lactose, LDC/ODC, mannitol, starch, sucrose, sulfur, VP
Staphylococcus aureus	Grape-like clusters (bunches), BAP: β-hemolysis with large light yellow colonies, facultative anaerobe, TSA: large white colonies
Found in urine, wounds, or food	**Positive tests:** β-lactamase, catalase, casein, coagulase, gelatin, glucose*, mannitol, MR, nitrate, salt, starch; **Negative tests:** sulfur
Staphylococcus epidermidis	Grape-like clusters (bunches), BAP: α-hemolysis with white to grey colonies, facultative anaerobe; **Positive tests:** catalase, glucose
Often a contaminant	**Negative tests:** casein, coagulase, mannitol, starch, sulfur
Lactococcus lactis	Cocci in pairs or chains, BAP: α- or γ-hemolysis, facultative anaerobe; **Positive tests:** glucose, lactose, mannitol, starch, sucrose, VP
Found in food or environment	**Negative tests:** catalase, coagulase, indole, gelatin, nitrate; Variable: casein, sucrose

Gram-Positive Bacilli (Non-sulfur producing)	
Bacillus	Large rods, BAP: β-hemolysis, facultative anaerobe, endospores, salt tolerant; **Positive tests:** β-lactamase, catalase, casein, citrate, DNA, gelatin, glucose, motility, nitrate, sucrose
Found in food or contaminant	**Negative tests:** indole, lactose, LDC/ODC, mannitol, PAD, VP; Variable: oxidase, starch
Clostridium sporogenes	Large rods, BAP: β-hemolysis, anaerobe, subterminal endospores; **Positive tests:** bile esculin casein, DNA, gelatin, glucose, lipase, motility
Found in food or soil	**Negative tests:** catalase, indole, lactose, mannitol, nitrate, sucrose, VP
Corynebacterium	Clubbing or palisades of small rods, BAP: α- or γ-hemolysis, facultative anaerobe; **Positive tests:** catalase, glucose
Found in the mouth/environment	**Negative tests:** coagulase, gelatin, motility Variable tests: citrate, starch, sucrose, mannitol
Listeria innocua	Medium rods, BAP: γ-hemolysis, facultative anaerobe; **Positive tests:** bile esculin, catalase, glucose, MR, VP
Found in food or environment	**Negative tests:** citrate, indole, LDC/ODC, mannitol, nitrate, oxidase, urease
Lactobacillus acidophilis	Medium rods, BAP: α- or γ-hemolysis, facultative anaerobe; **Positive tests:** bile esculin, glucose, lactose, sucrose, starch
Found in food or environment	**Negative tests:** catalase, casein, citrate, coagulase, gelatin, indole, mannitol, MR/VP, nitrate, oxidase, salt

Other	
Mycobacterium	Acid-fast bacillus in cords, BAP: α- or γ-hemolysis as rough colonies, obligate aerobe
Found in environment	**Positive tests:** catalase, citrate, nitrate, starch; **Negative tests:** motility, sulfur
Yeasts	Large Gram-positive coccus with budding, BAP: α-,β-,or γ-hemolysis, facultative anaerobe; **Positive tests:** catalase, glucose, maltose, oxidase, salt tolerant, sucrose
Found in the mouth/food/environment	**Negative tests**: bile esculin, motility, sulfur

*= positive is acid production; **O**= acid result plus gas is produced

Figure 20.2 Gram-positive microbe characteristics.

Gram-Negative Bacilli (catalase positive)	
Alcaligenes faecalis	Non-pigmented or transparent colonies, coccobacillus, aerobe: **Positive tests:** motility, nitrate, oxidase; Variable: citrate, ODC
Found in urine and wounds	**Negative tests:** DNase, gelatin, glucose, indole, lactose, mannitol, starch, sucrose, urea, LDC, MAC, XLD, HE, SS, PAD, TSI:K/K
Acinetobacter	Obligate aerobe, **Positive tests:** β-lactamase, citrate, glucose, LDC, ODC, MAC, XLD, HE, SS
Found in wounds	**Negative tests:** bile esculin, DNase gelatin, mannitol, motility, MR, nitrate, oxidase, sucrose, urease, VP, PAD TSI; K/K; Variable: lactose
Citrobacter	Facultative anaerobe; **Positive tests:** citrate, glucose, lactose, mannitol, motility, MR, nitrate, sucrose, sulfur, ODC, LDC, MAC, XLD, HE, SS, TSI:A/A+
Found in urine	**Negative tests:** indole, oxidase, urea, PAD, VP
Enterobacter	Facultative anaerobe, capsules, **Positive tests:** citrate, mannitol, motility, nitrate, VP LDC, ODC, MAC, XLD, HE, SS, TSI: TSI:A/@, glucose, sucrose, lactose (gas **O** for all)
Found in urine	**Negative tests:** indole, MR, oxidase, sulfur, urea
Escherichia coli	Facultative anaerobe; **Positive tests:** glucose, lactose, sucrose, mannitol (gas **O**), indole, motility, MR, nitrate, LDC, MAC, XLD, HE, SS, TSI:A/@
Found in urine or food	**Negative tests:** citrate, DNase ODC, oxidase, urea, VP, PAD
Klebsiella	Facultative anaerobe, capsules, mucoid colonies, **Positive tests:** bile esculin, catalase, citrate, glucose, lactose (gas **O** for both), mannitol, MR, TSI:A/@ nitrate, ODC, starch, sucrose, MAC, XLD, HE, SS
Found in urine	**Negative tests:** DNase, indole, lipase, motility, oxidase, sulfur, urea, VP, LDC, PAD
Morganella morganii	Cream colored, mucoid colonies, facultative anaerobe; **Positive tests:** indole, MR, nitrate, ODC, MAC, XLD, HE, SS, TSI: K/A
Found in urine and wounds	**Negative tests:** LDC, oxidase, urea, VP, PAD
Proteus vulgaris	Facultative anaerobe, can swarm over plates; **Positive tests:** indole, glucose (gas **O**), lactose, motility, MR, nitrate, sucrose, sulfur, urea, MAC, XLD, HE, SS, PAD, TSI:A/A+
Found in urine	**Negative tests:** citrate, LDC, ODC, oxidase, VP; Variable: gelatin
Providencia stuartii	Facultative anaerobe, **Positive tests:** glucose (gas **O**), indole, lactose, motility, MR, nitrate, sucrose, urea, MAC, XLD, HE, SS, PAD, TSI:A/A+
Found in urine	**Negative tests:** citrate, oxidase, LDC, ODC, sulfur, VP
Pseudomonas	Aerobe with blue-green or yellow pigments, smells like grapes; **Positive tests:** citrate, gelatin, indole, motility, MR, nitrate, oxidase, LDC, ODC, TSI:K/K
Found in urine and wounds	**Negative tests:** glucose, lactose, indole, urea, VP, MAC, XLD, HE, SS, PAD
Serratia	Facultative anaerobe, can be pigmented; **Positive tests:** citrate, gelatin, motility, glucose, lactose, mannitol, sucrose, VP, nitrate, ODC, LDC, MAC, TSI:A/A, XLD, HE, SS
Found in urine, wounds, soil	**Negative tests:** indole, MR, oxidase, sulfur, urea, ODC, PAD
Salmonella	Facultative anaerobe, **Positive tests:** citrate, glucose, motility, MR, nitrate, sulfur, LDC, ODC
Found in food	**Negative tests:** indole, oxidase, urea, VP, MAC, XLD, HE, SS, PAD, TSI:K/A+
Shigella	Facultative anaerobe; **Positive tests:** citrate, glucose, MR, nitrate, ODC
Found in food	**Negative tests:** motility, oxidase, sulfur, urea, VP, LDC, MAC, XLD, HE, SS, PAD, TSI:K/A Variable: indole
Gram-Negative Cocci (non-motile)	
Moraxella catarrhalis	Diplococci, BAP: γ-hemolysis, obligate aerobe; **Positive tests:** catalase, DNAse, nitrate, oxidase
Found in the mouth	**Negative Tests:** bile esculin, coagulase, colistin S, gelatin, glucose, lactose, maltose, sucrose, salt
Neisseria	Diplococci, BAP: α- or γ-hemolysis, obligate aerobe; **Positive tests:** catalase, oxidase
Found in the mouth	**Negative Tests:** bile esculin, coagulase, colistin S, gelatin, nitrate, salt TSI: K/K; Variable: glucose, lactose, maltose

Figure 20.3 Gram-negative microbe characteristics.

Name ______________________ Date ______________ Section ________

Partner ______________________ Unknown # ____ Source ______________

EXERCISE 20

Laboratory Report: Worksheet and Final Report: Unknown Identification

Gram stain of original specimen: ____

____

(Describe cell shape, arrangement, and Gram reaction)

Gram stains of TS agar subcultures: ____

____

(Describe cell shape, arrangement, and Gram reaction)

Test	Organism #1	Organism #2
Colony description		
(TS agar or blood)		
Hemolysis		
Gram stain		
Colony appearance macConkey (or EMB)		
Colony appearance mannitol salt agar		
Special stains		
Acid fast		
Capsule		
Endospore		
Lactose fermentation		
Glucose fermentation		
Sucrose fermentation		
Mannitol fermentation		
TSI slant		

Test	Organism #1	Organism #2
Sulfur production		
Indole production		
Motility		
Methyl red		
Voges–Proskauer		
Citrate utilization		
Urea hydrolysis		
Catalase test		
Coagulase		
Oxidase		
Oxygen requirements		
XLD		
Hektoen enteric (or SS)		
PAD		
ODC		
LDC		
Nitrate		

Consult the table compiled from the class results at the end of the second session in exercise 19 and figure 17.1 and figures 20.2 and 20.3 to help you identify your organisms.

Final Identification
Organism #1 = _____

Organism #2 = _____

1. What is the identification of your organisms? Discuss the process of identification (reasons for choosing specific tests, any problems, and other comments).

Organism #1:

Organism #2:

21

Chromogenic Agars and All-in-One Tests

Definitions

Chromogenic agar. An agar plate that allows for identification of bacteria based on the color the colony appears.

McFarland standard. A broth tube suspension (numbered 0.5, 1, 2, 3, etc.) that contains a standardized turbidity (and thus have a known spectrophotometer reading). A McFarland standard 3 equals approximately 9×10^8 CFU/ml.

Polymerase chain reaction (PCR). A method used to create millions of copies of a region of DNA in only a matter of hours. A synthetic way to copy a section of DNA by using the enzyme *Taq* polymerase to lay down individual amino acids into a complementary strand of DNA.

Presumptive. In a clinical setting, a presumptive microbiology test is an initial or preliminary test used to identify microorganisms, and the identity must be confirmed using a second testing method.

Turn around times (TAT). In a clinical setting, the time it takes from when a sample is given to a laboratory until a reliable result is reported to a physician or other health care provider.

Objectives

1. Describe the importance of Chromagars.
2. Identify bacteria using a Biolog, EnteroPluri-Test, or API®20E test strip.
3. Describe the importance of using a McFarland standard.

Pre-lab Questions

1. Why is a Chromagar only considered a presumptive test for many microorganisms?
2. What is the difference in creating an anaerobic state in the API®20E test strip versus the EnteroPluri-Test?

3. What reagents must still be added to the API®20E test strip and the EnteroPluri-Test after incubation?

Getting Started

In the previous labs, you have had an opportunity to inoculate and interpret results from many different types of agar and tube media. You may have noticed how time-consuming some of these methods are. Testing phenotypic traits has been the "gold standard" in identification of microbes (**table 21.1**). However, genotypic tests have become utilized more often as the technology has improved, giving reproducible results. Genotypic testing will be discussed in exercises 32–35. In order to facilitate quicker **Turn around times (TAT)** for identifying bacteria, some laboratories use **chromogenic agars** and all-in-one tests.

With chromogenic agars (also called Chromagars), one can identify a microbe to at least genus level based solely on appearance of the colony color. Some chromogenic agars are both selective and differential and can provide multiple colors for differentiation. Selective chromogenic agars usually grow either Gram-positive or Gram-negative bacteria, but are available for yeasts like *Candida* species. New chromogenic agars use multiple chromogens that turn color based on enzyme(s) production, making a rainbow of potential colony colors, allowing identification of many species of Enterobacteriaceae. Some chromogenic agars contain antibiotics and show a **presumptive**, or preliminary, identification to separate drug-resistant strains from drug susceptible strains that must be confirmed by another method. Methicillin-resistant *Staphylococcus aureus* (MRSA) appears dark blue, whereas methicillin-susceptible Staphylococci (MSSA) appear white (color plate 58). The identity is confirmed by latex agglutination test or PCR, to ensure an accurate identification.

Besides chromogenic agars, all-in-one tests can aid in identifying yeasts or bacterial species of Staphylococci, Streptococci, *Haemophilus*, Gram-negative

Table 21.1 Phenotypic and Genotypic Methods Used to Identify Microorganisms

Method	Comments
Phenotypic Characteristics	Most of these methods do not require sophisticated equipment and can easily be done anywhere in the world.
Microscopic morphology	Size, shape, and staining characteristics can give suggestive information as to the identity of the organism. Further testing, however, is needed to confirm the identification.
Culture characteristics	Colony morphology can give initial clues to the identity of an organism.
Metabolic capabilities	A set of biochemical tests can be used to identify a microorganism.
Serological testing	Proteins and polysaccharides that make up a microorganism are sometimes characteristic enough to be considered identifying markers. These can be detected using specific antibodies.
Protein profile	MALDI-TOF MS separates and sorts an organism's proteins by mass, generating a profile that provides a fast way to identify an organism grown in culture.
Genotypic Characteristics	These methods are increasingly being used to identify microorganisms. Even an organism that occurs in very low numbers in a mixed culture can be identified.
Detecting specific nucleo-tide sequences	Nucleic acid probes and nucleic acid amplification tests can be used to identify a microorganism grown in culture. In some cases, the method is sensitive enough to detect the organism directly in a specimen.
Sequencing rRNA genes	This requires amplifying and then sequencing rRNA genes, but it can be used to identify organisms that have not yet been grown in culture.

bacteria, or anaerobes. The tests vary in configuration utilizing different biochemical tests. Results are converted to a numerical sequence that is then looked up in a "code" book to determine the microbe present. Many all-in-one tests are packaged as a test strip or tray of some sort.

Some all-in-one tests that your instructor may include in this laboratory exercise are the bioMérieux API®20E, API® Staph-Ident, Biolog or Liofilchem® EnteroPluri-Test.

The API® (Analytical Profile Index) 20E is used to identify oxidase negative Enterobacteriaceae (**figure 21.1**), whereas the API® StaphIdent and Microgen® Staph are specific for Gram-positive Staphylococci. Each system is a plastic test strip containing dehydrated biochemical tests housed in individual cupules that are filled with a standardized turbidity (called a **McFarland standard**) of microbial suspension of the pure culture and incubated. In order to create an anaerobic state, several drops of sterile mineral oil are pipetted on top of the cupule. After incubation, each test result is read, and a numerical system identifies the bacterium present.

The biolog system uses a plate similar to the Microscan shown in color plate 59. Each well contains dehydrated biochemicals and a uniform turbidity is added to each well and incubated. Each well is read manually or automatically at 8 and 22 hours, depending on the system purchased.

The EnteroPluri-Test is an enclosed, bullet-shaped system in which biochemical media tests are housed in a see-through plastic tube (color plate 57). A sterile inoculating needle is touched to the colony of interest and then pulled through the housing system. As the needle is pulled through each test section, bacteria are deposited into the agars of each section. After incubation, the results of each test are read and a numerical system identifies the bacterium present.

For all identification systems, the unknown microorganism must be compared to known reference strains for a proper identification (**figures 21.1 and 21.2**). Regardless of which system is utilized, reagents usually still need to be added to the Voges–Proskauer, indole, and nitrate tests, etc.

McFarland standard tubes serve as an optical control for making turbid microbial suspensions. Each McFarland standard has a known turbidity range when placed in a spectrophotometer, which corresponds to an approximate microbial concentration. For example, a 0.5 McFarland standard is equal to approximately 1.5×10^8 colony forming units (CFUs)/ml. Isolated colonies being tested are aseptically added, usually with sterile swabs, to a sterile saline suspension held against a white paper with black lines until the desired turbidity is reached. This gives consistency to the concentration added to tests, ensuring test results are ready to read within a specified time frame. In addition, the

API®20E Test System Examples

Test	ONPG/β-galactosidase	ADH	LDC	ODC	CIT	H₂S	URE	TDA	IND	VP	GEL	GLU	MAN	INO	SOR	RHA	SAC	MEL	AMY	ARA	OX
Test Name	Ortho-Nitrophenyl-βD-Galactopyranosidase	Arginine Dihydrolase	Lysine Decarboxylase	Ornithine Decarboxylase	Citrate	Hydrogen Sulfide	Urea	Tryptophan Deaminase	Indole	Voges-Proskauer	Gelatin	Glucose	Mannose	Inositol	Sorbitol	Rhamnose	Saccharose	Melibiose	Amygdalin	Arabinose	Oxidase
Positive	Yellow	Red-orange	Red-orange	Red-orange	Blue-green	Black Diffusion	Red/orange	Reddish brown	Colorless/Lt yellow	Colorless	No diffusion/chunk	Yellow	Yellow	Yellow	Yellow	Yellow	Yellow	Yellow	Yellow	Yellow	Colorless
Negative	Colorless	Yellow	Yellow	Yellow	Yellow	Colorless/Grey	Yellow	Yellow	Pink	Pink/Red	Diffusion of black pigment	Blue-green	Blue-green	Blue-green	Blue-green	Blue-green	Blue-green	Blue-green	Blue-green	Blue-green	Purple
Reagent to add (if applicable)								TDA Reagent (FeCl)	James Reagent (Kovacs)	α-naphthol & KOH Wait 10 min											
Point Value	1	2	4	1	2	4	1	2	4	1	2	4	1	2	4	1	2	4	1	2	4
Escherichia coli	+	–	+	+	–	–	–	–	+	–	–	+	+	–	+	+	–	+	–	+	–
Total of 3 blocks 5144552		5			1			4			4			5			5			2	
Enterobacter cloacae	+	+	–	+	+	–	–	–	–	+	–	+	+	+	+	+	+	+	–	+	–
Total of 3 blocks 3305572		3			3			0			5			7			7			2	
Providencia stuartii	–	–	–	–	+	–	+	+	–	–	–	+	–	+	–	–	+	–	–	–	–
Total of 3 blocks 234220		0			2			3			4			2			2			0	
Klebsiella pneumoniae	+	–	+	–	+	–	+	–	–	–	+	+	+	+	+	+	+	+	+	+	–
Total of 3 blocks 5216773		5			2			1			6			7			7			3	
Citrobacter freundii	+	+	–	+	+	+	–	–	+	–	–	+	+	–	+	+	–	–	+	+	–
Total of 3 blocks 3744513		3			7			4			4			5			1			3	
Serratia marcescens	+	–	+	+	+	–	–	–	–	–	+	+	+	+	–	+	–	+	+	–	–
Total of 3 blocks 5306351		5			3			0			6			3			5			1	
Proteus vulgaris	+	–	–	–	+	+	+	+	–	–	–	+	–	–	–	–	+	–	–	–	–
Total of 3 blocks 0634020		0			6			3			4			0			2			0	

Figure 21.1 API examples. **Source:** API®

Test	Glucose	Gas	Lysine	Ornithine	H₂S	Indole	Adonitol	Lactose	Arabinose	Sorbitol	Voges–Proskauer	Dulcitol	Phenylalanine	Urea	Citrate
Positive	Yellow	lifted wax	Purple	Purple	Black	Pink/Red	Yellow	Yellow	Yellow	Yellow	Red	Yellow	Dark Brown	Purple	Blue
Negative	Red	wax intact	Tan	Tan	Tan	Tan	Red	Red	Red	Red	Colorless	Green	Green	Tan	Green
Reagent to add (if applicable)						Add 3-4 drops of Kovac's. Wait 10-15 sec.					Add 3 drops of α-napthol & 2 drops of KOH. Wait 20 min.				
Point Value	4	2	1	4	2	1	4	2	1	4	2	1	4	2	1
Escherichia coli	+	+	+	+	–	+	–	+	+	+	–	–	–	–	–
Total of 3 blocks		7			5			3			4			0	
75340															
Enterobacter cloacae	+	+	–	+	–	–	–	+	+	+	+	–	–	–	+
Total of 3 blocks		6			4			3			6			1	
64361															
Providencia stuartii	+	–	–	–	–	+	–	–	–	–	–	–	+	+/–	+
Total of 3 blocks		4			1			0			0			5	
41005															
Klebsiella pneumoniae	+	+	+	–	–	–	+	+	+	+	+	+	–	+	+
Total of 3 blocks		7			0			7			7			3	
70773															
Citrobacter freundii	+	+	–	+/–	+	–	–	+/–	+	+	–	+/–	–	+/–	+
Total of 3 blocks		7			2			1			4			1	
72141															
Serratia marcescens	+	+/–	+	+	–	–	–	–	–	+	+	–	–	–	+
Total of 3 blocks		5			4			0			6			1	
54061															
Proteus vulgaris	+	+	–	–	+	+	–	–	–	+	–	–	+	+	–
Total of 3 blocks		6			3			3			4			6	
63346															

Figure 21.2 EnteroPluri-Test examples. **Source:** API®

unknown test results will more likely match a known result in the database. McFarland standards are often used in commercial, industrial, research, and clinical laboratories.

The API®20E contains additional tests not discussed in previous exercises. The Ortho-Nitrophenyl-β-D-Galactopyranoside (ONPG) test detects the presence of the enzyme β-galactosidase. The ONPG serves as the substrate for detecting β-galactosidase production, which is one enzyme required for lactose fermentation; a permease is the other. The β-galactosidase breaks lactose into galactose and ortho-nitrophenol, but is not considered a lactose (*lac*) operon inducer. Another test utilizes tryptophan deaminase (TDA), which deaminates the amino acid tryptophan into indolepyruvic acid and ammonia. After incubation, ferric chloride is added, reacting with any pyruvic acid, and turning the section a reddish brown color. The arginine dihydrolase test determines if the bacterium can utilize the amino acid arginine as a sole carbon source. Additional sugars potentially used in fermentation include sorbitol, rhamnose, saccharose, melobiose, arabinose, and amygdalin. The tests will change from red to yellow when acid is produced and the pH decreases. Additional EnteroPluri-Test sections include the sugars of adonitol and dulcitol, which also detects sugar fermentation.

Some other automated identification systems include Vitek,® MicroScan, Phoenix, and Matrix-Assisted Laser Desorption/Ionization Time-Of-Flight (MALDI-TOF). With MALDI-TOF, the bacterium is added to a stainless steel plate, which is then identified within 3 seconds. The Vitek® and Microscan systems all require bacteria to be suspended in solution, which is added to testing wells, and then incubated. After incubation, reagents are added to some tests and results are read and compared to an electronic database. The database provides the most probable microbial match. All of these systems require a grown pure culture of the microorganism being tested, which is one reason the **polymerase chain reaction (PCR)** and quantitative real-time PCR (qPCR) will eventually replace conventional methods.

Larger laboratories often use an automated system where the identification and antibiotic sensitivity are performed simultaneously. Advantages are that once the microbe has been isolated, the system identifies the microbe and sensitivity, often in just a couple of hours. This allows treatments to be started more quickly.

Disadvantages of automated systems are that they are expensive to buy and maintain. In addition, the patient testing cards, trays, or wells are also very expensive—some cards and trays cost upwards of $900 each. Many of the database systems are updated frequently, but can cost $10,000–$25,000 or more per year for a subscription to access a database to identify cultured bacteria from specimens.

Materials

First Session
Per team

   Chromagar plate (a non-selective UTI Chromagar is recommended to test more bacteria than listed), 1

   Uninoculated Chromagar plate (negative control), 1

   Cultures (24 hours, 37°C) of:

   Escherichia coli

   Klebsiella pneumoniae

   Enterobacter cloacae

   Providencia stuartii

   Serratia marcescens

   Citrobacter freundii

   Proteus vulgaris (optional)

Second Session
Per Team

   EnteroPluri-Test, 1

   OR

   Biolog plate for Gram-negatives, 1

   OR

API®20E test strip, 1

API®20E plastic bottom tray holder and top cover, 1

Fine/thin sterile pipet and bulb, 1

Test tube containing 5 ml of sterile distilled H₂O, 1

Test tube containing 10 ml of sterile 0.85% saline solution, 1

McFarland standard 3, 1

Sterile mineral oil in tubes, 1

Sterile pipet for mineral oil transfer, 1

Inoculating fluid tubes for Biolog

Inoculation swab

Demonstration fluid tubes at 95% turbidity

Spectrophotometer

GEN microplate

Pipettor for 100 µl (multichannel for 8 tips is preferred)

Reservoir plate for inoculating fluid

Pipet tips for pipettor

Third Session
Per team

   Droppers of Kovac's reagent

   Barritt's A and Barritt's B

   Ferric chloride

   Pipets with bulbs or tips, 3

First Session

1. Label a Chromagar plate with the organism assigned, your name, date, temperature, and laboratory section.

2. Flame your loop and allow it to cool. Inoculate a Chromagar plate with a loopful of assigned bacteria and streak for isolation.

3. Incubate the plate at 37°C for 18–24 hours. Fresh growth is needed for the second session and older colonies may not give proper reactions.

Second Session

1. Obtain your Chromagar plate. Look at the colony color(s), size, shape, etc. Does the colony you were assigned match how it should appear? Record your results in the Laboratory Report.

API®20E System

1. Obtain a API®20E test strip and the top (no slots) and bottom (slots) plastic covers. Label the plastic edge overhang with the organism assigned, your name, date, temperature, and laboratory section.

2. Add about 5 ml of sterile water to the bottom of the tray, filling the bottom slots using pipets or water bottles. Do not overfill them. The test strip should not float on any water and any excess should be removed before the test strip is added.

3. Using a sterile swab, obtain several (3–4) medium sized colonies from the Chromagar plate.

4. Aseptically, add the sterile swab of organisms to a sterile 0.85% saline tube until it reaches a McFarland standard of 3. Tip: You can quickly swirl the swab in both directions to greatly increase the turbidity.

5. Discard the swab in the proper container.

6. If more turbidity is needed, obtain a new sterile swab and repeat steps 3–5.

7. Lay the test strip on the tray.

8. Obtain a sterile glass pipet with bulb, and hold the test strip and bottom tray vertical until the bottom cupule portion is filled.

9. **Slowly** pipet the saline solution containing bacteria into **ALL** of the cupules. Pipet very slowly to avoid bubbles forming in the cupules. Bubbles can be difficult to remove once more liquid is added to the cupules. A reaction cannot occur if the bacteria do not contact the reagents in the cupules. The larger the pipet tip, the more difficult it is to pipet very slowly.

10. Under fill the (underlined) labeled sections of ADH, LDC, URE, H_2S, and ODC. It should be filled to just the beginning of the curved section of the plastic in a U shape.

11. **Add** 3–4 drops of **oil** to fill up the ADH, LDC, ODC, H_2S, and URE tests to the top of the cups. This creates an anaerobic state.

12. Properly discard the pipets or your instructor may have you leave the pipet in oil for other groups.

13. Incubate at 37°C for 18–24 hours.

Biolog

1. Obtain a Gram-negative microplate and label with the organism, name, date, temperature, and laboratory section.

2. Choose a 3 mm colony to add to the inoculation fluid tubes. Use a vertical "stamping" motion to touch the colony with a sterile cotton-tipped inoculation swab.

Add the swab to the fluid and swirl the swab in the tube bottom to suspend the colonies.

3. Use the demonstration tube at 95% transmission to use as a turbidity guide.

4. Blank an uninoculated inoculation fluid tube.

5. Check that the fluid tube you inoculated shows 95% transmission and pour the fluid tube liquid into the multichannel pipet reservoir.

6. Pipet 100 µl of inoculation fluid containing the bacterial suspension into each well.

The suspension will form a gel-like consistency once inoculated into the wells.

7. Eject the tips into the proper disposal container.

8. Cover the plate with its lid and incubate at 33°C for 24 hours. At least three of the carbon source wells need to be positive before reading the plate.

EnteroPluri-Test

1. Obtain an EnteroPluri-Test. Label the test with the organism assigned, your name, date, temperature, and laboratory section.

2. Choose a couple of medium-sized isolated colonies to test, and circle them using a marking pen on the plate bottom.

3. Remove the blue-colored endcap first by twisting counter-clockwise. You will see a looped over metal handle. The inoculating needle you will use to pick up a bacterial colony will be under the white cover. Do **NOT** flame this needle.

4. Once you are ready to inoculate the test, remove the white endcap by turning counter-clockwise.

5. Gently press the needle to the desired isolated colonies, moving gently back and forth and rotating the needle so all sides are covered in bacteria. Be careful so as to not dig into the agar. The plastic from the tube should also not come into contact with the bacteria.

6. Using the handle end, gently pull the needle through each testing section. Once the needle comes out, reinsert the needle back into where you just pulled it out from, so the handle is still by the citrate end (and opposite the glucose).

7. You will see a small indentation on the wire. Line this indentation up with the plastic end of the housing, break the needle handle off at the notch, leaving the needle inside the tube.

8. Keep the broken off needle handle in your hands so it does not touch a surface and become contaminated.

9. Replace the two end covers by twisting the caps clockwise onto the housing unit.

10. Turn the tube so you can see the small U's in the test sections of the tube.

11. Using the broken off part of the needle handle, punch holes in the U's of the plastic film to the adonitol, lactose, arabinose, sorbitol, VP, dulcitol/PA, urea, and citrate sections. These sections require aerobic growth conditions.

12. Discard the handle wire piece in the proper container.

13. Incubate the tube at 37°C for 18–24 hours either on the flat side horizontally, or in a test tube rack with the glucose section pointed up into the air.

Third Session

API®20E System

1. Add reagents to the following tests before reading the test results.

2. Add 1 drop of Kovac's reagent to the IND test. Wait up to 2 minutes for a red ring to develop.

3. Add 1 drop of Barritt's A and 1 drop of Barritt's B to the VP test. Wait up to 10 minutes for a red color to develop.

4. Add 1 drop of ferric chloride to the TDA test.

5. Place a positive or negative sign (+ or −) under the block for each test.

6. Add up the points of the positive tests in the blocks of three.

7. Each positive result is worth either a 4, 2, or 1 so the number entered into the block will be a 0 (all negative tests), 1, 2, 3, 4, 5, 6, or 7.

8. Place the number for the first block as the 1st number in the sequence, the 2nd, etc. to determine the number. This number can then be looked up in the API®20E reference or quick reference book.

9. Record your results in the Laboratory Report.

Biolog

1. Obtain your microplate and make sure at least three carbon source wells appear positive.

2. Purple wells are considered positive.

3. If reading manually, record the positive wells in the Laboratory Report and then look up the results in the book.

4. If the microplate is read spectophotometrically, a plate reader can be used if it is set to 750 nm. Record your results in the Laboratory Report.

EnteroPluri-Test

1. Add 3–4 drops of Kovac's reagent to the H$_2$S/IND test by using a pipet tip to puncture the plastic film and add the reagent. Wait 10–15 seconds for a pink or red color to develop.

2. Place the tip in the proper disposal container.

3. Add 3 drops of Barritt's A and 2 drops of Barritt's B to the VP test by using a pipet tip to puncture the plastic film and add the reagent. Wait up to 20 minutes for a red color to develop.

4. Place the tip in the proper disposal container.

5. Place a positive or negative sign (+ or −) under the block for each test.

6. Add up the points of the positive tests in the blocks of three. Record your results in the Laboratory Report.

7. Each positive result is worth either a 4, 2, or 1 so the number entered into the block will be a 0 (all negative tests), 1, 2, 3, 4, 5, 6, or 7.

8. Dispose of the plate or tubes in the proper disposal container.

EXERCISE

21

Laboratory Report: Chromogenic Agars and All-in-One Tests

Results and Questions

The API and EnteroPluri strips are meant to be used with 24 hours, oxidase-negative Gram-negative bacteria grown in pure cultures.

Oxidase Result? + or –   Bacterium given? _____

Chromagar Results

1. What is the color of the uninoculated Chromagar? _____ 2. Bacterium used? _____
3. Colony appearance on Chromagar (size, shape, color, etc.)? _____
4. Does the colony morphology match the bacterium you were given?   Yes   No

Biolog Results

	1	2	3	4	5	6	7	8	9	10	11	12
A	A1 Negative Control	A2 Dextrin	A3 D-Maltose	A4 D-Trehalose	A5 D-Cellobiose	A6 Gentiobiose	A7 Sucrose	A8 D-Turanose	A9 Stachyose	A10 Positive Control	A11 pH 6	A12 pH 5
B	B1 D-Raffinose	B2 α-D-Lactose	B3 D-Melibiose	B4 β-Methyl-D-Glucoside	B5 D-Salicin	B6 N-Acetyl-D-Glucosamine	B7 N-Acetyl-β-D-Mannosamine	B8 N-Acetyl-D-Galactosamine	B9 N-Acetyl Neuraminic Acid	B10 1% NaCl	B11 4% NaCl	B12 8% NaCl
C	C1 α-D-Glucose	C2 D-Mannose	C3 D-Fructose	C4 D-Galactose	C5 3-Methyl Glucose	C6 D-Fucose	C7 L-Fucose	C8 L-Rhamnose	C9 Inosine	C10 1%Sodium Lactate	C11 Fusidic Acid	C12 D-Serine
D	D1 D-Sorbitol	D2 D-Mannitol	D3 D-Arabitol	D4 myo-Inositol	D5 Glycerol	D6 D-Glucose-6-PO4	D7 D-Fructose-6-PO4	D8 D-Aspartic Acid	D9 D-Serine	D10 Troleandomycin	D11 Rifamycin SV	D12 Minocycline
E	E1 Gelatin	E2 Glycyl-L-Proline	E3 L-Alanine	E4 L-Arginine	E5 L-Aspartic Acid	E6 L-Glutamic Acid	E7 L-Histidine	E8 L-Pyroglutamic Acid	E9 L-Serine	E10 Lincomycin	E11 Guanidine HCl	E12 Niaproof 4
F	F1 Pectin	F2 D-Galacturonic Acid	F3 L-Galactonic Acid Lactone	F4 D-Gluconic Acid	F5 D-Glucuronic Acid	F6 Glucuronamide	F7 Mucic Acid	F8 Quinic Acid	F9 D-Saccharic Acid	F10 Vancomycin	F11 Tetrazolium Violet	F12 Tetrazolium Blue
G	G1 p-Hydroxy-Phenylacetic Acid	G2 Methyl Pyruvate	G3 D-Lactic Acid Methyl Ester	G4 L-Lactic Acid	G5 Citric Acid	G6 α-Keto-Glutaric Acid	G7 D-Malic Acid	G8 L-Malic Acid	G9 Bromo-Succinic Acid	G10 Nalidixic acid	G11 Lithium Chloride	G12 Potassium Tellurite
H	H1 Tween 40	H2 γ-Amino Butyric Acid	H3 α-Hydroxy-Butyric Acid	H4 β-Hydroxy-D,L-Butyric Acid	H5 α-Keto-Butyric Acid	H6 Acetoacetic Acid	H7 Propionic Acid	H8 Acetic Acid	H9 Formic Acid	H10 Aztreonam	H11 Sodium Butyrate	H12 Sodium Bromate

Source: GEN III MicroPlate™, Instructions for Use. Biolog, Inc.

1. What might be an explanation for a CHROMagar colony not matching the description for a given bacterium?

2. Why might you see different results on an all-in-one test than what a known microbe should theoretically test as?

3. Give one explanation of why might you see different results for the same bacterium on an API®20E test strip than on an EnteroPluri-Test?

4. What is the purpose of a McFarland standard, and how do you know if you have made your suspension properly?

INTRODUCTION to Microbial Genetics

In this section, three aspects of microbial genetics will be studied: selection of mutants, gene transfer, and gene regulation.

Microorganisms and their mutations can be identified using different methodologies (**figure I.22.1**). You have already utilized biochemical tests in exercises 16–20, and serological testing was also performed on Streptococci to differentiate various species. Serological testing can also be utilized on various serotypes of Salmonella to determine the specific type. Molecular typing and bacteriophage typing will be performed in exercises 28 and 32–36, respectively. Antibiotic susceptibility was also performed in exercise 14.

Selection of Mutants   Mutations are constantly occurring in all living things. The replication of DNA is amazingly error-free, but about once in every 100 million duplications of a gene, a change is made. There are three possible outcomes:

1. There is no effect. Perhaps the altered base did not lead to structural change in a protein and the cell remained functional.

2. The mutation has affected a critical portion of an essential protein, resulting in the death of the cell.

3. The mutation has enabled the cell to grow faster or survive longer than the other non-mutated cells. This outcome rarely occurs.

Gene Transfer   Bacteria can transfer genetic material to other bacteria in three ways. These are conjugation, transduction, and transformation (**figure I.22.2**).

Conjugation occurs during cell-to-cell contact and is somewhat similar to sexual recombination seen in other organisms. The transferred DNA can be either chromosomal or a small, circular, extra piece of DNA called a plasmid.

Transduction is the transfer of genes from one bacterial cell to another by a bacterial virus. These viruses, called bacteriophage or phage, package a bacterial gene along with the viral genes and transfer it to a new cell.

The third method of transferring genes is transformation, which is also called DNA-mediated transformation. (The word *transformation* is sometimes used to define the change of normal animal cells to malignant cells—a completely different system.) In bacterial transformation, isolated DNA is mixed with viable cells. It then enters the cells, which are able to express these new genes. Although it seems impossible for a large molecule such as DNA to enter through the cell wall and membrane of a living cell, this is indeed what happens.

Gene Regulation   Another aspect of genetics is the expression of genes. A cell must be economical with its energy and material and must not make enzymes or other products when they are not needed. On the other hand, a cell has to be able to "turn on" genes when they are required in a particular environment. Gene regulation is examined in exercise 24.

Method	Comment
Biochemical typing	Biochemical tests are most commonly used to identify various species of bacteria, but in some cases they can be used to distinguish different strains. A group of strains that have a characteristic biochemical pattern is called a biovar or a biotype.
Serological typing	Proteins and carbohydrates that vary among strains can be used to differentiate strains. A group of strains that have a characteristic serological type is called a serovar or a serotype.
Molecular typing	Gel electrophoresis can be used to detect restriction fragment length polymorphisms (RFLPs). Whole-genome sequencing (WGS) is increasingly being used to detect differences.
Phage typing	Strains of a given species sometimes differ in their susceptibility to various types of bacteriophages.
Antibiograms	Antibiotic susceptibility patterns can be used to characterize strains.

Figure I.22.1 Methods to characterize microorganisms.

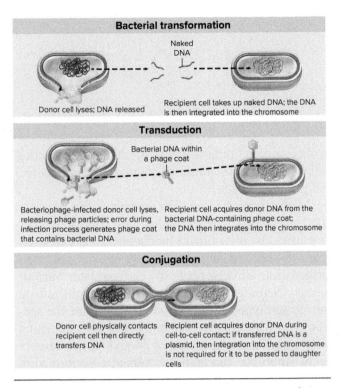

Bacterial transformation

Naked DNA

Donor cell lyses; DNA released

Recipient cell takes up naked DNA; the DNA is then integrated into the chromosome

Transduction

Bacterial DNA within a phage coat

Bacteriophage-infected donor cell lyses, releasing phage particles; error during infection process generates phage coat that contains bacterial DNA

Recipient cell acquires donor DNA from the bacterial DNA-containing phage coat; the DNA then integrates into the chromosome

Conjugation

Donor cell physically contacts recipient cell then directly transfers DNA

Recipient cell acquires donor DNA during cell-to-cell contact; if transferred DNA is a plasmid, then integration into the chromosome is not required for it to be passed to daughter cells

Figure I.22.2 Methods of horizontal gene transfer.

Definitions

Antibiotic. A chemical produced by certain molds and bacteria that kills or inhibits other organisms.

Mutation. A change in the nucleotide sequence of a cell's DNA that is then passed on to daughter cells.

Mutation rate. The number of mutations per cell division.

Sensitive. An organism killed or inhibited by a particular antibiotic.

Wild type. An organism that has the typical characteristics of the species isolated from nature.

Objectives

1. Explain the concept of selection and its relationship to mutation.

2. Explain that mutations are random events and that the cell cannot cause specific mutations to occur, no matter how advantageous they may be.

3. Calculate the number of streptomycin-resistant mutant bacteria that occur in an overnight culture of a sensitive strain.

Pre-lab Questions

1. Explain how mutations and horizontal gene transfer select for a large diversity of microorganisms.$^{ASM\ 2}$

2. Explain what types of genetic variations can impact microbial functions like drug resistance.$^{ASM\ 15}$

3. What unique cell structures do bacteria have that can be a target for antibiotics?$^{ASM\ 7}$

Getting Started

All the bacterial cells in a pure culture are derived from a single cell. These cells, however, are not identical because all genes tend to mutate and form mutant organisms. The spontaneous **mutation rate** of genes varies between 1 in 10^4 to 1 in 10^{12} cell divisions, and even though they are quite rare events, significant mutations are observed because bacterial populations are very large. In a bacterial suspension of 10^9 cells/ml, one could expect 10 mutations of a gene that mutated 1 in every 10^8 divisions.

Mutant bacteria usually do not grow as well as the **wild-type** normal cell because most changes are harmful, or at least not helpful. If, however, conditions change in the environment and favor a mutant cell, it will be able to outcompete and outgrow the cells that do not have the advantageous mutation. It is important to understand that the mutation is a random event that the cell cannot direct. No matter how useful a **mutation** may be in a certain situation, it just happens to the cell, randomly conferring an advantage or disadvantage to it.

In this exercise, you will select bacteria resistant to streptomycin. Streptomycin is an **antibiotic** that kills bacteria by acting on their ribosomes to prevent protein synthesis. (However, it does not stop protein synthesis in animals because eukaryotic ribosomes are larger than those of bacteria and therefore different.) **Sensitive** *Escherichia coli* cells can become resistant to streptomycin with just one mutation.

In this exercise, you will select organisms resistant to streptomycin by adding a large population of sensitive bacteria to a bottle of trypticase soy broth containing streptomycin. Only organisms that already had a random mutation for streptomycin resistance will be able to survive and multiply (**figure 22.1**).

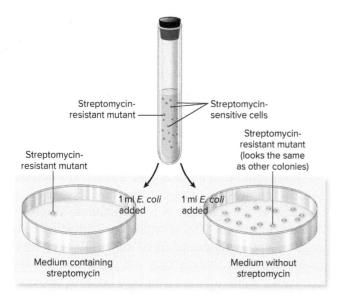

Figure 22.1 Selection of streptomycin-resistant *Escherichia coli* cells.

Labels in figure:
Streptomycin-resistant mutant
Streptomycin-sensitive cells
Streptomycin-resistant mutant (looks the same as other colonies)
Streptomycin-resistant mutant
1 ml *E. coli* added
1 ml *E. coli* added
Medium containing streptomycin
Medium without streptomycin

Materials

Per team

First Session

Flasks (or bottles) containing 50 ml trypticase soy broth, 2

Trypticase soy agar deeps, 2

Sterile Petri dishes, 2

1 ml pipets, 2 or 1 ml micropipettor with pipet tips

Overnight broth culture (~18 hours) of *E. coli* K12 (about 10^9 cells/ml), 5 ml

Streptomycin solution at 30 mg/ml

Second Session

Trypticase soy agar deeps, 2

Sterile Petri dishes, 2

1 ml pipets, 2

Tubes of 0.5 ml trypticase soy broth or sterile water, 2

0.1 ml streptomycin

First Session

1. Melt and place two trypticase soy agar deeps in a 50°C water bath. These will be used to make pour plates. Once tubes are removed from the water bath, you will need to work quickly to avoid the agar from solidifying.

2. Label one Petri plate and one flask "with streptomycin." Label the second flask and plate "Control-No streptomycin" (**figure 22.2**).

3. Add 0.3 ml (300 µl) streptomycin to the flask labeled "with streptomycin" and 0.1 ml (100 µl) to one of the melted, cooled agar deeps. Discard the pipet or pipet tip into the proper container.

4. Immediately inoculate the trypticase soy agar deep with 1 ml (1,000 µl) of the *E. coli* culture, mix well, and pour into the plate labeled "with streptomycin." Avoid moving the plate until the plate has completely solidified, around 15 minutes or so. Only organisms that already have mutated to streptomycin resistance will be able to grow on this plate. Discard the pipet or pipet tips into the proper container.

5. Add 1 ml (1,000 µl) of *E. coli* to the tube of melted, agar without streptomycin and pour into the plate labeled "Control-No streptomycin."

6. Add 1 ml (1,000 µl) of *E. coli* to each flask.

7. Incubate the plates and flasks at 37°C for 24–48 hours. If using bottles, lay them on their side to increase aeration. Make sure there are no holes in the sterile tin foil used to cover the flasks.

Second Session

1. Melt and cool two tubes of trypticase soy agar in a 50°C water bath.

2. Pour one tube of melted agar into a Petri dish labeled "without streptomycin" and let harden.

3. Add 0.1 ml streptomycin to the other tube of melted agar, pour into a Petri dish labeled "with streptomycin," and let harden.

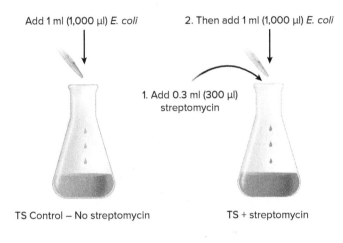

Add 1 ml (1,000 μl) E. coli

2. Then add 1 ml (1,000 μl) E. coli

1. Add 0.3 ml (300 μl) streptomycin

TS Control – No streptomycin

TS + streptomycin

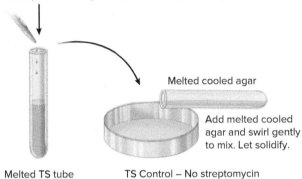

1. Add 1 ml (1,000 μl) E. coli to deep tube. Gently mix the tube and pour into plate.

Melted cooled agar

Add melted cooled agar and swirl gently to mix. Let solidify.

Melted TS tube

TS Control – No streptomycin

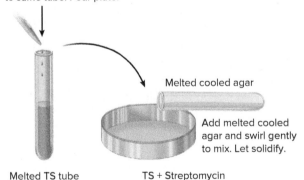

1. Add 0.1 ml (100 μl) streptomycin to deep tube.

2. Then add 1 ml (1,000 μl) E. coli to same tube. Pour plate.

Melted cooled agar

Add melted cooled agar and swirl gently to mix. Let solidify.

Melted TS tube

TS + Streptomycin

Figure 22.2 Inoculating media with and without streptomycin with a culture of *E. coli* (first session).

4. Examine the bottles and plates inoculated during the first session. Note whether there is growth (turbidity) or not in both of the bottles. Count the number of colonies growing in the pour plates. How many streptomycin-resistant mutants/ml were present in the original inoculum? (See exercise 8 for calculations.) Compare it to the growth of organisms in the control plate without streptomycin. If there are more than 300 colonies or the plate is covered by confluent growth, record as TNTC or "too numerous to count." Record the results.

5. Test the bacteria growing in the bottles and on the plates for sensitivity or resistance to streptomycin in the following way: divide both agar plates into four sections as diagrammed in **figure 22.3**, then take a loopful of broth from the bottle without streptomycin and inoculate the first quadrant sector of each agar plate. Do the same with the broth culture containing streptomycin, adding it to the second quadrant.

6. Suspend some organisms from the control plate in a tube of trypticase soy broth or sterile water (there will not be any isolated colonies) and inoculate the third quadrant. Dig an isolated colony out of the agar plate containing the streptomycin and suspend it in a tube of trypticase soy broth or sterile water. Use a loopful to inoculate the fourth quadrant of each plate. Incubate the plates at 37°C for 24–48 hours.

7. Predict which bacteria will be sensitive to streptomycin and which will be resistant.[ASM 28b & 29]

Third Session

1. Observe the growth on each sector of the plates and record the results. Were they as you predicted?

2. Occasionally, mutants not only will be resistant to streptomycin but also will require it. If you have one of these unusual mutants, be sure to show it to the instructor.

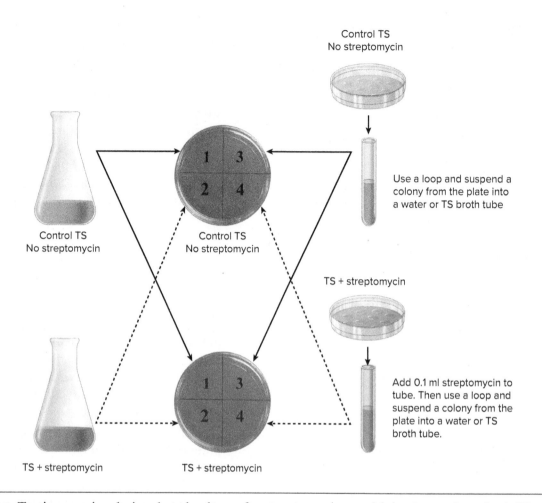

Control TS
No streptomycin

Use a loop and suspend a
colony from the plate into
a water or TS broth tube

Control TS
No streptomycin

Control TS
No streptomycin

TS + streptomycin

Add 0.1 ml streptomycin to
tube. Then use a loop and
suspend a colony from the
plate into a water or TS
broth tube.

TS + streptomycin

TS + streptomycin

Figure 22.3 Testing previously incubated cultures for streptomycin sensitivity (second session). The middle plates are poured the day of the second session. The plates on the right are what grew the first session. A colony is suspended in broth and then added to each plate. Courtesy of Anna Oller, University of Central Missouri.

EXERCISE

22

Laboratory Report: Selection of Bacterial Mutants Resistant to Antibiotics ASM 7&14

Results (Second Session)

Source	Growth / No Growth	Number of Colonies
TS broth (control)		
TS broth plus streptomycin		
TS agar plate (control)		
TS agar plate plus streptomycin		

Results (Third Session)

Source	Growth on TS Agar Plate	Growth on TS Agar Plate + Streptomycin
TS broth (control)		
TS broth plus streptomycin		
TS agar (control)		
TS agar plus streptomycin		

1. How many organisms/ml were streptomycin resistant in the original overnight culture of sensitive *E. coli*?

2. Two bottles of trypticase soy broth (with and without streptomycin) were inoculated in the first session with 1 ml of an overnight culture of *E. coli*. After incubation, why was one population sensitive to streptomycin and the other population resistant?

3. How were you able to estimate the number of streptomycin-resistant organisms already present in the overnight culture of *E. coli* growing in the trypticase soy broth?

4. Why should antibiotics only be used if they are necessary?

5. Which is correct?

   a. An organism becomes resistant after it is exposed to an antibiotic.

   b. An antibiotic selects organisms that are already resistant.

6. Why is streptomycin ineffective for treating viral diseases?

23

Transformation: A Form of Genetic Recombination

Definitions

Competent. In horizontal gene transfer, a physiological condition in which the cell is capable of taking up DNA.

DNase. An enzyme that cuts DNA, making it useless for transformation.

Lysate. The solution that contains cell contents like membranes, DNA, and proteins.

Naked DNA. DNA released from lysed, or disrupted, cells and no longer protected by an intact cell.

Objectives

1. Describe the process of transformation, and observe it in the laboratory.
2. Interpret the use of genetic markers.
3. Explain the importance of controls in an experiment.

Pre-lab Questions

1. Why is *Acinetobacter* used for the transformation instead of other genera of bacteria?
2. How do you lyse cells to release DNA?
3. What is the purpose of adding streptomycin to the TSY plate?

Safety Precaution: *Acinetobacter* can cause pneumonia in immunologically compromised individuals.

Getting Started

In this exercise, transformation is used to transfer the genes of one bacterium to another. It gives you a chance to see the results of what seems to be an impossible process—a huge DNA molecule entering an intact cell and permanently changing its genetic makeup.

Basically, the process involves mixing DNA from one strain of lysed (disrupted) cells with another strain of living cells. The DNA then enters the viable cells and is incorporated into the bacterial chromosome. The new DNA is expressed, and the genetic capability of the cell may be changed.

In order to determine whether the bacteria are indeed taking up additional DNA, the two sets of organisms (DNA donors and DNA recipients) must differ in some way. One strain usually has a "marker," such as resistance to an antibiotic or the inability to synthesize an amino acid or vitamin. In this exercise, a gene responsible for conferring resistance to the antibiotic streptomycin is transferred to cells that are sensitive to it (**figure 23.1**).

The antibiotic streptomycin will be used in this exercise. Antibiotics can be added to agar plates in numerous ways. You saw antibiotic discs being used by placing them on the agar surface in exercise 14. Most antibiotics need to be filter sterilized if not purchased in a sterile solution. The high heat of autoclaving will denature most antibiotics. Antibiotics can then either be aseptically added to cooled agar right before pouring plates or pipetted and spread on top of the agar surface and allowed to be absorbed.

The organism used in this exercise is *Acinetobacter* (a-sin-NEET-o-bacter), a short, Gram-negative rod found in soil and water. The prefix *a* means "without," and *cine* means "movement," as in cinema; thus, *Acinetobacter* is non-motile. This organism is always **competent**, which means it can always take up **naked DNA**. Most bacteria are not competent unless they are in a particular part of the growth curve or in a special physiological condition. Many times chemicals must be added to make a cell competent and able to take up DNA. In any event, for transformation to occur, the DNA must not be degraded (chemically broken down). If an enzyme such as **DNase** is present, it cuts the DNA into small pieces, preventing transformation.

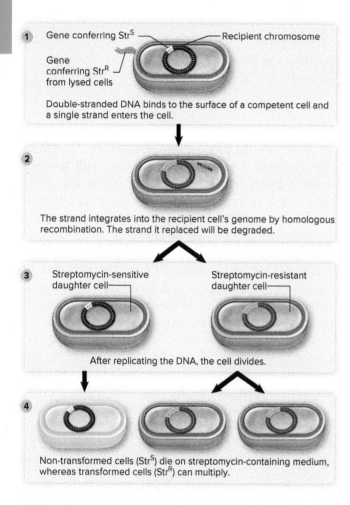

1 Gene conferring Str^S — Recipient chromosome

Gene conferring Str^R from lysed cells

Double-stranded DNA binds to the surface of a competent cell and a single strand enters the cell.

2 The strand integrates into the recipient cell's genome by homologous recombination. The strand it replaced will be degraded.

3 Streptomycin-sensitive daughter cell — Streptomycin-resistant daughter cell

After replicating the DNA, the cell divides.

4 Non-transformed cells (Str^S) die on streptomycin-containing medium, whereas transformed cells (Str^R) can multiply.

Figure 23.1 Transformation of cells with a gene conferring streptomycin resistance.

In this exercise, aseptic technique is imperative for the experiment to work properly. *Acinetobacter* that is resistant to the antibiotic streptomycin will be used as the source of donor DNA. When the bacterial culture is subjected to the detergent SDS (sodium dodecyl sulfate) with high heat, the *Acinetobacter* cells should lyse, releasing the DNA into the solution. When you use a loop to add the DNA lysate to the TSY plate, if all the bacteria were lysed, you should not see any bacterial growth. The broth culture of the streptomycin-susceptible *Acinetobacter* also serves as a control to make sure you have a pure culture of growth. The streptomycin-resistant *Acinetobacter* cells also serve as a control to ensure antibiotic resistance and that you had a pure culture of bacterial growth. You will then add a loopful of streptomycin-susceptible cells to

your agar plate. Make sure you have a healthy loopful of the cooled DNA lysate and add it on top of the *Acinetobacter* cells in the *d* sector. Lastly, add the streptomycin-susceptible *Acinetobacter* cells to the last sector. Next add a healthy loopful of DNase, which should degrade the bacterial DNA. Then add the loopful of the DNA lysate on top of those two items. Transformed cells need some time to make the newly inserted DNA permanent and allow for metabolic pathway changes, which is why TSY agar is used first. Then the bacterial growth from the first plate is transferred to a plate containing streptomycin to see if transformation actually occurred. Another genus of bacteria known to possess competence is *Neisseria* so you could utilize a resistant and susceptible culture of *Neisseria sicca* using the antibiotic colistin.

Materials

Per team

TSY (trypticase soy yeast extract) agar plate, 1

TSY agar plate with streptomycin (second session), 1

Broth culture of *Acinetobacter* Str^R (resistant to streptomycin), 1

Broth culture of *Acinetobacter* Str^S (sensitive to streptomycin), 1

Tube with 0.1 ml detergent SDS (sodium dodecyl sulfate) in 10× saline citrate, 1

Solution of DNase

1 ml pipet, 1

Class equipment

60°C water bath with test tube rack

PROCEDURE

First Session

1. Transfer 1 ml of Str^R *Acinetobacter* broth culture into the tube of SDS. Label the tube and incubate it in a 60°C water bath for 30 minutes. The detergent (SDS) will lyse the cells, releasing DNA and other cell contents into the solution.

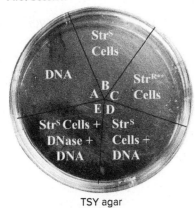

First Session

Str^S Cells

DNA

A B C Str^{R**} Cells

E D

Str^S Cells + DNase + DNA

Str^S Cells + DNA

TSY agar

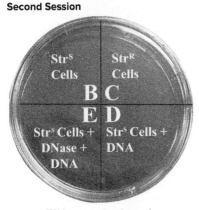

Second Session

Str^S Cells

Str^R Cells

B C
E D

Str^S Cells + DNase + DNA

Str^S Cells + DNA

TSY agar + streptomycin

Figure 23.2 Labeled sections of TSY plates. The ** denotes the resistant cells to add. Courtesy of Anna Oller, University of Central Missouri.

This is called the **lysate**. Any cells that are not lysed will be killed by the 30-minute exposure to 60°C water. Label the tube "DNA."

2. Divide the bottom of the trypticase soy yeast extract agar plate into five sectors using a marking pen. Label the sections "DNA," "Str^S (streptomycin sensitive)," "Str^R (streptomycin resistant)," "Str^S + DNA," and "Str^S + DNase + DNA" (**figure 23.2**).

3. Inoculate the TSY plate as indicated by adding a loop of the broth culture, DNA, or DNase in an area about the size of a dime to each sector. **Avoid cross-contamination of sectors**.

   a. DNA. The lysed mixture of Str^R cells is the source of DNA. It also contains RNA, proteins, and all the other cell components of the lysed cells that do not interfere with the transformation.

b. Str^S cells. Inoculate a loopful of the Str^S culture.

c. Str^R cells. Inoculate a loopful of the Str^R culture.

d. Str^S cells + DNA. Inoculate a loopful of Str^S cells and then add a loopful of the DNA (lysed Str^R cells) in the same area. **This is the actual transformation**. Str^R cells will grow here if transformed by the DNA.

e. Inoculate a loopful of Str^S cells as above, and in the same area add a loopful of DNase, then add a loopful of DNA. It is important to add these in the correct order or transformation will occur before the DNase can be added. The DNase should degrade the DNA and prevent transformation.

4. Properly label the rest of the plate and incubate the plates at room temperature for several days or at 37°C for 48 hours.

Second Session

1. Observe the plate you prepared in the first session. There should be growth in all sectors of the plate except the DNA sector *(a)*. If the DNA control sector shows growth, it indicates that your crude DNA preparation was not sterile and contained viable cells. If this has happened, discard your plates and borrow another student's plate after he or she is finished with it; there should be sufficient material for more than one team. Why is it so important that the DNA preparation be sterile?

2. Be sure you understand the purpose of each control. Which sector demonstrates that:

   a. the Str^S cells were viable and could grow on the agar plate?

   b. the Str^R cells were viable?

   c. the lysed mixture of Str^R cells contained no viable organisms?

   d. DNA is the component of the lysed cells responsible for transformation?

3. Divide the bottom of a TSY + streptomycin plate into four sectors and label them "Str^S," "Str^R," "Str^S + DNA," and "Str^S + DNase + DNA."

4. Streak a loopful of cells from the first plate to the corresponding sectors on the TSY + streptomycin plate. Lightly spread them in an area about the size of a dime. Cells growing on this plate must be streptomycin resistant.

5. Properly label the rest of the plate and incubate at room temperature for several days, or at 37°C for 48 hours, or until cells have grown.

Third Session

1. Observe the TSY + streptomycin agar plate inoculated last session and record the results. Did you transform the cells sensitive to streptomycin to cells that were resistant and could now grow on streptomycin?

2. Did the results from the controls confirm that naked DNA can transfer resistance to streptomycin?

EXERCISE 23

Laboratory Report: Transformation: A Form of Genetic Recombination

Results

Indicate growth (+) or no growth (−) in each sector. Refer to figure 23.2.

	Agar Section	A	B	C	D	E
First Session	TSY	DNA	Str^S cells	Str^R cells	Str^S cells + DNA	Str^S cells + DNase + DNA
		No antibiotic added	*Acinetobacter* DNA from SDS lysed cells	Streptomycin susceptible cells	Streptomycin resistant cells	Transformation	No transformation as DNase should degrade cells
Expected results		−	+	+	+	+
Growth? (+ or −)						
What do your results mean?						
Second Session	Agar Section	A	B	C	D	E
	TSY + streptomycin		Str^S cells	Str^R cells	Str^S cells + DNA	Str^S cells + DNase + DNA
What do you hypothesize the expected results to be?						
Actual results: Growth? (+ or −)						
What do your results mean?						

1. Were *Acinetobacter* Str^S cells sensitive to streptomycin?

2. Were *Acinetobacter* Str^R cells resistant to streptomycin?

3. Was the DNA (cell lysate) free of viable cells?

4. Did transformation take place?

5. Did the DNase prevent transformation?

Results of first session

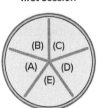

TSY agar

Results of second session

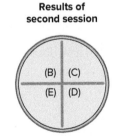

TSY agar + streptomycin

Questions

1. What two components were mixed together to show transformation?

2. What is the action of DNase?

3. What control showed that transformation—not conjugation or transduction—was responsible for the results?

4. If the Str^S cells placed in sector *(b)* had grown on the TSY + streptomycin agar plate, would you have been able to determine if transformation had taken place? Explain.

5. If you had used a DNA lysate containing viable cells, would it have been possible to determine whether transformation had taken place? Explain.

6. How does transformation differ from conjugation and transduction?

Definitions

Catabolite repression. The cell's mechanism for preventing the synthesis of enzymes needed to break down an energy source if a simpler energy source, such as glucose, is available.

Constitutive enzymes. Enzymes constantly produced by the cell.

Inducible enzymes. Enzymes produced only under certain environmental conditions.

ONPG (o-nitrophenyl-β-D-galactoside). A compound used for detecting the presence of the enzyme β-galactosidase.

Operon. A group of linked genes whose expression is controlled as a single unit.

Substrate. A substance on which an enzyme acts to form products.

Transcription. The process of copying genetic information coded in DNA into messenger RNA.

Objectives

1. Describe the concept of induction.
2. Explain the *lac* operon and the use of ONPG.
3. Explain the concept of catabolite repression and how it is tested.

Pre-lab Questions

1. What is an inducible enzyme?
2. What is the purpose of adding ONPG to your glucose, lactose, and glycerol tubes?
3. If you see a clear zone on your starch plate, what does it mean?

Getting Started

A bacterial cell has all the genetic information to produce and operate a new cell. This includes the enzymes necessary for obtaining energy and synthesizing the necessary cellular components.

A cell must work efficiently and carefully utilize available nutrients without investing energy in enzymes not needed. Some enzymes are utilized in the basic energy pathways of the cell and are continually synthesized. These are called **constitutive enzymes**.

Others, termed **inducible enzymes**, are only needed when their specific **substrate** is available. For example, *Escherichia coli* can break down the sugar lactose with an enzyme called β-galactosidase. If lactose were not present in the environment, it would be a waste of energy and of intermediate compounds to synthesize this enzyme. Therefore, these kinds of enzymes are called inducible enzymes because the presence of the substrate induces their synthesis.

How does the cell control inducible enzymes? In bacteria, this takes place on the level of **transcription**. Inducible enzymes can be found in an **operon**, which has a promoter and operator followed by the genes involved in the enzymatic activity (**figure 24.1**). The repressor binds to the operator gene, blocking transcription if the substrate for the enzyme is not present. When the substrate is present, it binds to the repressor, allowing the RNA polymerase to transcribe the genes for β-galactosidase. The genes involved with lactose utilization are in a cluster termed the *lac* operon.

Another method of control utilized by the cell is **catabolite repression**, which occurs when a cell has a choice of two sources of energy, one of which is more easily utilized than the other. An example is the presence of both glucose and starch. Glucose can immediately enter the glycolytic pathway, whereas starch must first be cleaved with amylase. Amylases and other enzymes cost the cell energy and materials to produce, so it is much more economical for the cell to utilize glucose if it is present. When glucose is present along with starch, the glucose represses the synthesis of amylase even though the enzyme would normally be induced in the presence of starch.

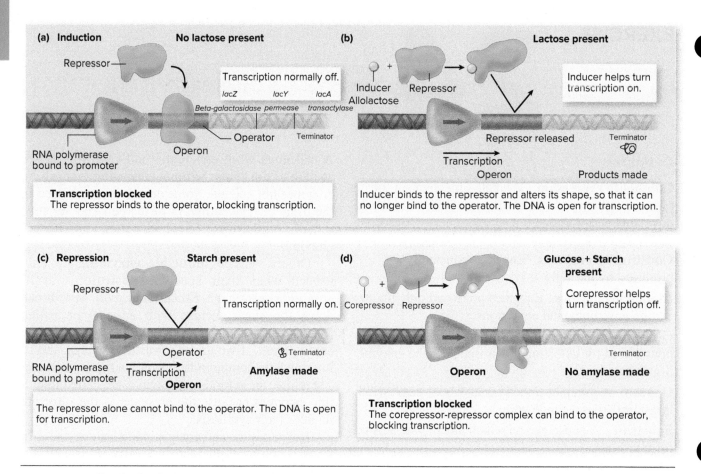

Figure 24.1 Catabolite induction and repression. (*a* and *b*) Lactose and the *lac* operon. (*a*) No lactose prevents transcription, whereas (*b*) lactose presence allows transcription. Note that only the genetic elements take part as in the regulation of a biosynthetic pathway. Repression is shown as the operon is normally turned on (*c*), and then the presence of another molecule prevents amylase production (*d*) when it would normally be made or synthesized.

The first procedure of this exercise will investigate the induction of β-galactosidase, a particularly important enzyme system used in recombinant DNA techniques as a measure of gene expression. The enzyme β-galactosidase cleaves the disaccharide lactose into glucose and galactose. It would be inefficient for the cell to synthesize this enzyme if lactose were not present; therefore, only the presence of lactose induces the synthesis of β-galactosidase.

However, when actually testing this reaction, it is difficult to detect the presence of β-galactosidase, a colorless protein. This problem has been solved by adding **ONPG (o-nitrophenyl-β-D-galactoside)**. Most enzymes are specific for just one substrate or reaction, but β-galactosidase can also cleave ONPG, a synthetic compound, forming a yellow dye. Therefore, a positive test for β-galactosidase is a yellow color in the presence of ONPG.

The second procedure of this exercise will test for the catabolite repression of amylase. Amylase is an exoenzyme; it is excreted outside the cell because the large starch molecule may be too large to pass through the cell membranes. The presence of starch is detected by adding iodine to the plate, which turns starch purple (figures 18.2 and 24.2, color plate 33).

Induction

Materials

Per pair of students

Mineral salts + 0.2% glucose broth, 5 ml/tube

Mineral salts + 0.2% lactose broth, 5 ml/tube

Mineral salts + 0.2% glycerol broth, 5 ml/tube

3 ml of ONPG

1 ml sterile pipets, 3

Cultures of the following:

Overnight trypticase soy broth cultures of
E. coli (cultures of *Enterobacter* and
Klebsiella may also be tested)

PROCEDURE

First Session

1. Label each tube very carefully, being sure
   to keep the sugars identified as all tubes will
   appear identical. Add the bacterium, your
   name, temperature, date, and lab section.

2. Inoculate each tube (glucose, lactose, and
   glycerol) with a drop or loopful of *E. coli*.

3. Incubate at 37°C until the next laboratory
   period or for at least 48 hours.

Second Session

1. Examine the tubes for growth—all the tubes
   should be turbid.

2. Add 1 ml of ONPG to each tube. The
   indicator compound ONPG is cleaved
   by β-galactosidase into a yellow product.

3. Incubate at room temperature for
   30 minutes.

4. Examine the tubes to determine if the
   broth has turned yellow—an indication
   of the presence of the induced enzyme
   β-galactosidase.

5. Record the results.

Catabolite Repression

Materials

Per pair of students

Nutrient + starch agar plate, 2

Nutrient + starch + glucose agar plate, 2

Cultures of the following grown on TS slants:

Escherichia coli (negative control)

Bacillus

Gram's iodine

PROCEDURE

First Session

1. Label two plates (one starch and one starch +
   glucose plate) with *Bacillus*, media type, your
   name, temperature, date, and lab section. See
   figure 24.3.

2. Inoculate the middle of each plate type (starch
   and starch + glucose) with *Bacillus* in a square
   area of a few mm, or in an "S" streak.

3. For the second set of plates (one starch and one
   starch + glucose plate), divide each plate in half.
   Label one half as Neg Control-Uninoculated.
   Do not inoculate this half of the plate.

4. Label the other half of the plate with *Escherichia*, media type, your name, temperature,
   date, and lab section. Incubate plates at 30°C

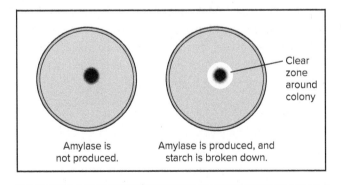

Figure 24.2 Appearance of plates after flooding
with Gram's iodine. See color plate 33.

for 24–48 hours. Try not to let the *Bacillus* grow over more than a third or half the plate. You may have to refrigerate the plates if the colony becomes too large.

Second Session

1. Flood the agar plates with Gram's iodine. If starch is present, it will turn purple. If the starch has been broken down with amylases, a clear zone will appear around the colony (figure 24.2, color plate 33).

2. Record the results.

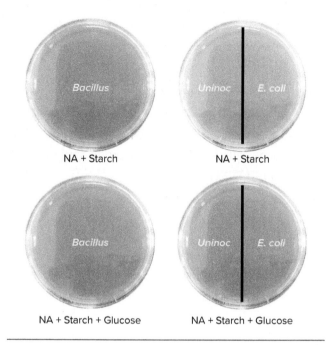

Figure 24.3 Labeled starch and starch + glucose plates. Courtesy of Anna Oller, University of Central Missouri.

EXERCISE

24

Laboratory Report: Gene Regulation: Induction and Catabolite Repression

Results

Results of Induction

Organism: _____	Glucose	Lactose	Glycerol
Amount of growth? (1+, 2+, etc.)			
Color after ONPG added?			
β-galactosidase present?			

Results of Catabolite Repression

	Starch			Starch + Glucose		
	Bacillus	*E. coli*	Uninoculated	*Bacillus*	*E.coli*	Uninoculated
Zone of clearing present?						
Amylase present?						

Questions

1. Which substrate induced β-galactosidase?

2. What chemical reaction produced the yellow color?

3. What results would you expect if β-galactosidase were a constitutive enzyme?

4. In the catabolite repression exercise, did the *Bacillus* have the capacity to synthesize amylase (amylase +)? How did you determine your answer?

5. Did you observe catabolite repression when glucose was added to the starch? How did you determine your answer?

6. What results would you expect if amylase were a constitutive enzyme?

INTRODUCTION   to the Other Microbial World

The phrase *the other microbial world* refers to organisms other than bacteria, the major organisms of study in other parts of this manual. The organisms included here, with the exception of viruses, are eukaryotic organisms, and many are of medical importance. Yeasts, molds, protozoa, algae, intestinal animal parasites, and helminths are all discussed. Viruses that infect both prokaryotes (bacteria) and eukaryotes (animal and plant cells) are also introduced, but only bacterial viruses will be studied.

Mycology, the study of fungi, is the subject of exercise 25. Fungi lack chlorophyll, differentiating them from algae, and they contain membrane-bound organelles and are larger than bacteria. The yeast *Saccharomyces* is vital to bread, beer, and wine production; the filamentous fungus *Penicillium chrysogenum* produces the antibiotic penicillin. Some fungi spoil bread and jelly, and *Aspergillus* produces food toxins (aflatoxins). Many filamentous fungi, like mushrooms, puffballs, toadstools, and molds, are visible with the naked eye.

Antibiotics effective against prokaryotes are often ineffective against eukaryotes. If an antifungal agent kills a fungus, it may damage human cells since human cells are also eukaryotic. Thus, eliminating fungal infections is often more difficult than controlling bacterial infections, even with emerging bacterial resistance to antibiotics. Fortunately, many fungal infections are opportunistic[1] infections that healthy individuals rarely acquire, other than cutaneous fungal infections like athlete's foot.

Fungi are cultivated in a laboratory in a similar manner as bacteria. Physiologically, all fungi are heterotrophs (they require an organic source of carbon, such as glucose). Many fungi are aerobes, some are facultative anaerobes, and a few are obligate anaerobes (found in cow rumens (stomachs)). Most fungi grow optimally at 20–30°C, but some grow well at 45–50°C (such as *Aspergillus fumigatus*, an opportunistic filamentous fungus known to cause pulmonary aspergillosis).

Protists, including algae, are discussed in exercise 26. Protists are eukaryotes that are not animals, plants, or fungi. Many are motile using flagella, cilia, or pseudopodia. Algae are photosynthetic and usually have a cell wall comprised of cellulose. Protists serve as decomposers in soil and water ecosystems, and some are food sources for larger predators. Parasitic diseases are still a worldwide public health problem, both in developed and developing countries. In developing countries, parasitic diseases are prevalent due to poverty, malnutrition, lack of sanitation, and lack of health care. Effects on humans range from nutritional loss and minor discomfort (pinworm infections), to debilitating and life-threatening infections (malaria).

Parasitic infections are also seen in industrialized countries. Fecal contamination of drinking water by wild animals, such as deer and beavers, has caused outbreaks of giardiasis and cryptosporidiosis (intestinal diseases) in parts of the United States. The causative agent of giardiasis is the flagellated protozoan *Giardia lamblia* (see figure 29.3), which produces **cysts** that can pass through faulty water supply filters. Cryptosporidiosis is caused by a coccidian protozoan called *Cryptosporidium parvum*, which is also small enough to pass through water supply filters. The flagellated protozoan *Trichomonas vaginalis* is sexually transmitted, can cause vaginitis in women, and infections have increased in retirement communities. In addition, pinworm infections commonly occur in children. Exercise 27 provides insights on parasites and techniques used to diagnose them.

Virology, the study of *viruses* (the word for "poison" in Greek), is the subject of exercise 28. Tobacco plant leaves often had a yellow mottled appearance, which affected the quality of the leaves. Some thought the disease was caused by a virus, which was too small to be seen with a light microscope. Infected tobacco leaf sap was passed through a small filter that removed bacteria, and the

---

[1] Opportunistic infections are associated with debilitating diseases (such as cancer) and use of cytotoxic drugs, broad-spectrum antibiotics, and radiation therapy, all of which can suppress the normal immune response.

clear infiltrate retained its infectious properties. In the mid-1930s, viruses were finally seen using an electron microscope. In 1935, Wendell Stanley succeeded in crystallizing tobacco mosaic virus (TMV), allowing him to observe structural differences from living cells (**figure I.25.1**).

Bacterial viruses, known as **bacteriophages**, were first described by Twort (1915) and later by d'Herelle (1917). d'Herelle observed their filterable nature and their ability to form **plaques** on an agar plate seeded with a lawn of the host bacterium (see figure 28.2). Both Twort and d'Herelle worked with coliform bacteria isolated from the intestinal tract.

The agar medium provided a fast, easy way to recognize, identify, and quantify bacteriophages. Viruses that infect mammalian cells also form structures analogous to plaques when cultivated on mammalian cell culture media. Rather than plaques, the viruses caused cell morphology changes, termed **cytopathic effects (CPEs)** (**figure I.25.2**). The CPEs observed are dependent upon the nature of both the host cell

and the invading virus. Mammalian cells require complex growth media, including blood serum, as well as 1–4 days incubation with the virus before CPEs are visible. Therefore, we will work with an *Escherichia coli* bacteriophage that can be obtained from a pure culture collection (exercise 28).

Definitions

Bacteriophage. A virus that infects bacteria. Frequently termed *phage*.

Cyst. A dormant resting protozoan cell characterized by a thickened cell wall.

Cytopathic effects (CPEs). An observable change in a cell *in vitro* caused by a viral action such as cell lysis.

Mycology. The study of fungi.

Plaque. A clear area in a monolayer of cells caused by phages infecting and lysing bacteria.

Virology. The study of viruses.

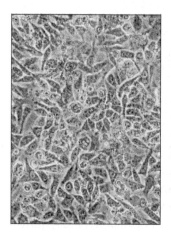

(a) Monolayer

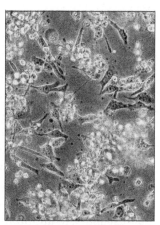

(b) Rounded & Dying Cells

Figure I.25.2 Cytopathic effects (CPEs). (*a*) A monolayer of cells. (*b*) The same cell monolayer incubated with a virus for a couple of days showing morphological changes such as irregularly shaped and dying cells.

Source: CDC

Modified from *"The Genome Sequence of Lone Star Virus, a Highly Divergent Bunyavirus Found in the Amblyomma americanum Tick,"* PLoS ONE, April 2013, Volume 8, Issue 4, e62083, with permission from University of California, San Francisco and the Centers for Disease Control and Prevention.

0.5 µm

Figure I.25.1 Tobacco mosaic virus. Electron micrograph (magnification is × approximately 70,000). Compare the length and width to a rod-shaped bacterium.

Omikron/Science Source.

25

Identification of Fungi ^{ASM34}

Definitions

Aflatoxin. A toxin produced by the genus *Aspergillus* that can cause cancer or even death at high concentrations.

Ascospore. Sexual spore characteristic of the fungus class Ascomycetes.

Ascus. An elongated finger-like shaped sac containing ascospores.

Aseptate. Also called non-septate or coenocytic. Fungi made up of long filaments of hyphae which do not contain any crosswalls or septa.

Basidiospore. A sexual spore characteristic of the fungal class Basidiomycetes.

Budding. In asexual reproduction, when a cell divides and the daughter cell (bud) separates from the mother cell. Genetically identical to the parent cell.

Chitin. A polymer derived from glucose that makes a fungal cell wall rigid and strong.

Chlamydospore. A resistant hyphal cell with a thick wall; it eventually separates from the hyphae and functions as a spore.

Coenocytic. Hyphae lacking septa, or cross walls, so repeated nuclear division is not accompanied by cell division. Long filaments of hyphae have multiple nuclei and no septa.

Columella. A swelling of the sporangiophore at the base of the sporangium that acts as a support structure for the sporangium and its contents.

Conidia (singular, conidium). Asexual spores produced from either the tip or side of the conidiophore.

Conidiophores. Specialized hyphae that support the conidia and produce asexual spores at the tip or side.

Conidiospores. The individual asexual spores produced at the tip or side of the conidiophore and make up the conidium.

Dermatomycosis (plural, dermatomycoses). A disease of the skin caused by an infection with a fungus (called a dermatophyte), usually from the genera *Trichophyton*, *Microsporum*, or *Epidermophyton*.

Dimorphic (dimorphism). The ability of pathogenic fungi to grow in two different forms. Often a mold phase containing hyphal filaments, and a yeast phase containing single cells are seen. The phase exhibited may be nutrient or temperature dependent.

Ergosterol. A sterol used in the cell wall that makes the cell rigid and strong. It functions similarly to cholesterol in human cells.

Ergotism. A disease that causes hallucinations due to toxins produced by the fungus *Claviceps purpurea*.

Fomite. An inanimate object, such as a desk or door handle, that may harbor microorganisms and infect a new host.

Germ tube. A tube-like outgrowth from an asexual yeast cell that develops into a hypha. Produced by *Candida albicans* grown under certain conditions.

Hypha (plural, hyphae). A thread-like fungal filament. Many filaments form a mycelium.

Lichen. A symbiotic relationship between one or two fungi and an alga. Discussed in exercise 26.

Meiosis. A process that reduces the chromosome number from diploid (2n) to haploid (1n).

Metula. A branch (or branches) at the end of the conidiophore that supports the sterigma.

Mitosis. In eukaryotes, a nuclear division process in which daughter cells receive the same number of chromosomes as the original parent.

Mold. A filamentous fungus, producing hyphae that can appear woolly.

Mycelium. A fungal mat made of tangled hyphae.

Mycoses (singular, mycosis). Diseases caused by fungi.

Onychomycosis. A fungal infection of the toenails or fingernails. Also known as tinea unguium.

Pseudohyphae. Short hyphal projections, elongated buds, seen in fungi grown under certain conditions.

Saprophyte. An organism that obtains nutrients from decaying organic matter.

Septum (plural, septa). A cross wall formed in hyphae that divides the hyphae into cells.

Spherule. A large, thick-walled circular structure filled with fungal endospores.

Sporangiospore. An asexual reproductive spore, usually circular in nature, found in Zygomycetes.

Sporangium. A sac that contains asexual spores.

Sterigma (plural, sterigmata). A specialized hypha that supports either a conidiospore or a basidiospore.

Superficial cutaneous mycosis. A fungal infection of the skin or hair that usually remains at the epidermal skin layer.

Subcutaneous mycoses. Fungal infections that penetrate the subcutaneous layer of the skin.

Systemic. An infection that has spread throughout the body to various tissues or organs, and can cause death.

Toadstool. A large, filamentous, fleshy fungus with an umbrella-shaped cap.

Yeast. A circular, unicellular, non-filamentous, fungus that is not a chytrid. It is found in nature on fermenting fruits.

Zygospore. Sexual spore characteristic of the fungus class Zygomycota.

Objectives

1. Describe what fungi are and how to distinguish them from one another, including how to distinguish members of the two major groups of fungi (non-filamentous and filamentous).

2. Explain how culturing yeasts and fungi differs from culturing bacteria.

3. Interpret macroscopic and microscopic colonial and vegetative cell morphology in order to identify *Saccharomyces*, *Rhizopus*, and *Penicillium*.

Pre-lab Questions

1. What is the difference in shape between the asexual structures of *Rhizopus* and *Penicillium*?

2. Which fungal division does not contain septa?

3. Name a disease *Candida albicans* can cause.

Getting Started

When you hear the word "fungus," what does it suggest? Mushrooms? **Toadstools?** Moldy fruits? Yeasts? Mushrooms and toadstools are macroscopic fungi, which do not require a microscope for identification, whereas yeasts and molds require microscopy for identification. However, fungi are considerably larger than bacteria, so they are easier to identify.

Fungal cell walls contain **chitin**, similar to what makes up insect exoskeletons, which makes them rigid and strong. Fungal membranes contain a sterol called **ergosterol**, which differs from cholesterol in animal membranes. This is why many antifungals inhibit ergosterol synthesis—it is one of the few differences between animal and fungal cells. Targeting a different cell structure is an attempt to minimize harmful drug effects on humans.

Fungi are divided into two categories, **yeasts** or **molds**, based upon cell characteristics. Yeasts (**figure 25.1**) are oval, unicellular, replicate by **budding**, grow at temperatures between 35°C and 40°C, and do not possess filaments (non-filamentous). *Schizosaccharomyces pombe* is a **fission** yeast used in research because they reproduce similarly to bacteria (**figure 25.2**). Molds (**figure 25.3**) are multicellular, replicate asexually by **mitosis** or sexually by **meiosis**, grow at temperatures <30°C, and possess long filaments called **hyphae**. Many hyphae make up a **mycelium**, which is visible to the naked eye. *Rhizopus nigricans* (color plate 60, and figures 25.3, **25.4**) is a mold contaminant of bread. The long filaments allow for nutrient absorption over a large surface area. Hyphae can contain cross walls, or **septa**, that separate filaments into individual cells. One or more nuclei are present within these cells. Hyphae that contain a septum are called septate, whereas hyphae that do not have a septum are called either **aseptate**, non-septate, or **coenocytic** (figure 25.3). Aseptate hyphae are long filaments that contain multiple nuclei.

Currently fungi are classified into phyla or divisions—Zygomycota, Ascomycota, Basidiomycota, and Chytridiomycota (**table 25.1**). The terms Zygomycota and Zygomycete are used interchangeably, just as Ascomycete, Basidiomycete, and Chytridiomycete are used. Zygomycota are differentiated from other phyla by their aseptate (coenocytic) hyphae.

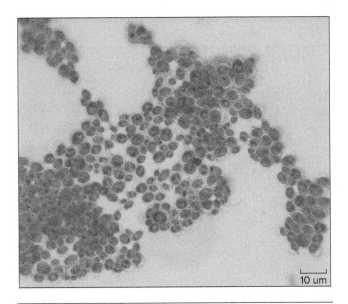

Figure 25.1 The yeast *Saccharomyces cerevisiae* containing ascospores and sexual asci. Each oval is a yeast cell. Courtesy of Anna Oller, University of Central Missouri.

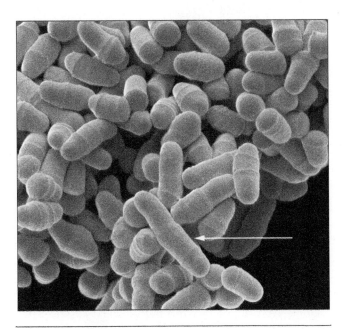

Figure 25.2 *Schizosaccharomyces pombe*, one yeast known to multiply vegetatively by fission (arrow). Steve Gschmeissner/Alamy Stock Photo.

Further, chytrids are found in aquatic environments parasitizing frogs and salamanders, usually killing them. Fungal classification is changing due to DNA sequencing and new technologies.

Fungi produce spores that are involved in asexual and sexual reproduction, whereas bacterial endospores

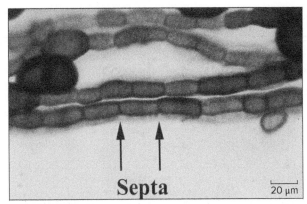

(a) Septate hyphae

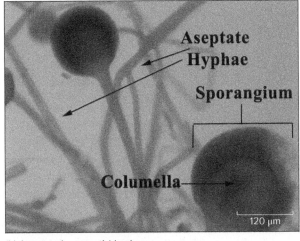

(b) Aseptate (coenocytic) hyphae

Figure 25.3 The two major hyphal types found in fungi. (*a*) Septate hyphae and (*b*) aseptate or non-septate hyphae, also called coenocytic. Courtesy of Anna Oller, University of Central Missouri.

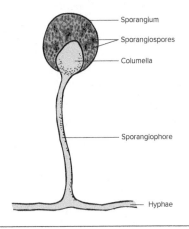

Figure 25.4 Intact asexual reproductive structure of the zygomycete *Rhizopus nigricans*. Note aseptate (coenocytic) hyphae and sporangiophore.

Table 25.1 Classification of Fungi

	Division			
	Zygomycetes	**Ascomycetes**	**Basidiomycetes***	**Chytridiomycetes**
Mycelium	Aseptate (coenocytic)	Septate	Septate	Aseptate (coenocytic)
Sexual spores	Zygospore (not in a fruiting body) found in terrestrial forms	Ascospores, borne in an ascus, usually contained in a fruiting body	Basidiospores, borne on the outside of a club-like cell (the basidium), often in a fruiting body	Flagellated gametes of + and − fuse together
Asexual spores	Sporangiospores, non-motile, contained in a sporangium	Budding, conidia, conidiospores, non-motile, formed at the end of a specialized filament, the conidiophore	Same as Ascomycetes (often absent)	Motile zoospores; sporangiospores
Habitat and common representatives	Terrestrial bread mold (*Rhizopus stolonifer* (*nigricans*)), fruit mold (*Mucor indicus*)	Terrestrial Yeasts (*Saccharomyces*), *Penicillium, Aspergillus,* morels (*Morchella esculenta*), ergot (*Claviceps purpurea*)	Stinkhorns, puffballs, mushrooms (*Agaricus bisporus*), *Cryptococcus neoformans*	Water, soil, herbivore guts; *Allomyces*

*Some of these fungi will no doubt form sexual spores in the right environment. In this event, they would need to be reclassified.

are considered dormant, and waiting for the return of a favorable environment. **Lichens** are often placed with the fungi for convenience because they are composed of one or two fungal species, and a green alga or cyanobacteria. They will be discussed in exercise 26.

A mushroom with its basic structures labeled is shown in **figure 25.5**. The main parts of the mushroom are the cap at the top of the fungus, which looks like an umbrella. On the underside of the cap are the gills, which open up and release spores. The stalk connects the mushroom to the soil and is more rigid than the cap.

Some mushrooms like Chanterelles (*Cantharellus*) and Morels (*Morchella*) are edible and sought after, whereas the "death cap mushroom," *Amanita*, are toxic and lethal if ingested. The white button mushrooms you buy in the store are in a genus called *Agaricus*. You will view a slide of a cross-section of the mushroom *Coprinus* (**figure 25.6**), called an "inky cap," because it leaks a blue-black liquid containing spores. It appears inky as the fungus ages because it makes enzymes to autodigest itself.

Fungi are important decomposers, and some produce unique enzymes, produce toxins, or cause infections. Some fungi produce life-saving antibiotics. Most fungi are **saprophytes**, meaning they live off of and decompose dead matter. *Sordaria fimicola* is a dung fungus found growing on herbivore feces,

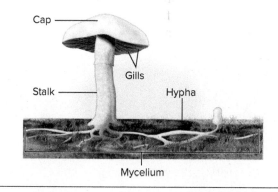

Figure 25.5 The structures of a mushroom.

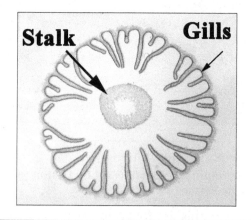

Figure 25.6 A cross-section of a mushroom. The cap has been removed, allowing the stalk to be visualized. The protrusion of the gills where spores are released can be seen. Courtesy of Anna Oller, University of Central Missouri.

and is often used in genetic studies to cross types of the fungus and view offspring colony colors. Fungi also produce enzymes that degrade cellulose and lignin, which is one reason why they grow on or around trees or stumps. *Claviceps purpurea* grows on wet grains like rye, and produces toxins that cause hallucinations (called **ergotism**), or even death at high concentrations, if ingested. *Armillaria*, called a honey mushroom, is a bioluminescent fungus that can cause root rot in oak trees. It is thought that the luminescence attracts insects that will then spread its spores in the environment. *Armillaria* contains hispidin, the precursor for luciferin, which facilitates the bioluminescence.

Why so much interest in fungi? Wheat rusts like *Puccinia triticina* are emerging worldwide in plant fungal infections, reducing crop yields, with no known control methods. *Trichoderma viride* is a fungus used to coat plant seeds to make them more resistant to other soil fungi.

Fungi are used on an industrial scale to make various enzymes vaccines, and antibiotics. Fungi may be a source of new antibiotics needed to save lives due to antibacterial resistance. Besides making the antibiotic penicillin, *Penicillium* species (color plate 62 and **figure 25.7**), are also used to make cheeses, like blue cheese. **Kombucha**, a fermented tea, and probiotics have become popular. Kombucha is made by dissolving sugar in boiling tea to which a thick gelatinous mat of bacteria and yeast (called a scoby) is added. A cloth covers a jar opening and the tea ferments for a couple of weeks.

Yeasts are used to make bread rise, and ferment wine and beer. Organic bread and food artisans, along with microbreweries looking for new yeast strains, have generated new interest in fungi.

Fungi usually reproduce both sexually and asexually (table 25.1). Zygomycetes produce **zygospores** as the sexual spores, but **sporangiospores** and **conidia** are used in asexual reproduction. Zygomycetes have aseptate hyphae and include *Rhizopus* and *Mucor* genera. *Mucor* is a fast-growing grayish mold, often seen covering strawberries or blueberries overnight. *Rhizopus nigricans* structures are labeled on color plate 55 and figure 25.4.

Ascomycetes form sexual spores (**ascospores**) within an **ascus**, which is a finger-like projection seen in *Morchella* in color plate 63. Ascomycetes make conidia asexually. *Penicillium, Aspergillus,*

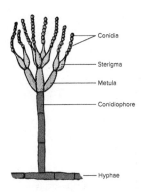

Figure 25.7 Intact asexual reproductive structure of the septate ascomycete *Pencillium chrysogenum*. Notice the absence of a columella and foot cell. Metulae are also symmetrically attached to the conidiophore.

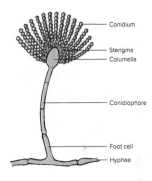

Figure 25.8 Intact asexual reproductive structure of the ascomycete *Aspergillus niger*. Note the presence of a foot cell, a columella, a septate conidiophore, and hyphae.

Candida, and *Saccharomyces* are all examples of Ascomycetes. *Penicillium,* shown in color plate 62 and figure 25.7, are commonly seen on orange rinds or other fruits. *Aspergillus* (**figures 25.8** and **25.9**) is a soil fungus that contaminates fruits, grains, and nuts. Some species are used to make saké and citric acid, and some produce a carcinogenic **aflatoxin**. Some colonize the sinuses or lungs (causing fungus balls to form) of immunocompromised people. As diabetes and other predisposing conditions increase globally, the number of immunocompromised individuals has increased.

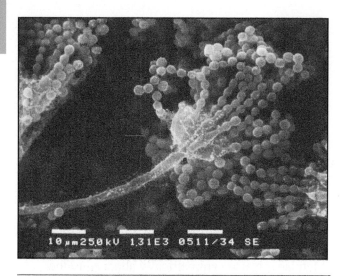

Figure 25.9 Scanning electron microscopy photo of *Aspergillus* showing conidia.
Source: Janice Haney Carr/CDC

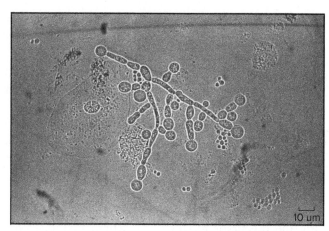

Figure 25.10 *Candida albicans* chlamydospores (large circular cells at the ends of hyphae) in a sputum sample from a patient with invasive pulmonary candidiasis.
Source: CDC/ Dr. Brinkman

Basidiomycetes reproduce sexually, producing basidia and **basidiospores** (color plate 64), but asexually produce conidia. Many mushrooms and puffballs are classified under the Basidiomycota. *Rhodotorula rubra* is a pigmented environmental yeast. Many fungi produce pigments to protect cells from oxidative stress and UV light.

Chytrids are usually found in water (*Allomyces macrogynus*), or the guts of herbivores. They produce flagellated spores called zoospores, and are therefore motile. Chytrids decompose pollen and cellulose. Chytrids can be parasitic to frogs and salamanders. In 1999, the chytrid *Batrachochytrium dendrobatidis* emerged, parasitizing frogs and salamanders, causing the outer skin layer to thicken. The fungus prevents water and nutrient absorption by frogs and suffocates salamanders. It is at least partially responsible for the amphibian population decline.

Candida albicans (**figure 25.10**) is the most common fungus diagnosed in humans, and common ailments are listed in **table 25.2** (color plate 65). In 2016, a multidrug resistant strain of *Candida auris* emerged in hospitalized patients in the United States, with an initial mortality rate >50%. This strain is different from *C. albicans* because it can demonstrate resistance to all three antifungal drug classes, forms biofilms and is viable for months on **fomites**, and is difficult to eradicate from patient rooms (www.CDC.gov/fungal/diseases).

Some pathogenic yeasts exhibit **dimorphism**, growing as a yeast at 37°C or forming mycelia at 25°C

(**figure 25.11**). Dimorphism can be induced by nutrient availability or growing the fungus at different temperatures. *Candida* form **chlamydospores** when grown on cornmeal agar or when isolated from the lungs (figure 25.10). *Candida* can also form elongated buds called **pseudohyphae** (**figure 25.12a**), when grown on Sabouraud's dextrose agar, or even **germ tubes** (**figure 25.12b**).

Fungal infections (**mycoses**) are divided into three categories based on the body tissues affected—**superficial cutaneous** (the outer dermis and epidermis layers of the skin), **subcutaneous** (the deeper hypodermis layer), and **systemic** (spread throughout the body) infections. Superficial cutaneous **dermatomycoses**, caused by fungi referred to as dermatophytes and often called ringworms, are fungal infections of the dermal skin layers. The ringworm is not a worm at all, but a fungus that causes circular patches of inflammation and hair loss (color plate 61), or thickened toenails, a condition known as **Onychomycosis**.

Dermatophyte infections are named for the body location where they occur: tinea pedis: feet; tinea capitis: head; tinea corporis: body; tinea barbae: beards; tinea cruris: groin; and tinea unguium: toenails or fingernails. Topical antifungals, like miconazole and tolnaftate, are often used for dermatophyte infections, but oral drugs like terbinafine or griseofulvin are prescribed for onychomycoses. They are commonly caused by *Trichophyton rubrum* (**figure 25.13**), *Microsporum canis*, or *Epidermophyton*. *Microsporum canis* is commonly found on animals, and

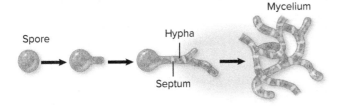

Figure 25.11 Development of a mycelium from a spore.

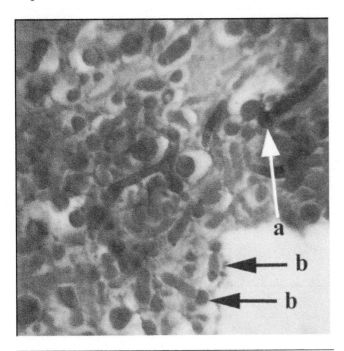

Figure 25.12 Pseudohyphae (*a*) and germ tubes (*b*) formed by *Candida albicans*. Courtesy of Anna Oller, University of Central Missouri.

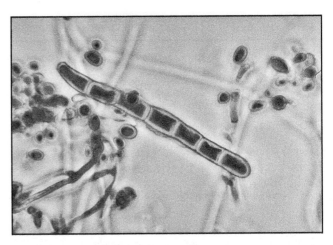

Figure 25.13 *Trichophyton rubrum*, a common dermatophytic fungus that causes ringworm. **Source:** CDC/ Dr. Libero Ajello.

humans can acquire this infection by direct contact with the fungus (i.e., petting). Avoid petting areas on animals with visible hair loss, and wash your hands well. Onychomycosis affects toenails or fingernails and can also be caused by *Candida albicans*, making it difficult to successfully treat with topical antifungals previously. To prevent athlete's foot, tinea pedis, avoid walking barefoot in public showers, pools, and motels as any fungal cells left on a floor have the potential to enter any breaks of the skin of your foot (color plate 61).

Subcutaneous infections caused by *Sporothrix schneckii* are seen when dirt on thorns or splinters enter puncture wounds, and is most commonly seen

in gardeners or florists. Potassium iodide is one treatment option. Wearing gloves while gardening around thorny bushes can reduce the likelihood of acquiring this fungus. Another fungus called *Malessezia furfur* and *Malessezia globosa* cause skin discolorations, are commonly acquired from tanning beds, and are aptly termed tinea versicolor. The fungus gives a characteristic "spaghetti and meatballs" appearance microscopically on skin scraping slides. Selenium sulfide or ketoconazole are common treatments.

Fungi that can cause systemic infections in immunocompromised individuals include *Blastomyces dermatitidis*, *Paracoccidioides brasiliensis*, *Histoplasma capsulatum*, and *Coccidioides immitis* (table 25.2). Systemic infections often originate in the lungs and then spread to other body locations. For example, *Coccidioides* arthroconidia from soil are inhaled, and a few days later a **spherule** is made in the lungs, allowing the fungus to quickly spread throughout the body. *Coccidioides immitis* infection can be diagnosed in humans using an ELISA test (exercise 36). Most people exposed to these fungi are asymptomatic, as the immune system prevents a symptomatic infection.

Yeasts are often initially grown on malt extract agar, whereas filamentous fungi are grown on Sabouraud's dextrose agar (SDA). Other commonly used agars include potato dextrose agar (PDA) and yeast potato dextrose (YPD). Depending on the source being tested, antibiotics may be added to the agar to inhibit bacterial growth. The incubation temperature is from 20°C to 25°C for filamentous fungi (except

Table 25.2   True Fungal (Dimorphic) Pathogens and Important Yeasts

Organism	Morphology	Ecology and Epidemiology	Diseases	Treatment
Cryptococcus neoformans, C. gattii	Single budding cells, encapsulated	Found in soil and pigeons' nests, opportunistic	Pneumonia, meningoencephalitis	Amphotericin B
Candida albicans	Budding cells, pseudohyphae, chlamydospores	Mouth, reproductive tract, intestines of humans, opportunistic, STI*	Thrush, vulvovaginal candidasis, cutaneous (skin), onychomycosis (nails), systemic	Fluconazole, miconazole, nystatin, terbinafine, tolnaftate, amphotericin B
True Fungal Pathogens				
Blastomyces dermatitidis	37°C: single large budding cells 20°C: mold with conidia	Soil in Midwest US, Inhaled	Pneumonia, blastomycosis (AKA Gilchrist disease)	Itraconazole, amphotericin B
Paracoccidioides brasiliensis	37°C: single and multiple budding cells 20°C: mold with white aerial mycelium	Soil in Central and South America, Inhaled	Pneumonia, paracoccidioidomycosis	Itraconazole, amphotericin B
Histoplasma capsulatum	37°C: single small budding cells 20°C: mold with chlamydospores	Soil in Midwest US, bird and bat droppings in caves, chicken coops, Inhaled	Pneumonia, histoplasmosis (cutaneous, systemic) (AKA Ohio valley fever)	Itraconazole, amphotericin B
Coccidioides immitis	37°C: thick-walled, endospore-filled spherical cells 20°C: mold with arthrospores	Soil in southwest US, such as San Joaquin Valley, Inhaled	Pneumonia, meningitis, osteomyelitis, coccidioidomycosis (AKA valley fever)	Itraconazole, amphotericin B

* STI= Sexually Transmitted Infection

Aspergillus fumigatus, which can grow at 45°C), but dimorphic yeasts will also grow at 30–37°C.

Yeasts can often be identified to species level based upon the ability to ferment specific sugars as the sole carbohydrate source (see exercise 19). Yeast is added to broth tubes made with a specific sugar, and the broth tube contains a Durham tube for detecting gas production. Following incubation, gas (CO_2) in the Durham tube constitutes a positive test for fermentation, which is an anaerobic process. The presence of yeast sediment or only a pH change is not indicative of fermentation.

Fungi are studied by several methods. Staining and microscopy can show cell and spore morphology, and sugar fermentation and biochemical tests can identify yeasts. Fungal DNA can now be isolated and PCR performed (exercise 32). In this exercise you will culture fungi, determine specific sugar fermentation of fungi, and view slides microscopically to see special structures.

Materials

Per team of 2–4 students
Sabouraud's dextrose broth cultures (48 hours, 25°C) of:
    Saccharomyces cerevisiae or
    Rhodotorula rubra
Sabouraud's or potato dextrose agar cultures (3–5 days, 25°C) of:
    Mucor indicus
    Penicillium chrysogenum (syn. *P. notatum*)
    Rhizopus stolonifer (syn. *R. nigricans*)
    Sordaria fimicola
    Trichoderma viride
    Armillaria mellea (on bread crumb agar)
Glucose fermentation broth tube, 1
Maltose fermentation broth tube, 1

Lactose fermentation broth tube, 1

Sucrose fermentation broth tube, 1

Sterile droppers, 4

Coverslips

Glucose acetate yeast sporulation agar, 1 plate

Sabouraud's dextrose agar (Sab or SDA), 1 plate

Potato dextrose agar (PDA), 1 plate

Yeast potato dextrose agar (YPD), 1 plate

Bread crumb agar, 1 plate

Slides for wet mounts

Demonstration slide of:

Aspergillus niger

Morchella

Coprinus

Schizosaccharomyces

Polyporus

Dissecting microscope(s)

Ruler divided in mm

Forceps

Methylene blue dropper bottle

Scotch™ tape (must be clear)

PROCEDURE

First Session

Fungal Culturing

1. Suspend the broth culture of *Saccharomyces cerevisiae*, *Rhodotorula rubra*, or another assigned microbe.

   Sugar fermentation study. Inoculate each fermentation tube (glucose, lactose, maltose, and sucrose) with a loopful of your assigned microbe. Place properly labeled tubes in a rack to incubate at 25–30°C for 48 hours.

2. **Yeast colony and vegetative cell morphology study**. Properly label a Sabouraud's dextrose agar (Sab/SDA) plate. With a sterile dropper, inoculate the agar surface with a *small* drop of yeast. Allow the inoculum to soak into the agar

before incubating upside down in the 25–30°C incubator for 48 hours.

3. **Yeast sexual sporulation study**. With a sterile dropper, inoculate the center of the sporulation agar plate with a small drop of the yeast broth culture. Allow the inoculum to soak into the agar before incubating right side up in the 25–30°C incubator for 48+ hours. Cultures freshly isolated from nature generally sporulate faster than laboratory-held cultures.

4. **Filamentous fungi inoculation**. Your instructor may have you learn culturing techniques. Working quickly using aseptic technique, use a loop to streak your assigned organisms over one-third of a Sab/SDA, YPD, and PDA plate. Using gently flamed forceps, add a sterile cover slip to the first area of the plates. Alternatively, your instructor may use scotch tape preps during the second session instead. Properly label your plates and incubate the plates at 25–30°C for 48+ hours.

5. **Bioluminescent study**. Obtain a bread crumb plate and streak one-third of the plate using *Armillaria mellea* or your assigned microbe. Properly label your plate and incubate the plate upside down at 22–25°C for 48+ hours.

6. **Colony characteristics of grown cultures** of *Mucor*, *Penicillium* (color plate 9), *Rhizopus*, *Sordaria*, *Trichoderma*, and *Armillaria*. Visually examine each Petri dish culture noting the following:

   a. Colony size. With a ruler, measure the diameter in millimeter.

   b. Colony color. Examine both the upper (front) and lower (reverse) surfaces.

   c. Presence of soluble pigments in the agar medium. (Are there any colors seen?)

   d. Colony texture (such as cottony, powdery, or woolly).

   e. Colony edge (margin). Is it regular or irregular?

   f. Colony convolutions (ridges). Are they present?

   g. Bioluminescence. View the plate in a dark room to see if any light is emitted. Is any glowing seen? If so, what color is it?

Enter your findings in table 25.5 of the Laboratory Report.

Note: Never smell fungal cultures—spore inhalation could cause infection. Keep air currents to a minimum.

Second Session

1. **Sugar fermentation study**. Examine fermentation tubes and record your results in **table 25.3** of the Laboratory Report:

   a. Presence or absence (+ or −) of growth (cloudiness).

   b. Presence or absence (+ or −) of gas in the Durham tube.

   c. Change of pH indicator (+ or −) to yellow (acid was produced).

   Note: Gas production indicates fermentation occurred. To detect false-negative results caused by super saturation, tubes with an acid reaction should be lightly shaken and the cap vented as a rapid release of gas can occur. All positive fermentation reactions are also positive for assimilation of that sugar, evidenced by clouding of the broth; however, sugars may be assimilated without being fermented.

2. **Yeast colony and vegetative cell morphology study**.

   a. Colony characteristics. If possible, observe Sabouraud's agar plates over a 5–7 days incubation period. Make note of the following in **table 25.4** of the Laboratory Report: For colony consistency (soft, firm)— probe the colony with a sterile needle to determine.

   b. Vegetative cell morphology. Remove a loopful of surface growth from each colony and prepare wet mounts. Observe with the high–dry objective lens, noting cell size and shape, and the presence or absence of pseudohyphae (figure 25.2*b*). Prepare and label a drawing in part 2*b* of the Laboratory Report.

3. **Sexual sporulation study**. With a sterile loop, touch the colony on the glucose–acetate sporulation agar plate and prepare a wet mount as above. Observe with the high–dry objective

lens, looking for asci containing one to four or more ascospores (figure 25.1). Prepare and label drawings in the Laboratory Report.

Note: If you do not find asci, re-incubate the plate up to one week, and re-examine periodically. Some yeast strains take longer than others to produce sexual spores.

4. **Filamentous fungi**. Describe the colonies seen on the agars in **table 25.5**.

5. Obtain a clean microscope slide and prepare a wet mount by adding one drop of methylene blue to a slide. Using flamed forceps, carefully lift the coverslip from your culture grown on agar and place on top of the methylene blue on the slide. Alternatively, your instructor may have you view a scotch tape prep to see hyphae and spores. Draw and label what you see in the Laboratory Report.

6. Observe a *Coprinus* and *Morchella* slide and record your drawings in the Laboratory Report.

7. **Morphological structures** can be viewed on grown agar plates and from slides. Draw any hyphae or spores seen in the Laboratory Report.

   a. On agar plates, use a dissecting microscope to view structures. Turn on both lights, the one under the stage and the one on the stereo head. Place a covered Petri dish right side up on the microscope stage. Using 4× magnification, use the course focus to move the magnification up or down to bring any structures into view. When you first observe structures like sporangiospores, stop moving the Petri dish.

   Note: First search for spores in both the inner and outer colony fringes.

   b. Draw and label any asexual reproductive structures that are seen using figures 25.4, 25.7, and 25.8 as reference. *Rhizopus* asexual structures will be on the underside of the Petri dish lid, and it may take 4–5 days of incubation before sporangiospores are seen. Covered slide cultures may be needed if conidia cannot be found for *Penicillium*.

8. **Morphology study** of asexual spores can be viewed on grown agar plates and from slides. Draw them in the Laboratory Report.

a. View your Petri dish on a dissecting microscope if slides are not available. The growth density can make it difficult to see intact asexual reproductive structures.

b. View any sporangiospores (*R. stolonifer*, color plate 60), conidia (*Penicillium* [color plate 62], *Mucor*), or chlamydospores by preparing a wet mount by adding one drop of methylene blue to a slide. Remove some aerial growth with a properly flamed loop and add it to the methylene blue. Add a coverslip. Observe the slide with the low-power and high–dry objective lenses of the microscope. Chlamydospores are in both surface and submerged mycelium. They are elongated and have thick walls.

9. In *Penicillium*, branching at the **conidiophore** tip can be symmetrical (figure 25.7) or asymmetrical, depending where the **metulae** attach to the conidiophore. When all of the metulae are attached at the conidiophore tip, branching of the sterigmata will appear symmetrical. If one of the metulae is attached below the conidiophore tip, asymmetrical branching occurs. This is an important diagnostic feature to differentiate *Penicillium*.

Note: If an ocular micrometer is available and time permits, measure the various structures, such as asexual spore sizes of different filamentous fungi.

EXERCISE

25

Laboratory Report:
Identification of Fungi ASM34

Results (Non-filamentous Fungi)

1. Fermentation study. Examine the tubes and record the results (+ or −) in table 25.3. A space has been left to record results in if your instructor assigned another microbe.

Table 25.3   Sugar Fermentation Study

Microbe	Glucose			Maltose			Lactose			Sucrose		
	Cloudy	Gas	Acid	Cloudy	Gas	Acid	Cloudy	Gas	Acid	Cloudy	Gas	Acid
Saccharomyces cerevisiae												

2. Yeast colony and vegetative cell morphology study.

   a. Colony characteristics. A space has been left to record results in if your instructor assigned another microbe.

Table 25.4   Colony Morphology on Sabouraud's Dextrose Agar

Microbe	Colony Morphology				
	Colony Color	Consistency	Diameter (mm)	Surface Appearance	Margin/Edge
Saccharomyces cerevisiae					

   b. Vegetative cell morphology. Draw what you see on 40× and record cell size, shape, presence or absence of pseudohyphae.

3. Sexual sporulation study. Draw what you see on 40× and record the presence or absence of asci or ascospores.

4. Describe the agar plate fungal colony characteristics in table 25.5.

Table 25.5   Colony Morphology of Fungi Cultured for _____ Days on _____ Agar

	Mucor	Penicillium	Rhizopus	Sordaria	Trichoderma	Armillaria
Colony size: diameter (mm)						
Colony color (forward)						
Colony color (reverse)						
Soluble pigments in agar						
Colony texture						
Colony margin						
Colony convolutions						
Bioluminescence						

5. Did you observe any differences in colony morphologies given by a fungus on different agar types? If so, explain them.

6. Draw and label the parts of a *Coprinus* and *Morchella* mushroom.

7. Morphological structures. Draw and label the asexual reproductive structures.

Mucor   *Penicillium*   *Rhizopus*   *Trichoderma*   *Aspergillus*   *Schizosaccharomyces*

8. Morphological structures. Draw and label the asexual spores.

Mucor   *Penicillium*   *Rhizopus*   *Trichoderma*   *Aspergillus*   *Morchella*

Questions

1. Explain the physiological differences between yeast fermentation and yeast assimilation of glucose.

2. Why would the growth of a pellicle, or film, on the surface of a broth growth medium be advantageous to the physiology and viability of a yeast?

3. What are some ways in which you might be able to differentiate *R. stolonifer* from *Mucor* simply by visually observing a Petri dish culture? What ways could you differentiate *Rhizopus* from *Sordaria*? From *Trichoderma*?

4. What problems might you have in identifying a pathogenic fungus observed in a blood specimen? What might you do to correct such problems?

5. In what ways can we readily distinguish
   a. fungi from bacteria?

   b. fungi from actinomycetes?

6. Define *opportunistic fungus*. Provide some examples. Are all medically important fungi opportunistic? Discuss your answers.

7. Name three pathogenic fungi that exhibit dimorphism and describe how causing the fungus to grow in a particular way might be used to target new drugs.

26

Identification of Protists: Algae, Lichens, Slime and Water Molds

Definitions

Algae. Photosynthetic eukaryotes found in aquatic environments that usually contain cellulose in their cell walls.

Cellular slime mold. A slime mold that possesses amoeboid-like cells that congregate; form fruiting bodies and spores that are released into the environment. *Dictyostelium discoideum* is an example.

Diatom. An algae that has symmetry, often seen at the bottom of ponds.

Lichen. A symbiotic relationship between a fungus and cyanobacteria or an alga.

Oomycete. A water mold. A protist that lives in wet or moist conditions.

Planaslo. A thickened solution added to wet mounts that slows down motile organisms so they can be viewed more easily.

Plasmodial slime mold. A terrestrial slime mold on trees and rocks that form a large, often colorful, moving plasmodium.

Protist. An unicellular or multicellular eukaryote that is not an animal, plant, or fungus.

Protozoa (singular, protozoan). A unicellular heterotrophic protist that is not a fungus, slime, or water mold. Often classified by method of locomotion.

Rotifer. A multicellular organism that is part of the pseudocoelomate with an intact digestive tract.

Water mold. An oomycete. A protist that lives in wet or moist conditions.

Objectives

1. Differentiate organisms into algae or protozoa.
2. Explain how algae and protozoa help clean water.
3. Why are invertebrates like *Rotifers* important to water quality?

Pre-lab Questions

1. How does *Paramecium* differ from *Euglena* in locomotive structures?
2. Why are water molds classified as protists?
3. What do algae require in order to live?

Getting Started

This exercise examines the diversity of the organisms that make up the microbial world of **Protists**. Different kinds of **algae**, which are photosynthetic, and **protozoa**, which lack cell walls, are studied. In this exercise we view the Underworld—microbes associated with water: the algae, protozoa, and water molds. The protozoan parasites will be discussed in exercise 27.

Algae can be either unicellular or multicellular and inhabit both fresh water and marine environments. All algae also contain chlorophyll, making them photosynthetic in nature. They are aptly named green, red, or brown, but most of what you will view will consist of green algae. Their cell walls are composed primarily of cellulose, but they lack transport mechanisms or complex reproductive structures. Some red algae are used to make the agar that is added to microbiological media to make Petri dishes solidify. Algae usually do not cause human diseases. However, they can overgrow at times, causing algal blooms, and some produce toxins harmful to humans and fish. Macroscopic algae that you can see and become fairly large include seaweed, lettuce, and kelp. Algae also make up a large portion of plankton.

Cladophora (**figure 26.1**) is a genus of common environmental alga that contains branches off from a main filament. It has been known to wash up onto beaches and shores in large masses, such as at Lake Michigan, with the decomposition of this large mass causing foul odors. It blooms if high phosphorous or nitrogen concentrations are present, such as with agricultural run-off seen in recent years.

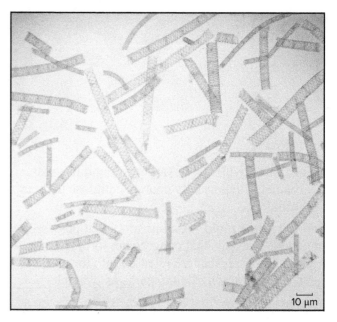

Figure 26.2 The algae *Spirogyra*. Courtesy of Anna Oller, University of Central Missouri.

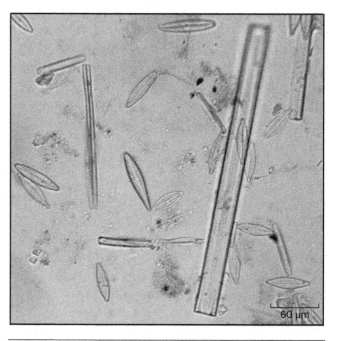

Figure 26.1 The algae *Cladophora* with branching filaments in pond water. Courtesy of Anna Oller, University of Central Missouri.

Figure 26.3 Many different Diatoms seen in pond water. Courtesy of Anna Oller, University of Central Missouri.

Spirogyra (**figure 26.2**) is another common environmental alga that is usually found in clean water environments. It can cause items in the water to become slimy. It contains the spirals or helices commonly associated with its name. The filaments are thin, but unlike *Cladophora*, it does not branch off from a filament. It contains filaments, and these filaments can simply break off during asexual reproduction. Conjugation tubes are involved in sexual reproduction of this alga.

Microscopic algae include the popular **Diatoms** (**figure 26.3**), which contain silicon dioxide in their cell walls. *Euglena* (**figure 26.4**) and *Volvox* (**figure 26.5**) are also common alga. All of these genera are commonly seen in pond water. *Diatoms* have

symmetry, just as humans have a left and right side. Diatoms can have many different shapes, but you will find symmetry in almost all of them. Diatoms fall to the bottom of oceans and ponds. They have many uses such as abrasives in cleaners, or being

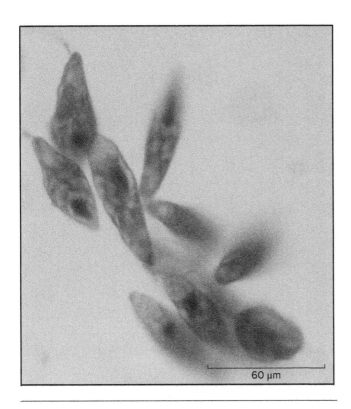

Figure 26.4 The algae *Euglena* showing elongated cells with a flagellum. Courtesy of Anna Oller, University of Central Missouri.

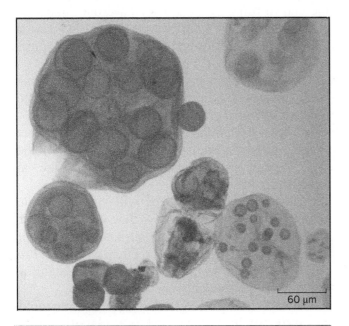

Figure 26.5 The algae *Volvox*, and some contain daughter cells. Courtesy of Anna Oller, University of Central Missouri.

mined as diatomaceous earth. Diatomaceous earth can be found in bags you can buy for planting your garden, and it has been used in cardiac catheter labs to determine how thin heparin has made a patient's blood.

Euglena is elongated and green with a flagellum. They move around the slide pretty quickly, which is one reason you add a thickened oil-like drop of **Planaslo**—so you can be fast enough to catch them! You can often see the flagellum if the light is turned down in your microscope or you have access to phase contrast.

Volvox appear as hollow spheres, but they can have daughter cells present within them. They will range in size as well. Most *Volvox* are considered to be freshwater species.

Lichens (color plate 66) are made up of one or two fungi and a cyanobacterium or algae. Lichens can appear as many different sizes, colors, and on many different surfaces. They grow very slowly, about 1 mm per year, and are thought to fragment (break off) for asexual reproduction. Only the fungi are thought to reproduce sexually. The fungal cells are thought to penetrate the cyanobacterium or algal cells. If you look at the tree trunks you can usually see green, blue, yellow, white, orange, or yellow crusty lichens. Lichens acquire most of their nutrients from the air, so the absence of lichens can indicate the presence of air pollution.

Protozoans are a very diverse group, containing unicellular, free-living organisms in aquatic environments that often serve as decomposers. In the food chain they often eat bacteria and algae, which are then eaten by larger organisms. Although they have previously been classified by locomotion, newer DNA technologies indicate that reclassification is forthcoming. Protozoan locomotion mechanisms include flagella (long projections), cilia (short projections), or pseudopodia (amoeboid-like).

Many protozoa seen in pond water samples include the Phylum Ciliophora, which contain many short, hair-like cilia on their surfaces that beat rhythmically by bending to one side. They contain two nuclei—a macro- and micronucleus. Ciliophora contains the following genera, which are all non-pathogens:

Paramecium (**figure 26.6**) is oval and around 100 μm in size. Cilia surround the organism and they are very active.

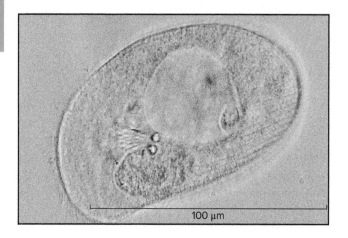

Figure 26.6 The protozoan *Paramecium* viewed at 40× in pond water. Courtesy of Anna Oller, University of Central Missouri.

Stentor (**figure 26.7**) appear as two funnels attached to filaments in which the heads have a circular whirring movement to bring in food.

Vorticella appear as an elongated funnel shape that have one head per very long stalk that may or may not be attached to plants. They also have cilia at their ends to bring in food.

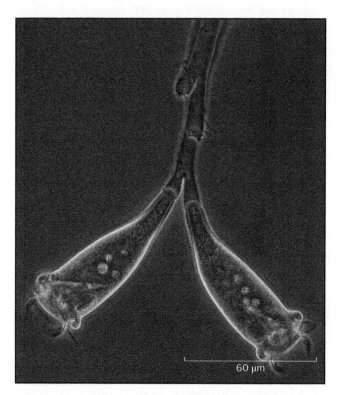

Figure 26.7 A *Stentor* viewed at 40× in pond water. Courtesy of Anna Oller, University of Central Missouri.

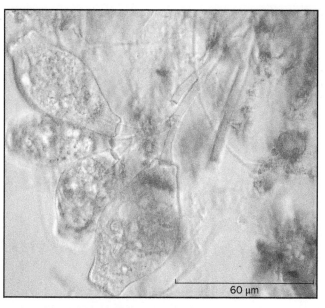

Figure 26.8 A *Carchesium* in pond water. Courtesy of Anna Oller, University of Central Missouri.

Euplotes are small, oval with a flagella tuft around their middle. They are around 60 μm in size and very active often appearing to go in circles.

Carchesiums (**figure 26.8**) resemble vorticella with the elongated funnel shape. They have cilia at the end to bring in food. They have one funnel per arm but have many funnels in a bunch (10+), similar to a "bouquet" of flowers. They are around 60 μm in size and can lunge and contract to obtain food.

Blepharisma appear oval in shape like a *Paramecium*, are around 80 μm, and are pink or red in color due to pigmented granules.

Coleps are around 80 μm in size, appear barrel shaped with cilia at one end and hooks at the other end. The squares that make up the "barrel" are calcium carbonate plates.

Most protozoans that you will view eat bacteria and/or yeasts. This is how they can organically clean up water. Some absorb oil, and some eat dirt and other particles in the water. Finally, *Amoeba proteus*, a unicellular protozoan, may be seen engulfing organisms with pseudopodia (**figure 26.9**).

Another group of organisms include **slime molds** and **water molds**, which used to be classified under fungi, but with technological advances, they are now both classified as protists. **Water molds** are called **oomycetes** and contain cellulose in their cell walls

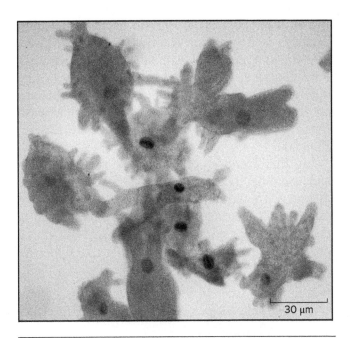

Figure 26.9 *Amoeba proteus* with pseudopodia.
Courtesy of Anna Oller, University of Central Missouri.

rather than the chitin seen in traditional fungal cells. This is one way they were distinguished from fungi.

The oomycete *Phytophthora infestans* was responsible for the Irish potato famine of 1846. Potatoes were about the only food people could afford to eat and about half of the country lived in poverty. Potatoes were the major crop people planted because they would grow in poor soil and in the cool, wet climates of Ireland and Scotland. Only one variety of potato plant was grown in Ireland, so all the plants were vulnerable to the same diseases. Excess moisture allowed the fungus to grow, and the fungus infected all parts of the plant: the stems, leaves, and tubers, making the potatoes inedible. Historians estimate that almost a million people died of starvation, and another million people fled from Ireland and Scotland to the United States.

Phytophthora ramorum is another oomycete responsible for causing a disease called sudden oak death, which began in California in 1995, when oak trees began to die. In 2000, the fungus was identified as *Phytophthora ramorum*. Since then it has continued to spread up and down the Pacific coast of the United States Once the infection was established, other countries started to ban US plant imports. In 2017, a *Phytophthora* research center was opened in the Czech Republic.

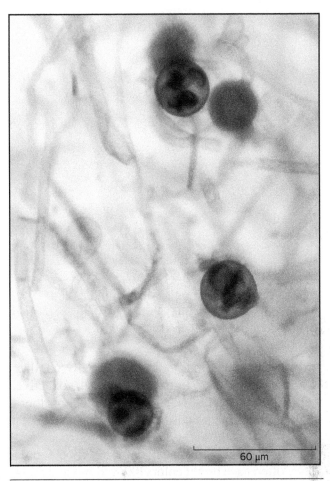

Figure 26.10 The oomycete *Saprolegnia*, with several oogonia containing eggs. Courtesy of Anna Oller, University of Central Missouri.

Saprolegnia (**figure 26.10**) is another oomycete that can cause infections in fish. In stagnant water, the oomycete can increase in numbers. When a fish gets an open wound, the oomycete will attach and cause secondary infections. It will usually lead to death. During sexual reproduction, a long hypha will have a rounded end with a sac, called an oogonium, that contains eggs. Asexual reproduction is by motile zoospores that travel in colonies together.

Slime molds are considered terrestrial, grow on trees and rocks, and are classified into two main types: **cellular slime molds** and **plasmodium slime molds**. Cellular slime molds have amoeba-like cells, and some species, like *Dictylostelium discoideum*, possess the ability to aggregate cells, form fruiting bodies via an elongated slug shape when they run out of food, and then release a spore into the environment.

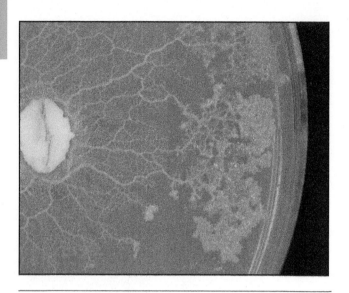

Figure 26.11 A plasmodial slime mold called *Physarum polycephalum*, a myxomycete.
Lisa Burgess/McGraw Hill Education.

Plasmodium slime molds are actually one large multinucleated cell (**figure 26.11**) and collectively called Myxomycetes. Haploid cells come together to make diploid cells and they divide repeatedly. The cytoplasm has the ability to shift and move to engulf bacteria, yeasts, and fungal spores as food sources. They often have bright colors like yellows and orange and in wet years can be very large "blobs" seen moving along trees or rocks.

They can be grown fairly easily. A small piece of tree bark is acquired from a tree and kept in a paper bag until the lab setup is ready. A large Petri dish with a moistened piece of filter paper is added to the Petri dish and then the small piece of tree bark is added to the moistened dish. The top lid is then added to cover the sample. Place in a darkened area like a cabinet and sometimes you can see growth in a day or two. Depending on the humidity level in your area, you may need to add more water to the paper periodically. If nothing is seen after a week, then no fruiting body was present on the bark sample in order to form a plasmodium.

Some other organisms you may see in your water slides, depending upon the source, are **Rotifers**, which are in the Phylum Rotifera. They are microscopic, multicelled pseudocoelomate animals that have cilia on their head to bring in food. Some are free-swimming out in the environment and some

live inside tubes. They bring in bacteria and other particles to help clean water as well. They have a bilateral symmetry too, similar to humans and diatoms. You may also see worms, which may or may not have segments. Members of the class Turbellaria are free-living planarians (flatworms), such as those found in the genera *Dugesia* (**figure 26.12**) and *Planaria*.

In this exercise, you will view prepared slides and examine living cultures of algae and protists in water. If any liquid ends up on your microscope, inform your instructor so any contamination can be properly disinfected.

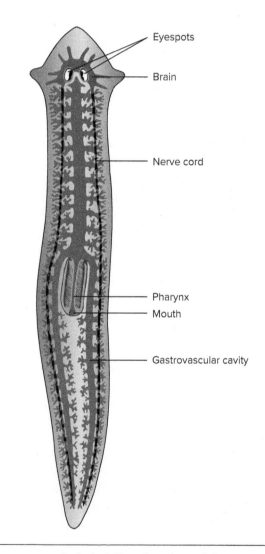

Figure 26.12 Labeled line drawing of the genus *Dugesia*, a free-living planarian in the class Tubellaria.

Materials

Cultures of the following:

Living cultures of a *Paramecium* species, *Amoeba proteus*, and a *Dugesia* or *Planaria* species

If available, a fresh sample of quiescent, stagnant pond water, which often contains members of the above genera. Students may wish to bring their own pond water.

Planaslo solution, 1 or more dropping bottles

Depression slides (hanging drop slides)

Coverslips

Vaseline

Toothpicks

PROCEDURE

Your study will consist of these procedures:

1. Observe the movements and structure of some living non-parasitic protozoa and worms often found in pond water.

   If pond water or prepared solutions of algae and protozoa are not available, then prepared slides of the following may be used:

   a. *Cladophora*
   b. *Spirogyra*
   c. *Diatoms*
   d. *Euglena*
   e. *Volvox*
   f. *Paramecium*
   g. *Amoeba proteus*
   h. *Saprolegnia*

2. Observe a prepared slide of a free-living flat-worm (*Dugesia* or *Planaria* species) with the low-power objective lens. Note the pharynx, digestive system, sensory lobes in the head region, and the eyespots.

Note: If the number of prepared slides is limited, these procedures may be performed in a different order to facilitate sharing.

1. Examination of free-living cultures.

   a. Pond water examination. Prepare a wet mount (exercise 3) using petroleum jelly to adhere a coverslip to a glass slide. Examine initially with the low-power objective lens and later with the high–dry objective lenses. Observe the mode of locomotion of any amoeboid or paramecium-like protozoa found. If their movements are too rapid, add a drop of Planaslo solution to slow them down. Describe their movements and prepare drawings in part 1*a* of the Laboratory Report.

   b. Examination of fresh samples of an amoeba (such as *Amoeba proteus*), para-mecium (e.g., *Paramecium caudatum*), and a free-living flatworm (such as *Dugesia* or *Planaria* species). Use wet mount slide preparations and examine as described in 1*a* for pond water. Record your observations in part 1*b* of the Laboratory Report.

EXERCISE

26

Laboratory Report: Identification of Protists: Algae, Lichens, Slime and Water Molds

Results

1. Examination of free-living cultures
   a. Pond water examination.
      Description of movements and drawings of any protozoa found in pond water.

   b. Examination of fresh samples of a free-living amoeba, paramecium, and flatworm.
      Description of movements and drawings with labels.

2. Draw pictures of the following microorganisms and record the magnification you viewed them under:

Cladophora   *Spirogyra*   *Diatoms*

Euglena   *Volvox*   *Paramecium*

Amoeba proteus   *Saprolegnia*   *Other:*_____

27

Identification of Parasites

Definitions

Acoelomate. Without a true body cavity. Platyhelminthes (flatworms) are an example.

Amoeba. Unicellular organisms with an indefinite changeable form and use pseudopodia to move.

Cercaria. A developmental stage for Schistosomiasis where a tail is used to propel the parasite in water to reach and penetrate the skin of a host.

Coelomate. With a true body cavity. Nematoda (roundworms) are examples.

Commensal. A relationship between two organisms in which one partner benefits from the association and the other is neither harmed nor benefits.

Cysts. Dormant, thick-walled cells. The stomach acids of many hosts allow the cyst to release a reproducing trophozoite.

Definitive host. The host in which the sexual reproduction or adult form of a parasite takes place.

Intermediate host. A host used by a parasite where asexual reproduction or an immature stage forms.

Merozoites. Schizont nuclei that become surrounded by cytoplasm and bud off as daughter cells.

Miracidium. A free-swimming ciliate seeking a snail species to penetrate and develop into a sporocyst.

Proglottid. A tapeworm neck segment formed by strobilation (transverse fission).

Pseudopodia. Cytoplasmic extensions that engulf particles and allow movement of amoeboid cells.

Schizont. An apicomplexan life stage in which the nucleus undergoes repeated nuclear division without corresponding cell divisions.

Scolex. The head of a tapeworm, used for attaching to the host's intestinal wall.

Sporocyst. A protozoan life cycle stage in which two or more of the parasites are enclosed within a common wall.

Trophozoites. Vegetative feeding form of some protozoa.

Objectives

1. Differentiate malarial and trypanosome parasites life history stages using stained blood smears.

2. Describe the morphology of some free-living, trophozoite, and cystic forms of intestinal parasites using prepared slides.

3. Explain the natural history and life cycle of an important human parasitic disease, schistosomiasis, using stained slides and a life-cycle diagram.

Pre-lab Questions

1. What is the common name for a trematode?

2. Provide the scientific name of a Mastigophoran.

3. What are the main steps in the life cycle of *Schistosoma mansoni*?

Getting Started

Because the natural histories of parasitic diseases differ from those of bacterial diseases, they merit a separate introductory laboratory experience with parasites, the diseases they cause, and techniques used to diagnose them.

The distinguishing features of parasitic life are the close contact of the parasite with the host in or on which it lives and its dependency on the host for life itself. This special association has led to the evolution of three types of adaptations not found in free-living relatives of parasites: loss of competency, special structures, and ecological ingenuity.

Parasites have become so dependent on their hosts for food and habitat that they now experience a *loss of competency* to live independently. With the exception of some viruses, they usually require a specific host, and many have lost their sensory and digestive functions; these are no longer important for their survival.

On the other hand, they have developed *special structures* and functions not possessed by their free-living relatives that promote survival within the host, such as hooklets and suckers for attachment. Parasites also have an increased reproductive capacity to compensate for the uncertainty in finding a new host. Tapeworms can produce up to 100,000 eggs per day.

Ecological ingenuity is demonstrated in the fascinating variety of infecting and transmitting mechanisms. This has led to complex life cycles, which contrast markedly with the relatively simple lifestyles of their free-living counterparts. Parasites possess diverse life cycles, ranging from species that pass part of each generation in the free-living state to others that require at least three different hosts to complete the life cycle. Some are simply transmitted by insects from one human host to a new host, or the insect may act as a host as well. Many protozoa develop resistant cysts, which enable them to survive in unfavorable environments until they find a new host. The eggs of flatworms and roundworms also have a protective coat.

These three strategies promote species survival and expansion by providing greater opportunities for finding and infecting new hosts, which is a continual problem for parasites. Successful interruption of these cycles to prevent their completion is an important feature of public health measures used to control diseases caused by parasites.

This exercise is designed to give you some practical experience with representative protozoan and helminthic parasites and with clinical methods used in their diagnosis and control. The following classification of parasites will serve as a guide to the examples you will be studying in this exercise. Many of the defined terms are in the exercise and not just in the introduction. It is not a complete listing.

Protozoa

Protozoa, members of the subkingdom of the kingdom Protista, are unicellular eukaryotic organisms. They usually reproduce by cell division and historically were classified according to the mode of locomotion; with new DNA information, many organisms are being reclassified (**table 27.1**).

Sarcodina (Amoeba)

This group includes amoebae that move and feed slowly by forming cytoplasmic projections called **pseudopodia** (false feet). A **trophozoite** (vegetative form) and **cyst** (resistant, resting cells) stage exist. It reproduces asexually by fission. Parasitic members include the **amoeba** *Entamoeba histolytica*, which causes amoebic dysentery in animals and humans (color plate 67). It ingests red blood cells and forms a four-nucleate cyst. Other amoeba species found in humans, like *Entamoeba gingivalis*, are harmless **commensals**.

Ciliophora (Ciliates)

Ciliophora possesses trophozoites that are motile using many short, hairlike cilia on their surface, reproduce by transverse fission, and usually have a mouth for feeding. They have two nuclei: a macronucleus and a micronucleus.

Balantidium coli (**figure 27.1**) is a common parasite in swine that has both a cyst and trophozoite

Table 27.1 Protozoa of Medical Importance

rRNA Classification	Disease-Causing Protozoan	Diseased Caused by Protozoan
Apicomplexan	*Plasmodium malariae, P. falciparum, P. ovale, P. vivax, P. knowlesi*	Malaria
	Toxoplasma gondii	Toxoplasmosis
	Cryptosporidium parvum	Cryptosporidiosis
	Cyclospora cayetanensis	Cyclosporiasis
Diplomonad	*Giardia lamblia (intestinalis)*	Giardiasis
Heterolobosean	*Naegleria fowleri*	Primary amebic meningoencephalitis
Kinetoplastid	*Trypanosoma brucei*	African sleeping sickness
Lobosean	*Entamoeba histolytica*	Amoebiasis
Parabasalid	*Trichomonas vaginalis*	Trichomoniasis

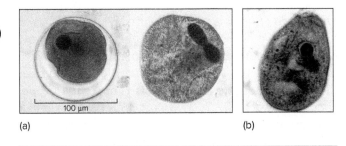

(a)   (b)

Figure 27.1 (a) *Balantidium coli* cysts viewed on 40×.
The macronucleus can be seen on the photo on the
right. (b) *Balantidium coli* trophozoite viewed on 40×,
showing the nucleus overlaid on the macronucleus.
(*a-b*) Courtesy of Anna Oller, University of Central Missouri.

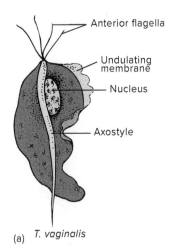

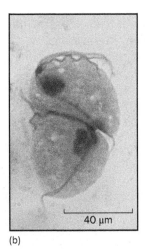

(a) *T. vaginalis*   (b)

Figure 27.2 (a) *Trichomonas vaginalis*.
Illustration of a typical mastigophoran protozoan.
(b) *Trichomonas vaginalis* viewed on 100×.
The flagella, nucleus, axostyle, and undulating
membrane can all be seen at the top. (b) Courtesy of
Anna Oller, University of Central Missouri.

form. It can infect humans who eat undercooked sausage, causing serious results.

Mastigophora

This group is motile using flagella, has a single nucleus, and many use longitudinal fission for reproduction. They usually have a cyst and trophozoite form. *Giardia lamblia* (*G. intestinalis*) is a diplomonad that has a cyst stage containing up to four nuclei, and a trophozoite stage that has a tapered flagellum for motility (color plate 68, figure 29.3). It causes a mild to severe diarrheal infection. *Trichomonas vaginalis* (**figure 27.2**) is a parabasilid found in the urogenital region. Most people are asymptomatic, but it can cause vaginitis in women. *Trichomonas* is considered a sexually transmitted infection and often diagnosed by accident in urine samples. *Trypanosoma gambiense* (color plate 69) infects the blood via tsetse fly bites, causing trypanosomiasis, or African sleeping sickness (**figure 27.3**), in cattle and humans. Cattle and other ungulates serve as reservoirs for this organism.

Trypanosoma cruzi is an emerging pathogen in the United States, causing American trypanosomiasis, or Chaga's disease. The life cycle is shown in **figure 27.4** The reduvid bug, or "kissing bug," feeds at night from capillaries near the mouth and eyes since they are close to the surface and easy to penetrate. The bug defecates near the wound where it fed and is wiped into the wound unknowingly. Most people are asymptomatic but over time it causes chronic heart failure. Opposums, dogs, and armadillos all serve as reservoirs.

Leishmania donovani (**figure 27.5**) is a kinetoplastid transmitted to humans by sandflies, which serve as an

intermediate host. It causes cutaneous lesions but can affect the liver in severe cases. In the United States, many cases are diagnosed in military personnel and travelers, especially those who have been to the Middle East.

Apicomplexa (Sporozoa) Sporozoa

Apicomplexa are obligate, non-motile parasites with complex life cycles: The sexual reproductive stage is passed in the definitive insect host, and the asexual phase is passed in the intermediate human or animal host. The genus *Plasmodium* (color plates 70 and 71) includes the malarial species (*malariae, ovale, vivax, falciparum, knowlsii*), in which the **definitive host** is the female *Anopheles* mosquito and the **intermediate host** is humans (**figure 27.6**). The genus *Coccidia* includes important intestinal parasites affecting fowl, cats, dogs, swine, sheep, and cattle. *Toxoplasma gondii* is a cat parasite that can harm the human fetus in an infected pregnant woman.

Helminths (Worms)

Helminths are multicellular eukaryotic organisms.

Phylum Platyhelminthes

Members of Platyhelminthes are flat, elongated, legless worms that are **acoelomate** and exhibit bilateral symmetry. This phylum contains three classes, Monogea, Trematoda, and Cestoda. Monogea includes the flatworms as seen in exercise 26.

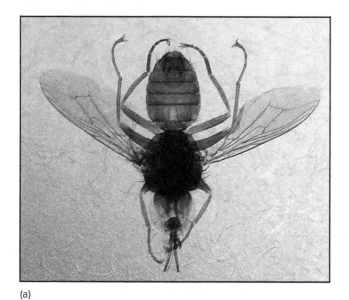

(a)

Trematoda (Flukes)

Flukes have an unsegmented body, and many have suckers to hold them onto the host's intestinal wall. Many flukes have complex life cycles that require aquatic animal hosts. The *Schistosoma* species (**figures 27.7, 27.8,** and **27.9**) are bisexual trematodes that cause serious human disease. They require fecal-contaminated water, snails, and contact with human skin for completion of their life cycles. *Opisthorchis* (formerly called *Clonorchis*), *Clonorchis sinensis*, and *Fasciola hepatica* (color plate 72) species are liver flukes acquired by eating infected raw fish and contaminated vegetables like watercress, respectively.

Cestoda (Tapeworms)

Tapeworms are long, segmented worms (**figure 27.10a**) with a small head (**scolex**) equipped with suckers and often hooklets (**figure 27.10b,** color plate 73) for attachment to the host's intestinal wall. The series of segments, or **proglottids,** contain the

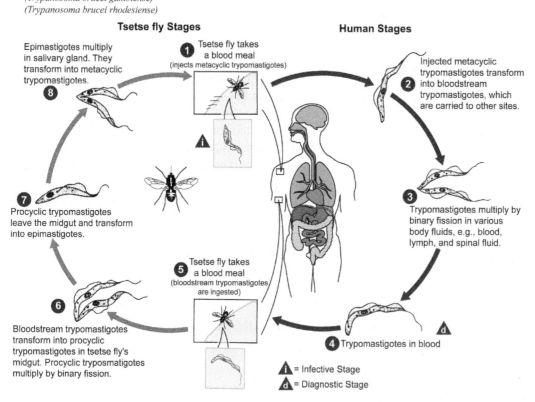

(b)

Figure 27.3 (*a*) A *Glossina* tsetse fly that can transmit the disease trypanosomiasis. (*b*) The life cycle for trypanosomiasis. (*a*) Courtesy of Anna Oller, University of Central Missouri; (*b*) CDC/Alexander J. da Silva, PhD, Melanie Moser.

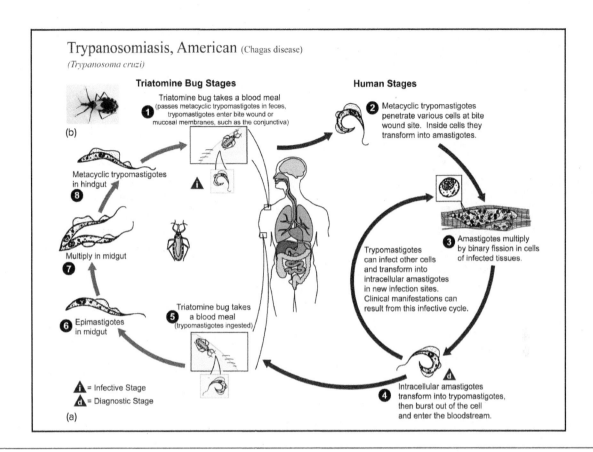

Figure 27.4 Chagas disease life cycle, seen in the United States. (*a*) CDC/Alexander J. da Silva, PhD, Melanie Moser; (*b*) schlyx/Shutterstock.

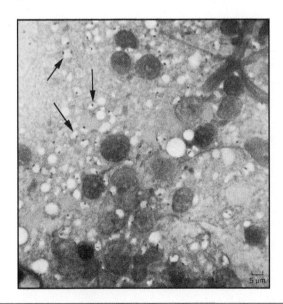

Figure 27.5 *Leishmania donovani* amastigotes in tissue. The small circular clearings are the parasite and the dots with the arrows pointing at them are part of their nucleus. Courtesy of Anna Oller, University of Central Missouri.

reproductive organs and thousands of eggs. These segments break off and are eliminated in the feces, leaving the attached scolex to produce more proglottids with more eggs. **Figure 27.11** illustrates the life cycle of the tapeworm in a human. The symptoms of *Taenia* tapeworm infection are usually not serious, causing only mild intestinal symptoms and loss of nutrition—not so for the *Echinococcus* tapeworm, which causes a serious disease. All tapeworm diseases are transmitted by animals.

Nematoda (Roundworms)

Members of Nematoda (roundworms) are present in large numbers in diverse environments, including soil, freshwater, and seawater. This phylum contains many agents of animal, plant, and human parasitic diseases. In contrast to the Platyhelminthes, these round, unsegmented worms are **coelomate** (have a body cavity) and have a complete digestive tract and separate sexes. Most require only one host and can pass part of their life cycle as free-living larvae in the soil. *Trichinella spiralis*

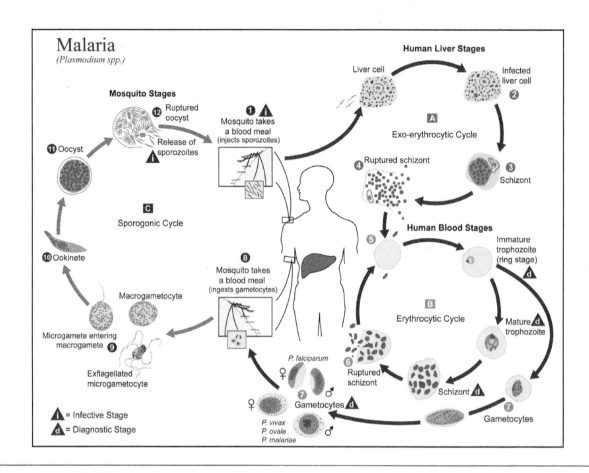

Figure 27.6 Malaria life cycle. CDC/Alexander J. da Silva, PhD, Melanie Moser.

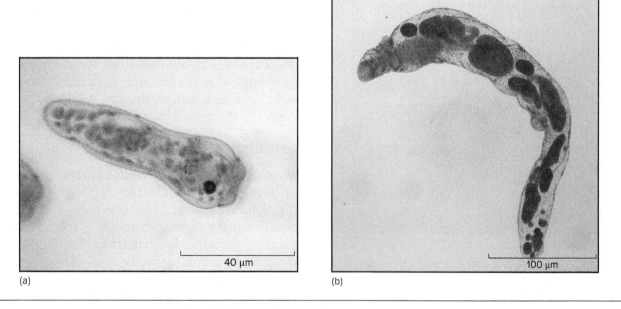

(a)

(b)

Figure 27.7 (*a*) A *Schistosoma* miracidium seeking a snail host to replicate in and (*b*) rediae containing sporocysts that are found in water viewed on 40×. (*a-b*) Courtesy of Anna Oller, University of Central Missouri.

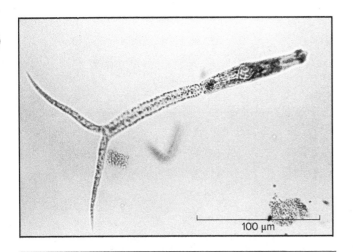

Figure 27.8 Schistosomal cercaria in the water seeking a host to penetrate and infect. Melissa Rethlefsen and Marie Jones, Prof. William A. Riley/Minnesota Department of Health, R.N. Barr Library/CDC.

(color plate 74) requires alternate vertebrate hosts. Humans become infected when they ingest inadequately cooked meat, such as pork or bear containing the larval forms in the muscles. *Ascaris lumbricoides* (**figure 27.12a** and **b**) is probably the most common human helminth worldwide. Larva migrate through the lungs and up the trachea, where they are then swallowed and mature in the intestines. *Enterobius vermicularis* causes pinworm (**figure 27.13**), a very common condition in children characterized by anal itching. *Trichuris trichura* (**figure 27.14**) is another common worm that causes whipworm in animals and humans. *Dirofilaria immitis* (color plate 75) is transmitted via mosquitoes and causes heartworms in animals. Left untreated, or treated with high doses of ivermectin, animals will perish. Animals must be treated with low doses of ivermectin over a period of time because a large number of dead worms can cause a lethal hypersensitivity or embolus.

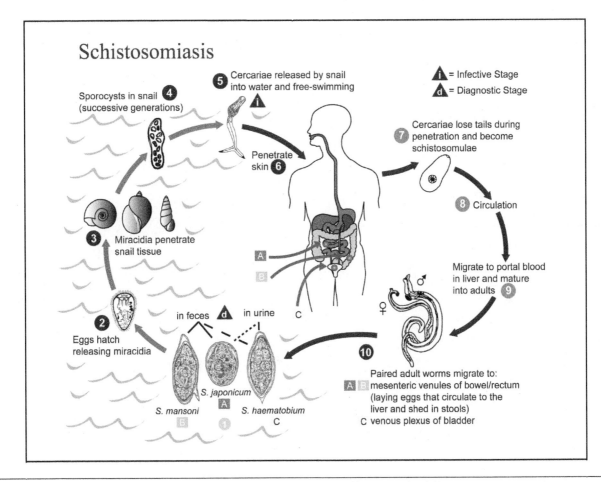

Figure 27.9 Life cycle of schistosomiasis. CDC/Alexander J. da Silva, PhD, Melanie Moser.

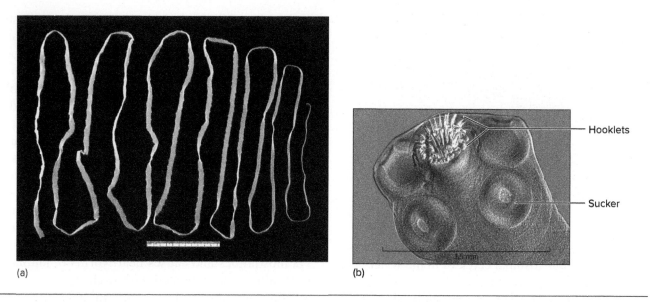

(a)

Hooklets

Sucker

(b)

1.5 mm

Figure 27.10 (*a*) *Taenia saginata* adult tapeworm. (*b*) A *Taenia solium* tapeworm scolex showing both hooklets and suckers for attachment to the intestine. *Taenia saginata* (beef tapeworm) lacks hooklets, whereas *Taenia solium* (pork tapeworm) has both. (*a*) Centers for Disease Control and Prevention; (*b*) Cultura Creative/Alamy Stock Photo.

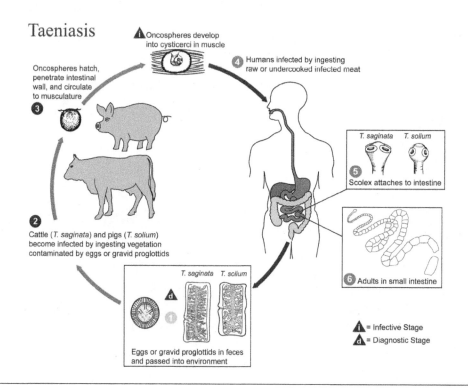

Taeniasis

🔺i Oncospheres develop into cysticerci in muscle

Oncospheres hatch, penetrate intestinal wall, and circulate to musculature

❸

❹ Humans infected by ingesting raw or undercooked infected meat

T. saginata   *T. solium*

❺ Scolex attaches to intestine

❷ Cattle (*T. saginata*) and pigs (*T. solium*) become infected by ingesting vegetation contaminated by eggs or gravid proglottids

T. saginata   *T. solium*

🔺d

❶

❻ Adults in small intestine

Eggs or gravid proglottids in feces and passed into environment

🔺i = Infective Stage

🔺d = Diagnostic Stage

Figure 27.11 Life cycle of *Taenia infection*. The adult tapeworm with scolex and proglottids develops from eggs in the human intestine. CDC/Alexander J. da Silva, PhD, Melanie Moser.

(a)

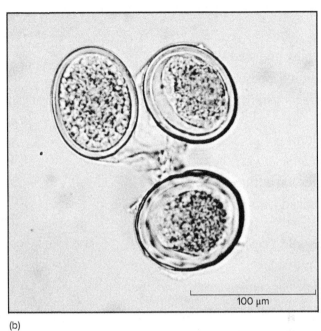

100 μm

(b)

Figure 27.12 (a) *Ascaris lumbricoides* worms. (b) *Ascaris lumbricoides* eggs viewed on 40×. (a) Source: James Gatheny/Centers for Disease Control; (b) Courtesy of Anna Oller, University of Central Missouri.

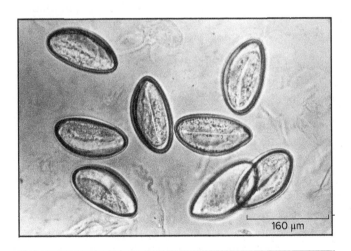

160 μm

Figure 27.13 *Enterobius vermicularis* eggs.
Centers for Disease Control and Prevention.

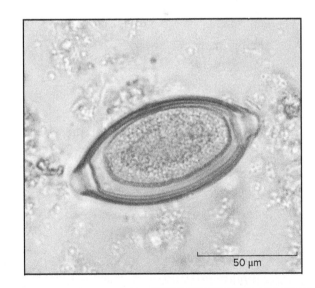

50 μm

Figure 27.14 Whipworm eggs in a fecal smear viewed at 100×. Notice the rounded end caps at each end. Courtesy of Anna Oller, University of Central Missouri.

Materials

Commercially prepared slides of:

Sarcodina (Lobosea)

Entamoeba histolytica trophozoite and cyst stages

Ciliophora (cilia)

Balantidium coli

Mastigophora (flagella)

Diplomonad *Giardia lamblia* trophozoite and cyst stages

Parabasalid *Trichomonas vaginalis*

Kinetoplastid

Trypanosoma gambiense

Leishmania donovani

Apicomplexa (Sporozoa)

P. falciparum ring, amoeboid schizont stages

Subkingdom Helminths (worms)

Phylum Platyhelminthes (flatworms)

Trematodes (flukes)

Schistosoma mansoni

    adult male

    adult female

    ovum (egg)

    ciliated miracidium

    infective ciliate cercaria

    sporocyst stage in snail liver tissue

Opisthorchis (Clonorchis) sinensis or *Fasciola hepatica*

Cestodes (tapeworms)

Taenia solium trophozoite or *T. saginata*

Nematodes (roundworms)

Ascaris lumbricoides

Enterobius vermicularis

Trichuris trichuris

Trichinella spiralis

Dirofilaria immitis

PROCEDURE

Your study will consist of these procedures:

1. Examine commercially prepared stained blood and fecal slides that contain human protozoan parasites.

2. Microscopically compare structures of parasitic worms to their free-living relatives.

3. Study the life cycles of the human parasitic diseases schistosomiasis, trypanosomiasis, and malaria.

Note: If the number of prepared slides is limited, these procedures may be performed in a different order to facilitate sharing.

1. Examination of stained slides for trophozoites and cysts in feces.

   a. Using the 10×, scan a *Balantidium coli* slide for trophozoites and cysts.

   b. Using the oil immersion objective lens, examine prepared slides of a protozoan, either the amoeba *Entamoeba histolytica* or the flagellate *Giardia lamblia*. In the trophozoite stage, observe the size, shape, number of nuclei, and presence of flagella or pseudopodia. In the cyst stage, look for an increased number of nuclei and the thickened cyst wall.

   c. Sketch an example of each stage, label, and record in the Laboratory Report.

2. Examination of protozoa present in stained blood slides.

   a. Use the oil immersion objective lens to explore a blood smear of *Plasmodium*, and locate blood cells containing the parasite. After a mosquito bite, the parasites are carried to the liver, where they develop into **merozoites.** Later, they penetrate into the blood and invade the red blood cells, where they go through several stages of development. The stages are the delicate ring stage (color plate 70), the mature amoeboid form, and the **schizont** stage (color plate 71), in which the parasite has divided into many infectious segments, causing the cell to rupture and release the parasites, which can then infect other cells. Sketch the red blood cells with the parasite inside them, and note any changes in the red blood cell shape, pigmentation, or the presence of granules due to the parasite. Identify and label the stage or stages seen, as well as the species, in part 2a of the Laboratory Report.

   b. Examine the trypanosome blood smear (color plate 69) with the oil immersion lens, and locate the slender flagellates between the red blood cells, noting the flagellum and undulating membrane. Sketch a few red blood cells along with a flagellate in part 2b of the Laboratory Report.

3. Examine a parasitic fluke such as *Opisthorchis* (*Clonorchis*) or *Fasciola*. Note the internal structure, especially the reproductive system and eggs if female, and the organs of attachment, such as hooklets or round suckers.

   a. Sketch each organism in part 3 of the Laboratory Report, and label the main features of each. Describe the main differences between the fluke and the free-living planaria.

   b. Examine the prepared slides of a tapeworm (*Taenia* species, figure 27.10), observing the head, or scolex, and the attachment organs—the hooklets or suckers. Locate the maturing proglottids along the worm's length. The smaller proglottids may show the sex organs better; a fully developed proglottid shows the enlarged uterus filled with eggs. Sketch, label, and describe its special adaptations to parasitic life in part 3 of the Laboratory Report.

   c. Observe a prepared slide of *Ascaris lumbricoides*, noting the differences between the flatworm, fluke, and tapeworm. Sketch in part 3 of the Laboratory Report.

   d. Observe a prepared slide of *Dirofilaria immitis* on 40×, noting the differences between the other classes of worms. Sketch in part 3 of the Laboratory Report.

4. Life cyle of *Schistosoma mansoni* and its importance in the control of schistosomiasis.

   a. Assemble five or six slides showing the various stages in the schistosoma life cycle: adult worm (male and female if available),

ova, ciliated miracidium, the sporocyst in snail tissue, and the infective cercaria.

Human schistosomiasis occurs when water is contaminated with human feces; this water is used for human bathing and wading, or irrigation of cropland; and snail species are present that are necessary as hosts for the sporocyst stage in fluke development and completion of its life cycle. The solution to this public health problem is very complex, not only because of technical difficulties in its control and treatment, but also because its life cycle presents an ecological dilemma. Many developing countries need food desperately, but the main sources now available for these expanding needs are fertile deserts, which have adequate nutrients but require vast irrigation schemes, such as the Aswan Dam in Egypt. However, due to the unsanitary conditions and the presence of suitable snail hosts, these projects are accompanied by an increase in the disease schistosomiasis, which is difficult to control and expensive to treat worldwide.

The **cercaria** larvae swim in contaminated water, penetrating the skin of workers. They migrate into the blood and collect in the veins leading to the liver. The *adults* develop there, mate, and release *eggs*. The eggs are deposited in the small veins of the large intestine, where their spines cause damage to host blood vessels. Some eggs die; however, others escape the blood vessels into the intestine and pass with the feces into soil and water. There they develop and then hatch into motile **miracidia**, which eventually infect suitable snail hosts and develop into saclike **sporocysts** in snail tissues. From this stage, the forktailed cercaria larvae develop; these leave the snail and swim in the water until they die or find a suitable human host, thus completing the complex life cycle involving two hosts and five separate stages.

   b. Now look at the prepared slides of all the *Schistosoma* stages discussed in the preceding description. Sketch each stage in the appropriate place in the life cycle diagram shown in part 4 of the Laboratory Report.

EXERCISE

27

Laboratory Report: Identification of Parasites

Results

1. Examination of stained slides for trophozoites and cysts. Draw and label accordingly trophs and cysts of:

   Entamoeba histolytica   *Giardia lamblia*   *Balantidium coli*

2. Examination of protozoa present in stained blood slides
   a. Examine blood smears of *Plasmodium species*.

   b. Examine blood smears of *Trypanosoma gambiense*.

3. Examine the following:

   a. *Opisthorchis (Clonorchis) sinensis* or *Fasciola hepatica* (parasitic).

   b. Study of a parasitic tapeworm (*Taenia* species).

   c. Study of *Ascaris lumbricoides*.

   d. Study of *Dirofilaria immitis*.

4. Life cycle of *Schistosoma mansoni* and possible methods of control

   a. For each space in this life cycle, sketch the appropriate stage, using the prepared microscope slides.

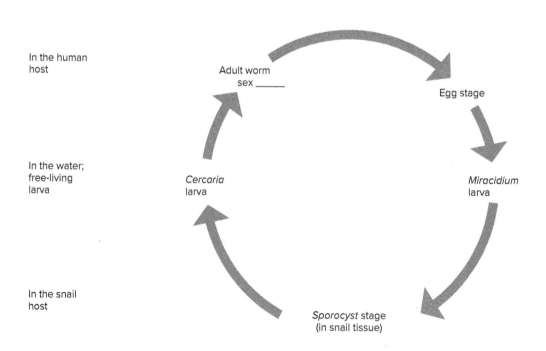

In the human host   Adult worm sex _____

Egg stage

In the water; free-living larva   *Cercaria* larva

Miracidium larva

In the snail host

Sporocyst stage (in snail tissue)

b. Propose a plan for public health control of schistosomiasis. Describe various strategies that might be developed by public health personnel to interrupt this cycle and thus prevent schistosomiasis. Show on a diagram where specific measures might be taken, and label. Explain each possibility and its advantages and disadvantages.

Questions

1. Which form—the trophozoite or the cyst—is most infective when found in a feces sample? Explain why.

2. In what ways are free-living and parasitic worms similar, such that they can be identified as closely related?

3. In what ways do the parasitic species differ from the free-living planaria in exercise 26? Use the chart to summarize your comparisons.

	Planaria	Fluke	Tapeworm
Outside covering			
Organs of attachment			
Sensory organs			
Digestive system			
Reproduction			

4. Estimate the length and width of a trypanosome. See color plate 69 for a clue. Show your calculations.

5. How is the *Echinococcus* tapeworm transmitted to humans? Does it cause a serious disease? What are two ways in which its transmission to humans can be prevented?

6. Room to draw other parasites.

28

Prokaryotic Viruses ^{ASM10, 18, 23, 34}

Definitions

Bacteriophage. A virus that infects bacteria; often abbreviated phage (rhymes with *rage*).

CRISPR (Clustered Regularly Interspaced Short Palindromic Repeats). Bacteria use CRISPR enzymes to recognize and remove CRISPR viral DNA sequences that have previously infected them.

Lytic virus. A virus that replicates within a host cell and causes it to produce phage, rupture, and die. See *virulent phage*.

Plaque. A clear or cloudy area in a bacterial lawn caused by phage infecting and lysing bacteria.

Plaque-forming units. A single phage that initiates the formation of a plaque. Abbreviated PFU.

Serial dilution. A dilution of a dilution, continuing until the desired final concentration is reached. (See exercise 8.)

Temperate phage. A phage that can either integrate into the host cell DNA or replicate outside the host chromosome, producing more phage and lysing the cell. See *lysogen*.

Titer. The concentration of virus in a sample (number/volume) (when used as a noun), or to determine the concentration (when used as a verb).

Virulent phage. A phage that always causes lysis of the cell following phage replication. Unlike a temperate phage, it cannot be integrated into the chromosome of the host. See *lytic virus*.

Objectives

1. Describe how to determine if a phage can or cannot infect a specific bacterium.
2. Calculate the concentration of a phage suspension.
3. Describe the steps involved in the phage life cycle.

Pre-lab Questions

1. What is a plaque?
2. How would you expect your plates to look beginning from the 10^{-1} going to the 10^{-4}?
3. What is the purpose of taking 1 ml from the 10^{-1} tube and adding to the next tube?

Getting Started

Viruses cannot replicate without infecting a living cell. To study a virus, you must also grow its host cells, so the easiest viruses to work with are those that infect bacteria. These are called **bacteriophages**, or simply phage (rhymes with *rage*). Many kinds of bacteriophages exist, but this discussion is limited to a well-studied DNA phage, such as lambda and T4. These phages first attach to the bacterial cell wall and then inject their DNA into the cytoplasm. There are two major outcomes of this injection, depending on whether the phage is lytic or lysogenic.

1. **Lytic**. The cell lyses about 30 minutes after infection, releasing approximately 100 virus progeny (**figure 28.1**).
2. **Lysogenic** (or **temperate**). The phage DNA integrates into the bacterial chromosome and is replicated with the bacterial DNA. In the future it may leave the chromosome when it directs the production of virus and lyses the cell. Bacteria that contain the DNA of a phage cannot be reinfected or lysed by the same type of phage due to **CRISPR** (Clustered Regularly Interspaced Short Palindromic Repeats).

Phage structures are shown in **figure 28.2**. Phage contain DNA within a protein coat, covered with a capsid, to make up the nucleocapsid. They possess tail spikes and fibers for attachment and the tail allows the DNA to be injected into the appropriate host cell.

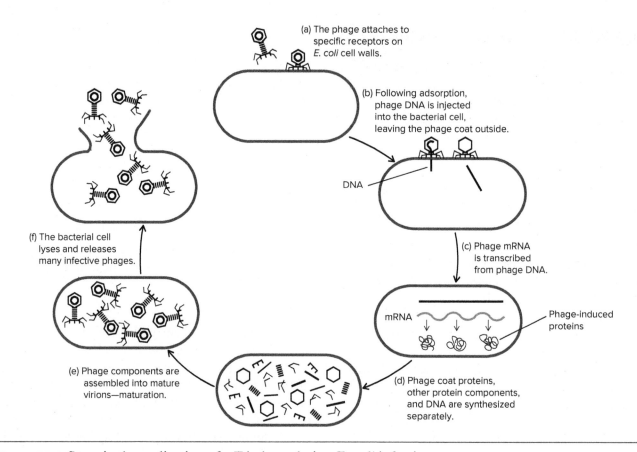

(a) The phage attaches to specific receptors on *E. coli* cell walls.

(b) Following adsorption, phage DNA is injected into the bacterial cell, leaving the phage coat outside.

DNA

(c) Phage mRNA is transcribed from phage DNA.

mRNA

Phage-induced proteins

(d) Phage coat proteins, other protein components, and DNA are synthesized separately.

(e) Phage components are assembled into mature virions—maturation.

(f) The bacterial cell lyses and releases many infective phages.

Figure 28.1 Steps in the replication of a T4 phage during *E. coli* infection.

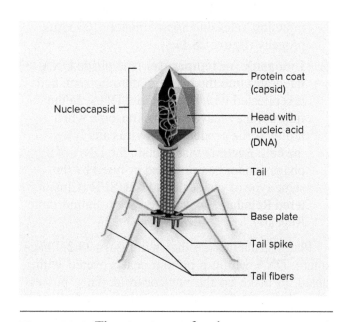

Nucleocapsid

Protein coat (capsid)

Head with nucleic acid (DNA)

Tail

Base plate

Tail spike

Tail fibers

Figure 28.2 The structures of a phage.

Phage are too small (about 200 nm) to be seen in a light microscope, but they can be detected if grown on a bacterial lawn. Phage and their host cells are mixed in a small tube of soft agar and then poured on top of an agar base plate. (Soft agar contains about half the concentration of standard agar so that the phage can diffuse more easily.) The plates are then incubated overnight at the optimum growth temperature for the host bacteria.

During incubation, bacteria multiply and produce a thick covering of bacteria, or bacterial lawn of growth, except in those places where phage have infected and killed the bacteria. The infected bacteria lyse, leaving clear areas called **plaques (figure 28.3)**. Because each plaque originated with one phage, the plaques can be counted—just like bacterial colonies—to determine the number of phage originally mixed with the soft agar. They are called **plaque-forming units (PFU)** instead of colony-forming units.

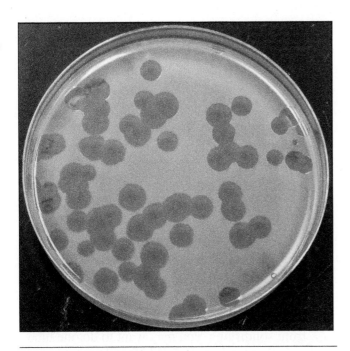

Figure 28.3 Bacteriophage plaques formed on agar medium seeded with a bacterial lawn. McGraw Hill Education/Lisa Burgess, photographer.

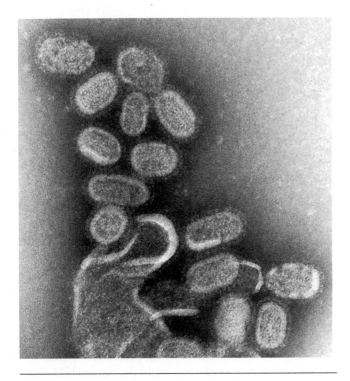

Figure 28.4 Electron microscopy of *influenza* viral particles. **Source:** CDC/Cynthia Goldsmith.

Although the appearance of the plaques can be influenced by many factors, in general, **virulent phage** produce clear plaques. **Temperate phage** produce cloudy plaques because many cells within the plaque were lysogenized instead of lysed and thus continue to grow and multiply. The plaques do not increase in size indefinitely because phage can replicate only in multiplying bacteria. The bacteria and the phage usually have stopped growing after 24 hours. Some bacteria may be resistant to a particular phage, so you may see a few bacterial colonies scattered throughout your plates. For example, *E. coli* C cannot be infected with the T4 phage.

It is important to study phage and to learn the techniques used to manipulate them for the following reasons:

1. Animal viruses, including human pathogens, are grown on tissue culture cells in the same fashion as phage on bacteria. Tissue culture cells are animal cells grown in bottles, tubes, or plates. The animal virus can form plaques by causing cells to degenerate or die. Tissue culture cells require a more complex (and more expensive) medium for growth, so it is convenient to learn viral technique with phage and bacteria.

2. Phage are used in recombinant DNA experiments and are used in studying bacterial genetics.

3. Phage identify different bacterial strains because one type of phage will only infect specific strains.

If you were to isolate phage that infect *Escherichia coli* from nature, sewage would be a good source. It contains high concentrations of *E. coli* and therefore a high concentration of *E. coli* phage. One can filter sewage samples to obtain just a filtrate of virus. Bacteria are removed by the filter, and the viral particles are small enough to pass through the filter. This is interesting to do, but the filtrate also contains animal viruses that could be a health hazard.

Filtration is frequently used to separate animal viruses from bacteria, as bacteria are about 10 times bigger than most animal viruses. They then can be studied without bacterial contamination. *Influenzavirus*, the causative agent of influenza, and Zika virus (color plate 76) are examples of an animal virus.

Materials

Per team

9 ml tryptone blank tubes, 8

4 ml overlay agar tubes, 10

Tryptone agar base plates, 10

Sterile 1 ml pipets (or micropipettors and tips), 10

Host *E. coli* bacteria for phage in late log phase (like *E. coli* B or K12), (Control)

Host suspensions of *E. coli* B and C labeled as Unknown numbers (1 or 2)

Escherichia coli phage suspension (like T4, or T4r)

Suspensions of the following:

Water bath (50°C)

In this exercise, you will **serially dilute** a suspension of phage so that you can count an appropriate number of plaques on a plate and calculate the phage titer. The number of phage/ml is the **titer**. You will also serially dilute a test bacterium to determine if the phage can infect the strain of *E. coli*. Sometimes phage are present but for some reason do not form a plaque.

Titering a Phage Suspension

PROCEDURE

First Session

Note: Small 1.7 ml centrifuge tubes may be used to make serial dilutions. You would use 900 µl solutions to which 100 µl of phage would be added.

1. Label four 9 ml tryptone blanks: 10^{-1}, 10^{-2}, 10^{-3}, 10^{-4} (**figure 28.5**) as Control.

2. Transfer 1 ml of the bacteriophage to the tube labeled 10^{-1} with a sterile 1 ml pipet. Discard the pipet. You must use fresh pipets each time so that you do not carry over any of the more concentrated phage to the next dilution.

3. Mix and transfer 1 ml of the 10^{-1} dilution to the 10^{-2} tube and discard the pipet.

4. Mix and transfer 1 ml from the 10^{-2} tube to the 10^{-3} tube and discard the pipet.

5. Mix and transfer 1 ml from the 10^{-3} tube to the 10^{-4} tube and discard the pipet.

6. Label five tryptone hard agar base plates: control, 10^{-1}, 10^{-2}, 10^{-3}, 10^{-4}.

7. Melt five tubes of soft overlay agar and place in a 50°C water bath. Let cool for about 10 minutes.

8. Add about 0.1 ml (or several drops) of *E. coli* broth to each tube of melted overlay agar.

   Note: Be careful to avoid contaminating the pipet between samples. (To prevent the media from solidifying, you may want to do one tube and plate from steps 8 and 10 and repeat for each dilution.)

9. Pour one of the *E. coli*-laden overlay agar tubes into the control plate.

10. Starting with the most diluted phage tube (10^{-4}), add 1 ml to the overlay agar and immediately pour on the tryptone agar base plate labeled 10^{-4}. Once you pour the plates, leave them to harden. Any movement of the plates can cause an uneven surface, making it difficult to obtain results.

    Note: Depending on the initial phage concentration, your instructor may have you add 0.1 ml of phage to the overlay agar.

11. Using the same pipet, add 1 ml of the 10^{-3} dilution to a tube of overlay agar and pour into the plate labeled 10^{-3}. You can use the same pipet because you started with the most dilute sample and therefore the phage carried over are minimal.

12. Repeat for the 10^{-2} and 10^{-1} phage dilutions.

13. Incubate the plates inverted at 37°C after the agar has hardened, about 15 minutes.

14. Repeat steps 2–13 labeling plates and tubes with the concentration and Unknown (U). Observe the results at 24 hours (or place plates in the refrigerator).

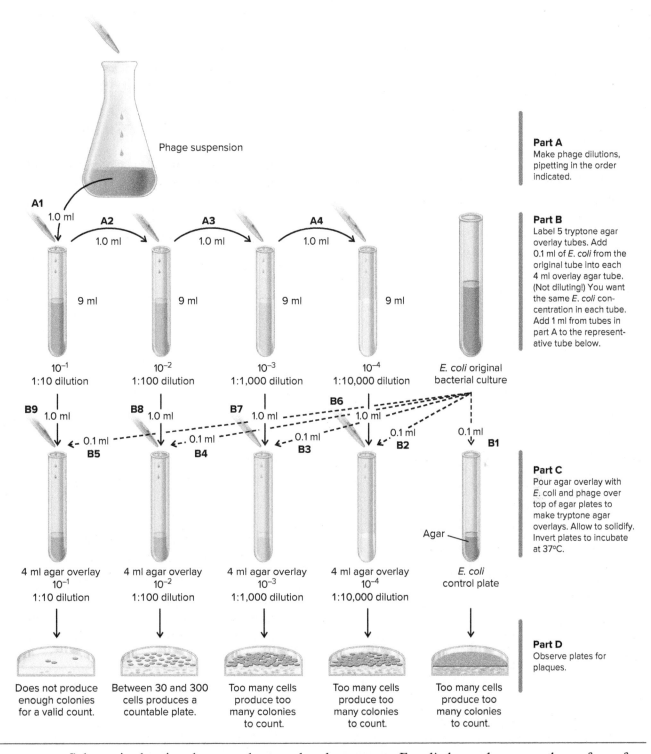

Phage suspension

Part A
Make phage dilutions, pipetting in the order indicated.

A1 1.0 ml   **A2** 1.0 ml   **A3** 1.0 ml   **A4** 1.0 ml

9 ml   9 ml   9 ml   9 ml

10^{-1}
1:10 dilution

10^{-2}
1:100 dilution

10^{-3}
1:1,000 dilution

10^{-4}
1:10,000 dilution

E. coli original
bacterial culture

Part B
Label 5 tryptone agar overlay tubes. Add 0.1 ml of *E. coli* from the original tube into each 4 ml overlay agar tube. (Not diluting!) You want the same *E. coli* concentration in each tube. Add 1 ml from tubes in part A to the representative tube below.

B9 1.0 ml   **B8** 1.0 ml   **B7** 1.0 ml   **B6** 1.0 ml

← 0.1 ml   ← 0.1 ml   ← 0.1 ml   0.1 ml   0.1 ml
B5   **B4**   **B3**   **B2**   **B1**

4 ml agar overlay
10^{-1}
1:10 dilution

4 ml agar overlay
10^{-2}
1:100 dilution

4 ml agar overlay
10^{-3}
1:1,000 dilution

4 ml agar overlay
10^{-4}
1:10,000 dilution

Agar

E. coli
control plate

Part C
Pour agar overlay with *E. coli* and phage over top of agar plates to make tryptone agar overlays. Allow to solidify. Invert plates to incubate at 37°C.

Does not produce enough colonies for a valid count.

Between 30 and 300 cells produces a countable plate.

Too many cells produce too many colonies to count.

Too many cells produce too many colonies to count.

Too many cells produce too many colonies to count.

Part D
Observe plates for plaques.

Figure 28.5 Schematic showing the procedure used to demonstrate *E. coli* phage plaques on the surface of agar plates. Plaques are represented by the light-colored areas. (These results show only one possibility.)

Second Session

1. Examine the plates. Select a plate containing between 30 and 300 plaques. As you count the plaques, place a dot with a marking pen under each plaque on the bottom of the Petri plates. These marks can be wiped off so that each team member can count the plaques.

2. Estimate the numbers on the other plates. They should vary by a factor of 10 as the dilution increases or decreases.

3. To determine the titer, use this formula:

   No. of plaques × 1/dilution × 1/ml of sample = **plaque-forming units**/ml

   Example: If 76 plaques were counted on the 10^{-4} dilution, then

   $$76 \times 1/10^{-4} \times 1/1 = 76 \times 10^4 \text{ pfu/ml}$$

4. Record the results of the Control and Unknown in the Laboratory Report.

EXERCISE

28

Laboratory Report:
Prokaryotic Viruses ASM10, 18, 23, 34

Results

1. Titering a phage suspension

Dilution	Control	10^{-1}	10^{-2}	10^{-3}	10^{-4}
Number of plaques (Control)					
Number of Plaques (Unknown)					

a. Which dilution of the control resulted in a countable plate? _____

b. Did the number of plaques decrease 10-fold with each dilution of the control? _____
_____

c. How many phage/ml were in the original control suspension? Show calculations (exercise 8).
_____

d. How many were in the Unknown suspension? _____

Questions

1. How would an agar plate appear after incubation
   a. if you forgot to add the bacteria to the phage/bacteria mixture?

   b. if you forgot to add the phage to the phage/bacteria mixture?

2. How can you distinguish a lytic phage from a temperate phage by observing their plaques?
   Which one did you titer?

3. Why can a plaque be considered similar to a bacterial colony?

4. If you continue to incubate the plates, will the size of the plaques increase?

5. What was the identity of the Unknown bacterium?

INTRODUCTION to Public Health

Communicable, or infectious, diseases are transmitted from one person to another. Transmission is either by direct contact with a previously infected person—for example, by touching—or by indirect contact with a previously infected person who has contaminated the surrounding environment.

A classic example of indirect contact transmission is an epidemic of cholera that occurred in 1854 in London. During a 10-day period, more than 500 people became ill with cholera and subsequently died. As the epidemic continued, John Snow and John York studied the area and proved by **epidemiological methods** only (the bacteriological nature of illness was not known at that time) that the outbreak stemmed from a community well on Broad Street known as the Broad Street Pump[*] (**figure I.29.1**). These researchers then discovered that sewage from the cesspool of a nearby home was the pollution source and that an undiagnosed intestinal disorder had occurred in the home shortly before the cholera outbreak. Neighboring people who abstained from drinking the pump water remained well, whereas many of those who drank the water succumbed to cholera.

Recently, Yemen and Haiti have had **epidemics** of cholera. Developed countries rarely have epidemics of cholera or typhoid fever, as they have implemented water, sewage, and food standards and regulations. (Public health sanitation is presented in exercise 29, which discusses water microbiology; food microbiology is discussed in exercise 30.) Other recent epidemics include Zika virus in Brazil, plague in Madagascar, and ebola in Africa.

Pandemics are a rise in the occurrence of a particular disease worldwide. The 1918 Spanish flu caused by a new *influenzavirus* strain killed an estimated 500 million people worldwide (www.cdc .gov). Besides the flu, the 2020 *Coronavirus* pandemic affected all parts of the world, shutting down many countries for a time. The chain of transmission (**figure I.29.2**) is an important component for understanding how to control or stop disease transmission.

Public health agencies are responsible for the prevention and control of communicable diseases. One such agency is the federal cabinet-level Department of Health and Human Services, which focuses on preventive medicine (vaccination) research, provides hospital facilities for military personnel, and gives financial assistance to state and local health departments. All state and local government agencies provide needed services to the community. Additionally, the Centers for Disease Control and Prevention (CDC) in Atlanta plays an important role in the prevention, reporting, and control of diseases. The CDC has recently investigated foodborne illnesses caused by *Listeria* in dairy products; *Salmonella* in hamburger, chicken, peanut butter, dogfood, and cantaloupe; and *Escherichia* in spinach, lettuce, and salad mixes.

Perhaps the most important international health agency is the World Health Organization (WHO), headquartered in Geneva, Switzerland. WHO distributes information, provides vaccines, and develops international regulations important for the control and eradication of epidemic diseases like Ebola in the Democratic Republic of the Congo and polio in Afghanistan and Pakistan. WHO was instrumental in eradicating smallpox through vaccination. Its current emphasis is on preventing the spread of HIV and the famine in parts of Africa. They estimate that about 35% of people worldwide do not have access to adequate sanitation.

The mission of another organization, UNICEF, is to help children in over 190 countries and territories (www.unicef.org). They advocate for children's rights and help those affected by poverty, violence, and diseases. This includes education about diseases, immunizing children against childhood diseases, and preventing HIV/AIDS transmission.

Finally, voluntary health organizations like the Bill and Melinda Gates Foundation focus on fighting diseases in developing countries.

[*]Snow, John. "The Broad Street Pump," in Roueche, Berton (ed.), *Curiosities of medicine.* New York: Berkley, Medallion, 1964.

In the following exercises, you will learn about public health by culturing microorganisms from water and food in exercises 29 and 30, and epidemiological methods (epidemiology) in exercise 31.

Definitions

Epidemic. The occurrence in a community or region of a group of illnesses of similar nature, clearly in excess of normal expectancy.

Epidemiological methods. Methods concerned with the extent and types of illnesses and injuries in groups of people and with the factors that influence their distribution. This implies that disease is not randomly distributed throughout a population but rather that subgroups within a given population differ in the frequency of susceptibility to different diseases.

HIV. Human immunodeficiency virus is acquired through sexual or blood contact.

Pandemic. The occurrence of a disease or illness across the world in multiple continents.

(a)   (b)

Figure I.29.1 (*a*) The John Snow pub in London where epidemiologists go to celebrate the heroics of John Snow's early epidemiological efforts to stem a cholera epidemic. (*b*) A replica of the pump, with pump handle attached, a monument dedicated to Dr. Snow in July 1992. At the time of the epidemic, he was so convinced that the disease was being carried by water from the pump that he had the pump handle removed. Robert Koch isolated and identified the cholera vibrio about 30 years later. (a) PC Jones pubs/Alamy Stock Photo. (b) Nathaniel Noir/Alamy Stock Photo.

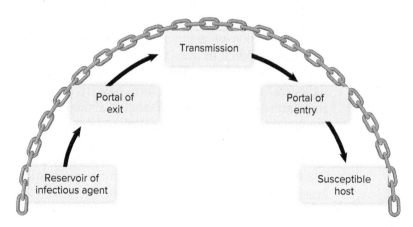

Figure I.29.2 Disease transmission shown as a chain of events, which if broken, will prevent or slow the spread of a disease.

Definitions

Aerobic bacteria. Microbes that grow and multiply in the presence of oxygen.

Anaerobic bacteria. Microbes that grow best or exclusively in the absence of oxygen.

Coliform bacteria. A collective term for bacteria that inhabit the colon, are Gram-negative, and ferment lactose.

Indicator organism. Microbes whose presence in an environment indicates possible fecal contamination.

Selective growth medium. A growth medium containing substances that inhibit the growth of certain organisms but not others.

Objectives

1. Explain the use of a multiple-tube fermentation technique for detecting the presence and number of coliform indicator organisms present in water samples.

2. Explain the membrane filter technique for detecting the presence and number of coliform bacteria in water samples.

Pre-lab Questions

1. In the United States, which bacterium is used as an indicator organism for coliforms?

2. Name an advantage of the membrane filter technique over the multiple-tube fermentation test.

3. What color will a colony testing positive on the Endo media exhibit?

Getting Started

Water, water, everywhere
Nor any drop to drink.

The Ancient Mariner, Coleridge, 1796

This rhyme refers to seawater, undrinkable because of its high salt content. Today the same can be said of freshwater supplies, polluted primarily by humans and their activities. A typhoid epidemic, dead fish on the beach, and red tides are all visible evidence of pollution. There are two primary causes of water pollution: dumping of untreated (raw) sewage and inorganic and organic industrial wastes, and fecal pollution by humans and animals of both freshwater and groundwater. In the United States, sewage and chemical wastes have been reduced as a result of federal and local legislation requiring a minimum of secondary treatment for sewage, as well as severe penalties for dumping of chemical wastes.

Fecal contamination can be difficult to control in developing countries, particularly when people drink water that is also used for bathing, swimming, and washing clothes. UNICEF estimates that 6,000 children die every day from waterborne diseases such as cholera, typhoid fever, bacterial and amoebic dysentery, and viral diseases like polio and infectious hepatitis (www.unicef.org).

Increased organic matter in polluted water allows **anaerobic bacteria** to increase their numbers in relation to the **aerobic bacteria** originally present (**figure 29.1**). *Sphaerotilus natans* is a nuisance bacterium that forms an external sheath (**figure 29.2**) that allows it to adhere to water pipes and reduce their carrying capacity.

Microbes are also beneficial in water purification. In smaller sewage treatment plants, raw sewage is passed through a slow sand filter, wherein microorganisms present in the sand degrade (metabolize) organic waste compounds before the effluent is discharged. Sewage effluent is then chlorinated, ozonated, or subjected to ultraviolet light to further reduce fecal microbial contaminants; many places now recycle their sewage into potable, or drinkable, water. Sewage treatment plants and the control of raw sewage discharge reduced the annual typhoid fever death rate in the United States from about 70 deaths per 100,000 population to

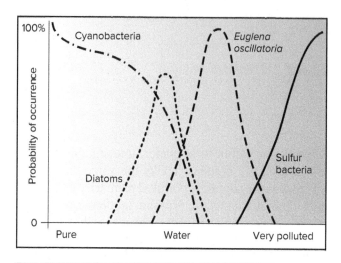

Figure 29.1 Diversity of microbes found in pure to very polluted water. Note the change from aerobic to anaerobic microbes as the water becomes more polluted. **Source:** Settlemire and Hughes, *Microbiology for Health Students*, Reston Publishing Co., Reston, Virginia.

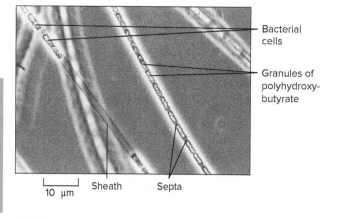

Figure 29.2 *Sphaerotilus* species, a sheathed bacterium that often produces masses of brownish scum beneath the surface of polluted streams. James T. Staley.

nearly zero. However, the potential danger of pollution is always present. Recently, Milwaukee, Wisconsin saw 403,000 cases of Cryptosporidiosis. Further, 86% of parasitic infections were caused by *Giardia lamblia* (*intestinalis*) (**figure 29.3** and color plate 68) from animal fecal contamination (www.ncbi.nlm.nih.gov/pmc/articles/PMC2901654/).

Conventional methods for detecting of fecal contamination are still widely employed, despite limitations. In general, these procedures are the same ones described by the American Public Health Association in *Standard Methods for the Examination of Water and Wastewater* (www.standardmethods.org).

The **indicator organisms** most commonly detected are **coliform bacteria**. Indicator organisms always occur in large quantities in feces, and are relatively easy to detect compared to waterborne

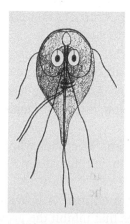

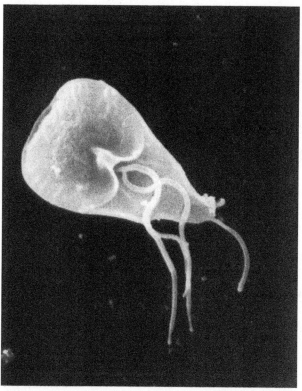

Figure 29.3 An illustration of a *Giardia lamblia* trophozoite and an electron microscope view of *Giardia lamblia*. *G. lamblia* is a waterborne protozoan pathogen that inhabits the intestinal tract of certain warm-blooded animals. **Source:** Janice Haney Carr/CDC

pathogens. Coliforms include facultative anaerobic, Gram-negative, non-spore-forming, rod-shaped bacteria that ferment lactose with gas production within 48 hours at 37°C, and include *E. coli* (6^{10}–9^{10} cells/g of feces) together with closely related organisms (exercise 19). Due to the diversity in microbial physiology, no single growth medium and no single set of cultural conditions can satisfy universal microbial growth. Non-coliforms are sometimes employed for water quality, primarily for confirmation, and include *Enterococcus faecalis* because it is found in the feces of warm-blooded animals.

Culturing *E. coli* or *Enterococcus faecalis* in water from water sources like reservoirs suggests inadequate disinfection methods. Drinking water standards state that it should be free of coliforms and contain no more than 10 other microorganisms per milliliter. Assuming pathogens are dying off faster than the indicator organisms, the absence of indicator organisms suggests the absence of bacterial pathogens. The diagnosis of pathogens is complicated and time-consuming, and thus less suited for routine investigations. Commonly used methods to detect coliforms are described below.

Plate Counts

Standard plate counts on nutrient agar incubated at two temperatures, 20°C and 35°C, provide a useful indication of the organic pollution load in water.

Multiple-tube Fermentation/Most Probable Number (MPN)

This technique employs three consecutive tests: first, a presumptive test; if the first test is positive, then a confirmed test; and, finally, a completed test (**figure 29.4**).

Presumptive Test Fermentation tubes are filled with a *selective growth medium* (lauryl lactose tryptose broth), which contain inverted Durham tubes for detection of fermentation and gas (figure 29.4). Lactose cannot be fermented by many bacteria, and sodium lauryl sulfate inhibits many bacteria, including spore formers, making this medium selective. A pH indicator dye, such as bromocresol purple, phenol red, or basic fuchsin, is added for detecting acid production. Tube turbidity indicates the presence of bacteria. The formation of gas in the Durham tube

within 24–48 hours constitutes a positive presumptive test for coliforms and the possibility of fecal contamination. The test is presumptive only; several other bacteria produce similar results.

The most probable number test (MPN) is useful for counting bacteria that reluctantly form colonies on agar plates or membrane filters but grow readily in liquid media. In principle, the water sample is diluted so some broth tubes contain one bacterial cell. The total viable count is then determined by counting the portion of positive tubes at the specific dilution out of the five tubes, and using a statistical MPN table used for calculating the total viable bacterial count (**table 29.1**).

Confirmed Test This test confirms the presence of coliforms when either a positive or doubtful presumptive test is obtained. A loopful of growth from a presumptive tube is inoculated into a tube of 2% brilliant green bile lactose broth containing a Durham tube and incubated at 35°C for 48 hours. This selective medium detects coliform bacteria in water and dairy food products by showing gas production in the Durham tube at 24–48 hours, which is a positive result. The correct concentration of the dye (brilliant green) and bile must be present. If the concentration is too high, coliform growth can be inhibited. Bile is naturally found in the intestine, where it encourages the growth of coliforms while discouraging other microbial growth.

Completed Test This test further confirms doubtful and positive confirmed test results. The test has two parts:

1. A plate of LES (Lawrence Experimental Station) Endo agar is streaked with a loopful of growth from a positive confirmed tube and incubated at 35°C for 24 hours. Typical coliform bacteria (*E. coli* and *Enterobacter aerogenes*) exhibit good growth on this medium and form red to black colonies with a sheen. *Salmonella typhi* exhibits good growth, but the colonies are colorless. *Staphylococcus* growth is inhibited.

2. Next, a typical coliform colony from an LES Endo agar plate is inoculated into a tube of 2% brilliant green bile broth and on the surface of a nutrient agar slant. They are then incubated at 35°C for 24 hours. A Gram stain is then prepared from growth present on the nutrient agar slant. The presence of gas in the brilliant green bile broth tube and Gram-negative,

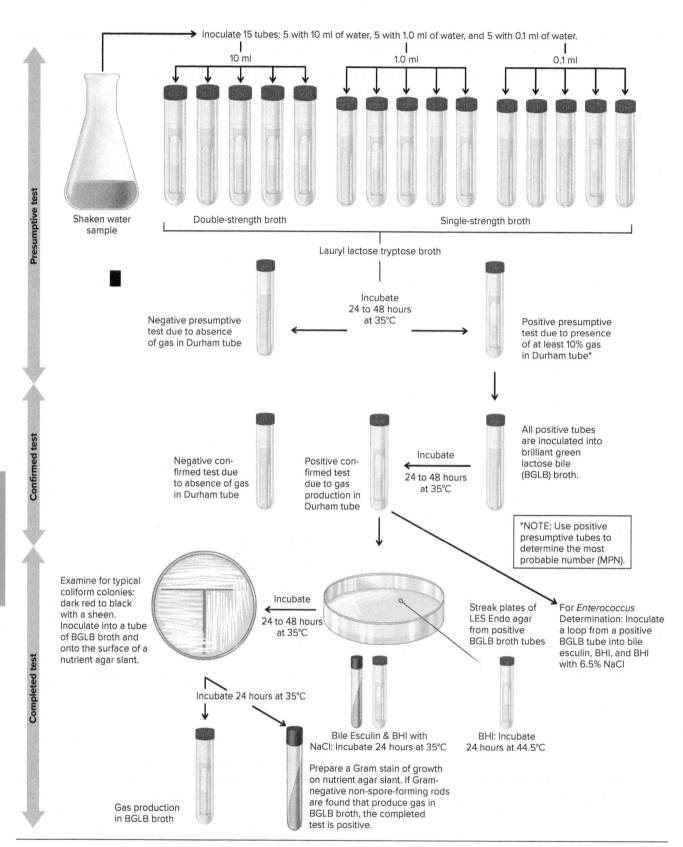

Inoculate 15 tubes: 5 with 10 ml of water, 5 with 1.0 ml of water, and 5 with 0.1 ml of water.

10 ml   1.0 ml   0.1 ml

Presumptive test

Shaken water sample

Double-strength broth   Single-strength broth

Lauryl lactose tryptose broth

Negative presumptive test due to absence of gas in Durham tube

Incubate 24 to 48 hours at 35°C

Positive presumptive test due to presence of at least 10% gas in Durham tube*

Confirmed test

Negative confirmed test due to absence of gas in Durham tube

Positive confirmed test due to gas production in Durham tube

Incubate 24 to 48 hours at 35°C

All positive tubes are inoculated into brilliant green lactose bile (BGLB) broth.

*NOTE: Use positive presumptive tubes to determine the most probable number (MPN).

Completed test

Examine for typical coliform colonies: dark red to black with a sheen. Inoculate into a tube of BGLB broth and onto the surface of a nutrient agar slant.

Incubate 24 to 48 hours at 35°C

Streak plates of LES Endo agar from positive BGLB broth tubes

For *Enterococcus* Determination: Inoculate a loop from a positive BGLB tube into bile esculin, BHI, and BHI with 6.5% NaCl

Incubate 24 hours at 35°C

Bile Esculin & BHI with NaCl: Incubate 24 hours at 35°C

Prepare a Gram stain of growth on nutrient agar slant. If Gram-negative non-spore-forming rods are found that produce gas in BGLB broth, the completed test is positive.

BHI: Incubate 24 hours at 44.5°C

Gas production in BGLB broth

Figure 29.4 Standard methods procedure for the examination of water and wastewater and for use in determining most probable number (MPN).

Table 29.1 MPN Index and 95% Confidence Limits for Various Combinations of Positive Results When Five Tubes Are Used per Dilution (10 ml, 1.0 ml, 0.1 ml)

Combination of Positives	MPN Index/ 100 ml	95% Confidence Limits		Combination of Positives	MPN Index/ 100 ml	95% Confidence Limits	
		Lower	Upper			Lower	Upper
0 0 0	<2	—	—	4-3-0	27	12	67
0-0-1	3	1.0	10	4-3-1	33	15	77
0-1-0	3	1.0	10	4-4-0	34	16	80
0-2-0	4	1.0	13	5-0-0	23	9.0	86
1-0-0	2	1.0	11	5-0-1	30	10	110
1-0-1	4	1.0	15	5-0-2	40	20	140
1-1-0	4	1.0	15	5-1-0	30	10	120
1-1-1	6	2.0	18	5-1-1	50	10	150
1-2-0	6	2.0	18	5-1-2	60	30	180
2-0-0	4	1.0	17	5-2-0	50	20	170
2-0-1	7	2.0	20	5-2-1	70	30	210
2-1-0	7	2.0	21	5-2-2	90	40	250
2-1-1	9	3.0	24	5-3-0	80	30	250
2-2-0	9	3.0	25	5-3-1	110	40	300
2-3-0	12	5.0	29	5-3-2	140	60	360
3-0-0	8	3.0	24	5-3-3	170	80	410
3-0-1	11	4.0	29	5-4-0	130	50	390
3-1-0	11	4.0	29	5-4-1	170	70	480
3-1-1	14	6.0	35	5-4-2	220	100	580
3-2-0	14	6.0	35	5-4-3	280	120	690
3-2-3	17	7.0	40	5-4-4	350	160	820
4-0-0	13	5.0	38	5-5-0	240	100	940
4-0-1	17	7.0	45	5-5-1	300	100	1300
4-1-0	17	7.0	46	5-5-2	500	200	2000
4-1-1	21	9.0	55	5-5-3	900	300	2900
4-1-2	26	12	63	5-5-4	1600	600	5300
4-2-0	22	9.0	56	5-5-5	≥1600	—	—
4-2-1	26	12	65				

"9221 MULTIPLE-TUBE FERMENTATION TECHNIQUE FOR MEMBERS OF THE COLIFORM GROUP (2017)", *Standard Methods For the Examination of Water and Wastewater.*

non-spore-forming rods constitute a positive completed test for the presence of coliform bacteria; this implies possible contamination of the water sample with fecal matter.

Enterococcus Determination

Using a positive brilliant green bile tube from the MPN confirmed test, a loop of bacteria is transferred to two BHI broth tubes, one containing 6.5% NaCl and the other lacking NaCl, and one bile esculin tube. The BHI tube containing NaCl and the bile esculin tube are incubated at 35°C for 48 hours, whereas the BHI tube without NaCl is incubated at 45°C for 48 hours. If a positive bile esculin test and growth in BOTH BHI tubes are seen, it is considered to confirm the presence of an *Enterococcus* species.

Membrane Filter Technique

For this technique, a known volume of water sample (100 ml) is vacuumed through a sterile polycarbonate or nitrocellulose acetate membrane filter. The filter is very thin (150 μm), and has a pore diameter of 0.45 μm; bacteria larger than 0.47 μm cannot pass through it and are caught on the top of the filter. Sample turbidity presents a serious obstacle if suspended matter clogs the filter before the required volume of water has passed through. Unknown samples often must be diluted.

Once the water is filtered, the filter disc is aseptically transferred using forceps to the surface of a wetted pad soaked with Endo broth contained in a Petri dish, and incubated at 35°C for 24 hours. To increase coliform specificity, Endo plates can be incubated at 44.5°C. Coliform colonies appear pink to dark red in color with a golden metallic sheen. The number of characteristic coliform colonies is counted and the total number of coliform bacteria present in the original water sample can be calculated. The membrane filter method provides accurate results if the coliform colony count is 30–300 organisms per filter disc. This procedure can be used in conjunction with the completed multiple-tube fermentation test for more specificity.

The advantages of the membrane filter technique over the multiple-tube fermentation test are: (1) reproducibility of results; (2) greater sensitivity, because larger amounts of water can be used; and (3) shorter time (one-quarter the time) for obtaining results.

Enzyme Detection

Enzyme-based tests have been approved by the Environmental Protection Agency (EPA) for detecting coliforms. The Colilert test contains two substrates, ONPG (ortho-nitrophenyl-β-galactoside) and MUG (4-methylumbelliferyl-β-D-glucuronide). The ONPG detects β-D-galactosidase production by coliforms by turning the solution yellow. *Escherichia coli* produces the enzyme β-D-glucuronidase, breaking down MUG into 4-methylumbelliferone, which fluoresces blue at 365 nm. Fluorescence without a color change is considered a negative test.

Newer methods of determining the total number of microorganisms include quantitative polymerase chain reaction (qPCR)/real-time PCR. qPCR uses fluorescent dyes to determine the presence and quantity of coliforms, so the target DNA sequence of the microbe(s) must be known. A disadvantage to this method currently is the cost, but it is cost-effective if many samples need processing. Environmental DNA (eDNA) is also becoming popular for detecting the fecal matter or other cells of specific (and sometimes endangered) organisms like turtles, but the target sequence must also be known.

In this exercise you will perform a combination of the testing methods described.

Materials

Per team (figure 29.4):

Sterile 150 ml screw-cap bottle for collecting water sample, 1

Membrane Filter Technique
Demonstration (**figure 29.5**)

Water sample for coliform analysis, 30 ml

1 liter side-arm Erlenmeyer flask, 1

Sterile membrane filter holder assembly, two parts wrapped separately in foil (figure 29.5, frames 2 and 3), 1 unit

A metal clamp for clamping filter funnel to filter base

Sterile membrane filters, 47 mm diameter, 0.45 μm pore size

Sterile forceps, 1 pair

Sterile 50 mm diameter Petri dishes, 3

Absorbent filter pads, 3

Tube containing 10 ml of sterile Endo broth (LES or MF), 1

Sterile 90 ml water blanks, 2

Erlenmeyer flasks containing 100 ml of sterile water, 6 (or a 1L flask, 1)

Sterile 10 ml pipets, 2

Vacuum pump or system

MUG

MUG broth (90 ml) or plates, 1

Sterile 10 ml pipets (if using broth), 1 OR

Micropipettor (100 μl) and micropipet tips if using plates, 1

Sterile spreader or rod, 1

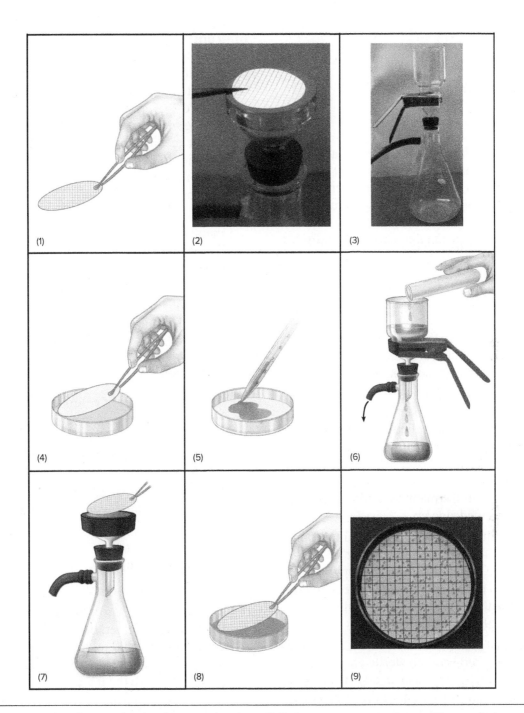

Figure 29.5 Analysis of water for fecal contamination. Cellulose acetate membrane filter technique.
(1) Sterile membrane filter (0.45 μm) with grid for counting is handled with sterile forceps. (2) The sterile membrane filter is placed on filter holder base with grid side up. (3) The apparatus is assembled.
(4) Sterile absorbent pads are aseptically placed in the bottom of three sterile Petri dishes. (5) Each absorbent pad is saturated with 2.0 ml of Endo MF broth. (6) A portion of well-mixed water sample is poured into assembled funnel and filtered by vacuum. (7) Membrane filter is carefully removed with sterile forceps after disassembling the funnel. (8) Membrane filter is centered on the surface of the Endo-soaked absorbent pad (grid side up) using a slight rolling motion. (9) After incubation, the number of colonies on the filter is counted. The number of colonies on the filter reflects the number of coliform bacteria present in the original sample.
(2, 3) Courtesy of Anna Oller, University of Central Missouri; (9) Lisa Burgess/McGraw Hill.

Membrane Filter Technique

First Session

1. Shake or swirl the water sample. Prepare two dilutions by transferring successive 10 ml aliquots into 90 ml blanks of sterile water (10^{-1} and 10^{-2} dilutions). **Note:** Re-swirl prepared dilutions before using.

2. Assemble the filter holder apparatus as follows (figure 29.5):

   a. Using aseptic technique, unwrap the lower portion of the filter base, and insert a rubber stopper.

   b. Insert the base in the neck of the 1-liter side-arm Erlenmeyer flask (figure 29.5, frame 2). One person should hold the filter folder onto the flask so it does not tip over.

   c. With sterile forceps (sterilize by dipping in alcohol and flaming them with the Bunsen burner), transfer a sterile membrane filter onto the glass or plastic surface of the filter holder base (figure 29.5, frames 1 and 2). Make certain the membrane filter is placed with the ruled side up.

   d. Aseptically remove the covered filter funnel from the foil, and place the lower surface on top of the membrane filter. Clamp the filter funnel to the filter base with the clamp provided with the filter holder assembly (figure 29.5, frame 3).

3. Prepare three plates of Endo medium by adding 2 ml of the tubed broth to sterile absorbent pads previously placed aseptically (using sterile tweezers) on the bottom of the three Petri dishes (figure 29.5, frames 4 and 5).

4. Remove the foil covering the filter opening, and pour the highest water dilution (10-2) into the funnel (figure 29.5, frame 6). Assist the filtration process by turning on the vacuum pump or Venturi vacuum system.

5. Rinse the funnel walls with two 100 ml aliquots of sterile water to recover any bacteria that adhered to the wall of the funnel. This ensures accurate results.

6. Turn off (break) the vacuum and remove the filter holder funnel. Using aseptic technique and sterile tweezers, transfer the filter immediately to the previously prepared Petri dish (figure 29.5, frames 7 and 8). Using a slight rolling motion, center the filter, grid side up, on the medium-soaked absorbent pad. Take care not to trap air under the filter, as this will prevent nutrient media from reaching all of the membrane surface (figure 29.5, frame 9).

7. Reassemble the filter apparatus with a new membrane filter, and repeat the filtration process first with the 10^{-1} water sample, and finally with a 100 ml aliquot of the undiluted water sample.

8. Label and invert the Petri dishes to prevent condensation from falling on the filter surface during incubation. Incubate plates for 24 hours at 37°C or 44.5°C.

Second Session

1. Count the number of coliform bacteria by using either the 4× objective lens of the microscope or a dissecting microscope. Count only those colonies that exhibit a pink to dark-red center with or without a distinct golden metallic sheen.

2. Record the number of colonies found in each of the three dilutions in **table 29.2** of the Laboratory Report.

MUG Determination

First Session

1. Using a sterile pipet, transfer 10 ml of the water sample to 90 ml of MUG broth. If plating onto a MUG plate, transfer 100 μl of the water sample to the plate.

2. Use a sterile spreader to spread the sample across the plate and let absorb for 10 minutes. Incubate the broth or plates at 37°C for 24 hours.

Second Session

1. Wearing eye protection, take the plate or broth into a dark room and shine a 360 nm UV light onto it. If you see fluorescence, it is considered a positive result. Record the result in the Laboratory Report.

Materials

Multiple-Tube Fermentation Technique

Water sample for coliform analysis, 55.5 ml

Test tubes (50 ml), containing 10 ml of double-strength lauryl sulfate (lauryl lactose tryptose) broth plus Durham tubes, 5

   Small test tubes containing 10 ml of single-strength lauryl sulfate broth plus Durham tubes, 10

   Sterile 10 ml pipet, 1

   Sterile 1 ml pipet calibrated in 0.1 ml units, 1

   Brilliant green bile 2% broth plus Durham tubes, 2

   LES Endo agar plate, 1

   Nutrient agar slant, 1

Enterococcus Determination

   Bile esculin slant, 1

   BHI tube, 1

   BHI tube containing 6.5% NaCl, 1

PROCEDURE

Multiple-Tube Fermentation Technique

Note: Your instructor may ask you to bring a 100 ml sample of water from home, a nearby stream, a lake, or some other location for analysis. When taking a tap sample, water should run for 5–10 minutes with the tap in the same position to prevent loosening of bacteria from inside the tap. Next, using aseptic technique, open a sterile bottle (obtained from the instructor beforehand) and collect a sample. If the sample cannot be examined within 1–2 hours, keep refrigerated until ready to test.

First Session (Presumptive Test)

1. Be sure to label all tubes before starting. Shake or swirl the water sample. Aseptically pipet 10 ml of the sample into each of the five large tubes containing 10 ml aliquots of double-strength lauryl sulfate broth. Next, with a 1 ml pipet, transfer 1 ml of the water sample into five of the smaller tubes, and then 0.1 ml into the remaining five small tubes of lauryl sulfate broth.

2. Incubate the test tubes for 24–48 hours at 37°C.

Second and Third Sessions (Presumptive and Confirmed Tests)

1. After 24 hours, observe the tubes for gas production by gently shaking/swirling the tubes. If gas is not evident in the Durham tube after shaking, re-incubate the tube for an additional 24 hours. Record any positive results for gas production in **table 29.3** of the Laboratory Report.

2. Observe the tubes for gas production and turbidity after 48 hours of incubation. If neither gas nor turbidity is present in any of the tubes, the test is negative. If turbidity is present but gas is not, the test is questionable; record as negative. If the tube shows gas production, the test is positive for coliform bacteria. Record your results in table 29.3 of the Laboratory Report.

3. MPN determination. Using your fermentation gas results in table 29.3, determine the number of tubes from each set containing gas. Determine the MPN by consulting table 29.1; for example, if you have gas in two of the first five tubes, in two of the second five tubes, and none in the third three tubes, your test readout is 2-2-0. Table 29.1 shows that the MPN for this readout is 9. Thus, your water sample contains nine organisms/100 ml water with a 95% statistical probability of there being between 3 and 25 organisms.

Note: If your readout is 0-0-0, it means that the MPN is less than two organisms/100 ml water. Also, if the readout is 5-5-5, it means the MPN is greater than 1,600 organisms/100 ml water. In this instance, what procedural modification is required to obtain a more significant result? Report your answer in Question 10 in the Laboratory Report.

4. The confirmed test should be administered to all tubes demonstrating either a positive or doubtful presumptive test. Inoculate a loopful of growth from each tube showing gas or dense turbidity into a tube of 2% brilliant green lactose bile broth. Incubate the tube(s) at 37°C for 24–48 hours.

Note: For expediency, your instructor may instruct you to inoculate only one tube. If so, for the inoculum, use the tube of lauryl sulfate broth testing positive with the least amount of water.

***Enterococcus* Determination:** If determining whether Enterococci are present, obtain a bile esculin slant, a BHI tube, and a BHI tube with 6.5% NaCl. Inoculate a loopful of growth from one tube with gas production seen in the Durham tube. Incubate the bile esculin and BHI tube with NaCl at 37°C for 48 hours and the BHI without NaCl at 45°C for 48 hours.

Fourth Session (Confirmed and Completed Tests)

1. Examine the 2% brilliant green lactose bile tube(s) for gas production. Record your findings in the confirmed test section of the Laboratory Report.

2. Streak a loopful of growth from a positive tube of 2% brilliant green lactose bile broth on the surface of a plate containing LES Endo agar. Incubate at 37°C for 24 hours.

3. Examine the bile esculin and two BHI tubes. Record your results in the Laboratory Report

Fifth Session (Completed Test)

1. Examine the LES Endo agar plate(s) for the presence of typical coliform colonies (dark red to black with a sheen). Record your findings in the completed test section of the Laboratory Report.

2. With a loop, streak a nutrient agar slant with growth obtained from a typical coliform colony found on the LES Endo agar plate. Also, inoculate a tube of 2% brilliant green lactose bile broth with growth from the same colony. Incubate the tubes at 37°C for 24 hours.

Sixth Session (Completed Test)

1. Examine the 2% brilliant green lactose bile broth tube for gas production. Record your results in the completed test section of the Laboratory Report.

2. Prepare a Gram stain of some of the growth present on the nutrient agar slant. Examine the slide for the presence of Gram-negative, non-spore-forming rods. Record your results in the completed test section of the Laboratory Report. The presence of gas and Gram-negative, non-spore-forming rods constitutes a positive completed coliform test.

EXERCISE

29

Laboratory Report: Water Quality Determination [ASM10, 28b, 34, 35]

Results

1. Membrane filter technique water sample ID: _____ Agar used: _____

Table 29.2   Number of Coliform Colonies Present in Various Dilutions of the Water Sample

Undiluted Sample	10^{-1} Dilution	10^{-2} Dilution

   a. Calculate the number of coliform colonies/ml present in the original water sample (show your calculations):

2. Multiple-tube fermentation technique

   a. Record the results of the presumptive test in table 29.3.

Table 29.3   Presumptive Test for the Presence or Absence of Gas and Turbidity in Multiple-Tube Fermentation Media

Water Sample Size (ml)	Presence of Gas and Turbidity*				
	Tube #1	Tube #2	Tube #3	Tube #4	Tube #5
10					
1					
0.1					

*Use a + sign to indicate turbidity and a circle (O) around the plus sign to indicate gas.

   b. Determine the MPN below:

     Test readout: _____ MPN: _____ 95% confidence limits: _____

   c. Confirmed test results (gas production in brilliant green lactose bile 2% broth):

     Sample Number   ***Gas (+ or −)***

                    24 hr   48 hr

   d. Appearance of colonies on LES Endo agar:

e. Completed test results:

Sample Number   *Gas (+ or −)*   *Gram-Stain Reaction*

3. *Enterococcus* determination   Record if the tubes contained growth using a + or − sign.
   Bile esculin
   BHI 44.5°C
   BHI with NaCl
4. MUG result: Fluorescence?_____
   Based off all of your results, is the water safe to drink (i.e., potable)? Why or why not?

Questions

1. What nutritional means could be used to speed up the growth of the coliform organisms when using the membrane filter technique?

2. Describe two other applications or uses of the membrane filter technique.

3. Why not test for pathogens such as *Salmonella* directly rather than using an indicator organism, such as coliform bacteria?

4. Why does a positive presumptive test not necessarily indicate that the water is unsafe for drinking?

5. List three organisms that usually give a positive presumptive test.

6. Describe the purpose of lactose and LES Endo agar in water quality testing.

7. Name at least two limitations of the membrane filter technique?

8. Define the term *coliform*.

9. Briefly explain what is meant by presumptive, confirmed, and completed tests in water analysis.

10. See Note 3, MPN determination, on page 261 for the question.

Definitions

Polymerase chain reaction (PCR). The process of replicating, or amplifying, a DNA sequence using a thermocycler.

Probiotic. A food or pill containing live bacteria that are ingested to increase the number of beneficial bacteria in the body.

Pulsed-field gel electrophoresis. The process of separating large DNA fragments (fingerprinting) using an electrical voltage that is applied in three directions.

Objectives

1. Explain the importance of fermentation.
2. Describe how cheese is made.
3. Explain the use of serial dilutions for quantifying bacteria in foods.

Pre-lab Questions

1. In fermentation, what is the importance of the balloon attached to the flask?
2. Why is rennet important in making cheese?
3. What is the importance of blending food for bacterial quantification?

Getting Started

Food microbiology encompasses the diverse microbes used to make foods like bread, wine, cheese, and yogurt, as well as the microbes that spoil food and cause illness, such as *Salmonella* or *Escherichia coli*. *Saccharomyces cerevisiae* is a common yeast that is used to make bread and wine, and fungi such as *Penicillium* are used to make blue cheese. *Lactobacillus acidophilus, Lactobacillus bulgaricus, Lactococcus lactis, Bifidobacterium bifidum*, and *Streptococcus thermophilus* are all bacteria commonly used in yogurts that contain live cultures. These bacteria are often considered **probiotics**, as they will grow in the intestinal tract and inhibit the growth of other bacteria that cause gas and diarrhea.

In an anaerobic environment, *S. cerevisiae* can convert sugars like glucose into alcohol and carbon dioxide during the process of fermentation. The juice pH and preservatives like ascorbic acid (vitamin C) can inhibit fermentation. Thus, finding a juice or cider with fewer preservatives is more likely to undergo fermentation. The balloon attached to a flask provides the anaerobic environment needed for fermentation to occur, and it also shows approximately the amount of carbon dioxide produced.

Cheeses can be produced using many different bacteria and fungi. Mozzarella cheese can be made quickly, whereas aged cheeses like Swiss and blue cheese take time for fermentation to occur. In nature, the viable bacteria left in milk convert milk sugars to acids, like lactic acid, that allow for curdling when the milk becomes warm. Curdling begins at around 31°C and stops at around 40°C, when it becomes too warm for the enzymes to work. The citric acid acidifies the milk, and the enzyme, rennet, then curdles the milk, creating the curds (solid) and whey (liquid). The citric acid affects the curd texture (too much = small firm curds), and salt provides flavor to the cheese.

Foods become contaminated in several ways. Fruits and vegetables may be contaminated while growing from the dirt, water, and animals in the environment. They can also be contaminated if fecal matter is used as fertilizer. Meats can be contaminated during slaughter or packaging. Contaminated surfaces or pasteurization failures can lead to contamination, such as in recent recalls of peanut butter. In addition, leaving food out at room temperature for over 2 hours allows bacteria to rapidly replicate. Touching contaminated food followed by improper hand washing and improper disinfection of the counter or sink is one of the leading causes of food poisoning.

The most commonly recovered bacteria from contaminated foods include *Escherichia, Salmonella, Shigella, Campylobacter, Listeria, Vibrio*, and *Yersinia*. A blender is often used to chop food up so the entire surface area of the food contacts water, thereby allowing the bacteria to be recovered on agar plates. Selective plates like XLD, MacConkey, or EMB can determine if common Enterobacteriaceae contaminants like *Escherichia* are present in the food. Viruses and parasites are also frequent contaminants, but they require detection methods like the **polymerase chain reaction (PCR)**. Bacteria recovered from stool samples in suspected foodborne illnesses are sent to the state department of health. In the lab, DNA fingerprinting is performed using either PCR or a technique called **pulsed-field gel electrophoresis**, in which genomic DNA is cut using enzymes. The DNA is then placed in a gel where an electrical voltage is applied in three directions. The DNA pieces created by the enzymes are separated into DNA bands of various sizes. The resulting banding pattern can then be compared to the national database, PulseNet, to see if the same bacterial strain was recently isolated. The Centers for Disease Control and Prevention (CDC) estimates that foodborne illnesses contribute to over 48 million illnesses and 3,000 deaths in the United States annually. The World Health Organization has implemented the Global Foodborne Infections Network for monitoring diseases worldwide.

In this exercise, you will have an opportunity to add microbes to food to see how wine and cheese are made, as well as examine foods for the presence of potential microorganisms. We recommend performing the food labs in a separate lab from where bacteria are used or not sampling the end products.

Fermentation

Materials

Per team
First Session
  500 ml sterile flask, 1
  100 ml apple or grape juice
  100 ml sterile graduated cylinder
  5 ml sterile pipet, 1
  5 g glucose
  Lead acetate paper, 1
  Masking tape
  Balloon, 1
  5 ml of a 24-hours *S. cerevisiae* culture grown in apple or grape juice
Second Session
  A small drinking cup, per person

PROCEDURE

First Session

1. Using a sterile graduated cylinder and aseptic technique, add 100 ml of juice to the 500 ml sterile flask.
2. Add 5 ml of *S. cerevisiae* to the 500 ml flask containing juice.
3. Add 5 g of glucose to the flask.
4. Tape a lead acetate strip to the inside lip of the flask. It should hang down into the flask. If the strip turns black, it indicates that hydrogen sulfide has been produced.
5. Stretch a balloon over the top of the flask to create an anaerobic environment.
6. Label the balloon with your group information.
7. Incubate at room temperature for about 7 days, or until the balloon begins to shrink.

Second Session

1. Record the balloon size in **table 30.1**.
2. Remove the balloon and record the smell of the juice.
3. Observe the color of the juice.
4. Observe the color of the lead acetate strip. Is it white? If it is black, you should not ingest the juice.
5. If the lead acetate strip is white, obtain a small cup for each person in the group.
6. Decant a small amount of juice into all of the cups at one time to avoid stirring up the yeast. What does it taste like?

Cheese Making (Mozzarella)

Materials

Per team
  Hot plate or burner, 1
  9 × 9 sterilized stainless steel pan, 1
  ½ gallon of 2% milk

Citric acid, 1 tsp.

Non-iodized (cheese) salt, 1 tsp.

Rennet, ¼ tablet

¼ cup sterilized water in small beaker

Sterilized scupula, 1

500 ml sterilized beaker, 1

Hot hand or potholder, 1

Gloves

Thermometer, optional

Cheesecloth, optional

PROCEDURE

1. Remove the foil from the sterilized scupula. The scupula should be held at all times.

2. Remove the foil from the sterilized water and add the rennet tablet to the water. Using the scupula, break up the rennet tablet as much as possible to help it dissolve.

3. Remove the foil from the pan and add ½ gallon of milk to the pan.

4. Turn the burner on to medium. Use the scupula to stir the milk so it does not burn.

5. Add the rennet solution and the citric acid to the pan and stir.

6. Add half of the salt and stir.

7. Once curds have completely formed, turn the burner off. Remove the foil from the 500 ml beaker.

8. One person wearing gloves will scoop out the curds into the sterile 500 ml beaker. Leave about 100 ml of whey in the beaker.

9. Add the rest of the salt.

10. Microwave the beaker containing the curds for 1 minute on high.

11. One person will use a hot hand and pour off the excess whey. You can use the scupula to help keep the curds in the beaker.

12. Once the curds have cooled enough to handle (usually 5–10 minutes), the person wearing gloves can stretch the curds and squeeze out the excess whey into the pan.

13. Once the texture appears smooth and the curd has cooled to room temperature, the person wearing gloves can tear off a piece of the curd for each team member for tasting.

Bacterial Recovery from Food

Materials

Per group:

    Blender, 1

    90 ml sterile water

    Balance, 1

    weigh boat, 1

    Sterile scupula, 1

    10 g of food (hamburger, chicken, spinach, lettuce, etc.)

    9.0 ml water blanks, 2

    10 ml pipets with bulbs, 1

    1 ml pipet with bulbs, 2

    Spreader or glass rod, 1

    XLD plates, 2 (can also use MacConkey or EMB plates)

PROCEDURE

First Session

1. Place a weigh boat on the balance and zero it.

2. Add 90 ml of sterile water to a disinfected blender.

3. Measure out 10 g of food with the scupula and place the food into the blender containing 90 ml of sterile water.

4. Blend the food for 5 minutes on high.

5. Label the water blanks and XLD plates as shown in **figure 30.1**.

6. Using a 10 ml pipet, transfer 1 ml from the blender to tube 1 and mix well. Properly discard the pipet.

7. Using a 1 ml pipet, transfer 1 ml from tube 1 to tube 2 and mix well.

8. Using the same pipet, transfer 1 ml from tube 2 onto the XLD plate.

10 ml of food + 90 ml sterile water
10^{-1}

Figure 30.1 Dilution scheme for determination of food quality. PhotoSpin, Inc/Alamy Stock Photo

Figure 30.2 A spreader used to spread colonies across the surface of plates. Courtesy of Anna Oller, University of Central Missouri.

9. Using the same pipet, transfer 1 ml from tube 1 onto the XLD plate. Properly discard the pipet.

10. Using a spreader (**figure 30.2**) or glass rod, spread the mixture across the entire surface of the agar plate, starting with plate 2, then plate 1. Properly discard the spreader.

11. Incubate the plates at 37°C for 24 hours.

Second Session

1. Observe the plates and record the number and the color of colonies seen in **table 30.2** in the Laboratory Report.

2. Calculate the number of organisms/milliliter (see exercise 8).

Alternative Experiment: Obtain a yogurt cup containing live organisms. Using a sterile loop, obtain a loopful of microbes from a stirred yogurt cup. Streak for isolation onto an MRS agar plate. Incubate at 30°C for 48 hours and then examine for small clear colonies.

EXERCISE

30

Laboratory Report: Microbial
Examination of Food

Table 30.1 Fermentation Results

	Results
Balloon size (mm)	
Color of lead acetate strip	
Smell	
Juice color pre-inoculation	
Juice color post-incubation	

Table 30.2 Food Quality Results

Type of Food	Hamburger		Chicken		Spinach		Lettuce	
	10^{-2}	10^{-3}	10^{-2}	10^{-3}	10^{-2}	10^{-3}	10^{-2}	10^{-3}
Total number of colonies								
Number of yellow colonies								
Number of red colonies								
Number of colonies producing hydrogen sulfide								
Is the food safe to eat raw?	Yes / No	Yes / No	Yes / No	Yes / No	Yes / No	Yes / No	Yes / No	Yes / No
Is the food safe to eat if cooked?	Yes / No	Yes / No	Yes / No	Yes / No	Yes / No	Yes / No	Yes / No	Yes / No
What was the concentration of bacteria?								

Questions

1. Why was a balloon added to the fermentation flask containing juice?

2. Do you think fermentation took place in the juice? Why or why not?

3. What was the texture of the cheese you made?

4. Why would contamination by *Escherichia* be a larger problem in spinach than in hamburger?

5. If you inoculated a loopful of yogurt to an MRS plate, did you see growth? Give one explanation of why the results most likely appeared as they did.

31

Epidemiology and Quorum Sensing ASM12, 15, 17, 21

Definitions

Biofilm. Many bacterial cells aggregate together, such as plaque in the oral cavity.

Communicable. An infectious disease that can be transmitted from one host to another.

Contagious. A communicable disease that is easily transmitted from one person to another.

Fomites. Inanimate objects that can act as transmitters of pathogenic microorganisms or viruses.

Index case. The first identified case in an outbreak.

Quorum sensing. The ability of microorganisms to sense or detect molecules produced by other bacteria, and when enough molecules are sensed, the bacteria change behaviors.

Objectives

1. Describe the common methods of communicable disease transmission.
2. Interpret data that simulate a disease outbreak and determine the index case.
3. Explain how quorum sensing can work in bacteria.

Pre-lab Questions

1. Is this lab a model of direct transmission or indirect transmission of a disease agent?
2. Why are we using the organism *S. marcescens* in this exercise?
3. What is the definition of *index case*?

Getting Started

Epidemiology is the study of factors that influence the spread and control of diseases within populations. Diseases such as colds, measles, and tuberculosis may be **communicable**, or **contagious**, when they are spread from one person to another. Communicable diseases are spread by either direct or indirect methods of transmission. Direct transmission occurs when individuals have person-to-person contact, such as with the transmission of fecal organisms from touching unwashed hands during handshaking or skin contact with a lesion generated from the herpes virus. Indirect transmission occurs when individuals transfer microorganisms to another person via respiratory droplets, when coughing or sneezing, or on **fomites**, inanimate objects such as doorknobs, pens, and toys that can harbor microorganisms.

In nature, many bacteria grow as a **biofilm**, or an aggregate of bacterial cells, to protect them from enzyme degradation, antibiotics, UV, etc. **Quorum sensing** (QS) is the ability of bacteria to sense molecules out in the environment, which causes a molecular or behavioral change (**figure 31.1**). Bacterial bioluminescence was one of the first characteristics attributed to quorum sensing; if enough bacterial cells are present they glow in the dark. Motility is another characteristic, as is pigment production, controlled by quorum sensing.

In this lab, we will simulate the transmission of a "disease" from one person to another and demonstrate how an epidemiologist would track the spread of a communicable disease back to the original source, or **index case**. Using a piece of candy to transfer the "disease" agent to gloved hands, you will shake hands with classmates and transfer the "disease" to others. Contact with the "disease" agent will be determined by touching gloved hands (which had contact with the candy) to the surface of a nutrient agar plate. The presence of red or purple colonies on the surface of the agar plate after 24 hours of incubation indicates contact with the "disease" agent. Further, you will then inoculate a bacterium onto different agars, one incubated aerobically and one anaerobically to see if you see a difference in the quantity of bacteria or pigment production.

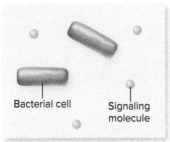

When few cells are present, the concentration of the signaling molecule is low.

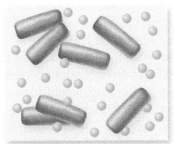

When many cells are present, the signaling molecule reaches a concentration high enough to induce the expression of certain genes.

Figure 31.1 Quorum sensing seen in bacteria.

Materials

Per student or team:

Disposable gloves (1 pair)

Petri dish or weigh boat with jelly bean (or other type of hard candy)

Nutrient agar plates, 3

Tryptic soy agar plates, 2

Culture

24-hours 37°C *Micrococcus roseus* (preferred), *Chromobacterium violaceum,* or *Serratia marcescens* (pigment producers) in trypticase soy broth

PROCEDURE

First Session

1. Each student will choose one weigh boat that contains a piece of candy covered in broth.

Note the ID number or letter of your weigh boat and write it here. _____

2. Place gloves on both hands and label your nutrient agar plate with your name and ID number. *Alternatively, you may use one glove and place this on your non-writing hand, so you may keep record of those you contact in the exercise.*

3. When instructed to start the exercise, pick up the candy with one of your gloved hands and roll the candy around the glove to spread the broth on your palms and fingers. (Pick up the candy with your non-writing hand, so you may keep a record of those you contact in the exercise.) Be careful not to drip any of the broth on your lab benches or clothes as it may be contaminated with live organisms.

4. Shake hands with four students, keeping track of their ID numbers and the order in which you contact them. Record your ID and your contacts in the space provided in the Laboratory Report.

5. Place your contaminated gloved hand on the surface of the NA plate. Be sure to rub the palm and fingers of the glove on the surface of the plate. Try not to break the surface of the agar.

6. Make sure most of the class has encountered enough contacts before removing your glove(s). Remove your gloves by turning them inside out as you pull them off of your hands. Dispose gloves in the BIOHAZARD waste containers along with your candy and weigh boat.

7. Incubate your nutrient agar plates at 30°C or room temperature overnight.

8. Obtain two nutrient and two tryptic soy agar plates. Label one NA plate as aerobic and one as anaerobic. Label one TSA plate as 25°C and one as 37°C. Then label with your name, quorum sensing, bacterium used, section, temperature, and date.

9. Use a loop to inoculate the TSA plates and streak for isolation.

10. Use a loop to inoculate the NA plates and streak for isolation.

11. Incubate both NA plates at 25°C, but place one in an anaerobe jar and incubate the other aerobically for 24 hours.

12. Incubate one TSA plate at 25°C and one at 37°C for 24 hours.

Second Session

1. Observe epidemiology plates for the presence of red- or purple-pigmented colonies. *Serratia marcescens* is a Gram-negative rod that produces prodigiosin, a type of red pigment that can range from orange to brick red in color. If this organism is present on your plate, you were infected with the "disease" agent. *Micrococcus roseus* is a Gram-positive coccus that produces a reddish carotenoid pigment (related to carotene found in carrots), whereas *Chromobacterium violaceum* is a Gram-negative facultative rod that produces the purple pigment violacein.

2. Record the data in **table 31.1** of all individuals who have red-pigmented colonies on their plates. Be sure to include the ID number of the positive plates as well as the ID numbers of all the individuals they shook hands with.

3. Based on the data, try to identify the index case. It may only be possible to narrow down the likely "suspects" to two individuals, the index case and the first person students shook hands with because this initial handshake will have a high concentration of bacteria. To narrow your possible choices for the index case, you may make the assumption that a likely "suspect" will have all four contacts contaminated or will be positive for pigmented growth on their plates.

4. Observe the four quorum sensing plates and record any differences you see between colony counts and pigment quantities/intensity between the plates incubated at different temperatures, oxygen requirements, and nutrient availability. Record your observations in the Laboratory Report.

EXERCISE

31

Laboratory Report: Epidemiology and
Quorum Sensing ASM12, 15, 17, 21

My ID _____ My contact #1 _____ My contact #2 _____ My contact #3 _____ My contact #4 _____

Table 31.1 Epidemiology Data Results

Name of People or Groups with Red or Purple Pigmented Colonies on Plate	Contact #1	Contact #2	Contact #3	Contact #4

Quorum Sensing Results

Record any differences in colony color/pigmentation or quantity of growth seen.

Temperature	TSA		Oxygen	NA		Nutrients
25°C			Aerobic		TSA	
37°C			Anaerobic		NA	

Questions

1. In your class, who was the index case? Briefly explain how you determined the index case.

2. In this lab, we are assuming that everyone who comes into contact with the disease agent is infected. Is this a model of what happens in "real life"? Explain your answer.

3. On your agar plates, there may be other bacteria growing on them that are not pigmented. What is the likely source of the non-pigmented bacteria?

4. For quorum sensing, what factors seemed to affect pigment production? Why might this be?

INTRODUCTION to Immunology and Biotechnology

Watson and Crick first proposed the structure of DNA in the early 1950s. In less than 50 years, the field had mushroomed: Geneticists had isolated DNA, transferred specific genes to another organism, and determined the sequence of DNA bases in specific genes, as well as determine entire bacterial genomes (**figure I.32.1**). These sequences have been used to identify or classify organisms, and to determine evolutionary relationships. **Restriction enzymes** have been used to insert DNA segments into another organism, so that the gene's function can be studied further. One instrumental use of this technology was cloning the human insulin gene into bacterial (*Escherichia coli*) plasmid DNA. These microorganisms were then grown in huge quantities, and the insulin was purified to treat diabetics. This standardized the batches of insulin that were given by injection.

New immunology techniques and testing methods have increased dramatically in the past few years. Biotechnology advances continue to amaze even the scientific community. DNA segments can be synthesized in the lab, which is what the polymerase chain reaction (PCR) does in exercise 32. We can then detect the DNA by running it on a gel as you will experience in exercise 33. Quantitative PCR (qPCR) will also quantitate the amount of DNA. The main drawback currently is the cost of the qPCR kits.

DNA, including ribosomal DNA, can be sequenced and then submitted to a database to determine if any existing matches exist, as in exercise 34. DNA microchips and microarrays (exercise 35) are allowing us to see the genes cancer cells turned on, and allow for customized therapy regimens. Enzymes can be used to cut DNA into segments, which can determine if the food poisoning you had is caused by the same microbe that caused another person's food poisoning thousands of miles away. It can also help to detect GMOs (genetically modified organisms). Enzyme-linked immunosorbent assays (ELISAs) in exercise 36 have been used for many years to detect viral and bacterial infections in humans, as well as autoimmune disorders. Their sensitivities have been lowered so large quantities of samples are no longer needed. Blood samples, which are still used in many tests and used to diagnose different conditions like leukemias, are seen in exercise 37. CRISPR now allows us to synthesize different cell functions and interfering RNAs (RNAi) can turn off genes. The advances in technology make one wonder, what will be the technologies of the future?

Definitions

Restriction Enzyme. An enzyme that recognizes and cleaves (cuts) a specific sequence of DNA.

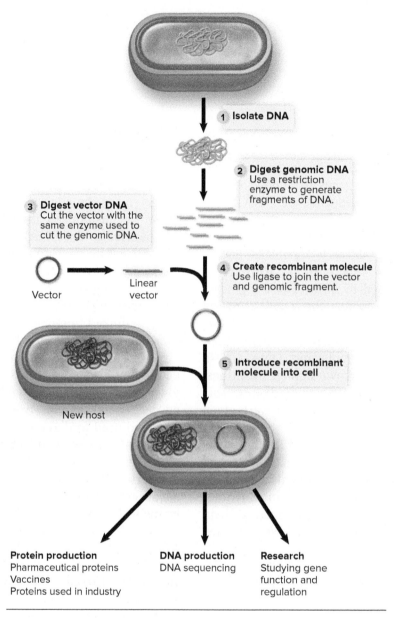

1 **Isolate DNA**

2 **Digest genomic DNA**
Use a restriction
enzyme to generate
fragments of DNA.

3 **Digest vector DNA**
Cut the vector with the
same enzyme used to
cut the genomic DNA.

Vector

Linear
vector

4 **Create recombinant molecule**
Use ligase to join the vector
and genomic fragment.

5 **Introduce recombinant
molecule into cell**

New host

Protein production
Pharmaceutical proteins
Vaccines
Proteins used in industry

DNA production
DNA sequencing

Research
Studying gene
function and
regulation

Figure I.32.1 The steps in cloning DNA.

32

Identifying DNA with Restriction Enzymes and the Polymerase Chain Reaction (PCR) ASM34, 36

Definitions

Amplify. In PCR, to make more DNA of a specific sequence and length.

Buffer solution. A salt solution formulated to maintain a particular pH.

Cell lysate. The liquid portion containing lysed cells.

DNA polymerase. Enzyme that synthesizes DNA; it uses one strand as a template to generate the complementary strand.

DNA sequencing. Determining the sequence (order) of nucleotide bases on a strand of DNA.

dNTPs (deoxyNucleoside TriPhosphates). The DNA bases used to make a DNA sequence, that is, A, T, C, and G.

Electrophoresis. Technique that uses an electrical current to separate either DNA or RNA fragments or proteins.

Invert. To gently mix a small tube. Place a thumb on the bottom of the tube and index finger on the top of tube and gently tip upside down.

Nucleotides. Comprised of a nitrogenous base, the nucleoside, a five carbon sugar like deoxyribose, and a phosphate group. The monomers for making both DNA and RNA.

Primer. A short (10–30 bp) DNA sequence that determines the segment of DNA that will be amplified.

Restriction enzymes. Enzymes isolated from bacteria that cut DNA at a specific sequence.

Supernatant. The liquid portion above the pellet.

***Taq* polymerase.** Heat-stable DNA polymerase of the thermophilic bacterium *Thermus aquaticus*.

Target DNA. In the PCR procedure, the region to be amplified.

Objectives

1. Explain the use of restriction enzymes to cut DNA into segments.
2. Explain the steps in PCR and how the reagents allow the steps to work.
3. Explain the use of DNA fingerprinting to identify DNA.

Pre-lab Questions

1. What will you use to transfer the liquids into the centrifuge tubes?
2. What are the three major steps in PCR?
3. Name the main reagents for PCR.

Getting Started

DNA from one organism can be distinguished from the DNA of another organism by using restriction enzymes (REs) and/or the polymerase chain reaction (PCR). The location of the specific sequences varies from species to species, and only identical strands of DNA are cut or amplified into the same number and size of fragments. This is the basis for comparing DNA from different organisms, one that has many applications. For example, in forensic (legal) investigations, the DNA from bloodstains can be compared with the DNA of a suspected murderer. In epidemiology investigations, the DNA from an *Escherichia coli* outbreak can be compared to another *E. coli* strain to see if the same food (like hamburger) caused the disease. Diseases like lyme disease and whooping cough are now diagnosed by PCR.

Enzymes are found in bacteria. The bacteria use enzymes to degrade foreign DNA that might enter their cell. Each cell methylates its DNA by adding a methyl group at a particular site, thereby preventing its own DNA from being degraded by its own restriction enzymes. Foreign DNA entering the cell does not have this specific pattern of methylation, and so the cell cleaves it, restricting its expression in the cell. Therefore, these enzymes are called **restriction enzymes** (**figure 32.1**). They can be isolated from bacteria and used in the laboratory for studying and manipulating DNA. In the laboratory, DNA samples to be compared are mixed with a restriction enzyme and incubated until the enzymes have cleaved the DNA at the recognition

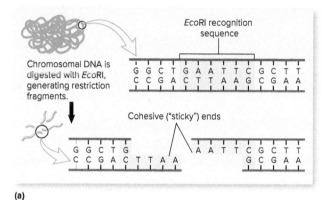

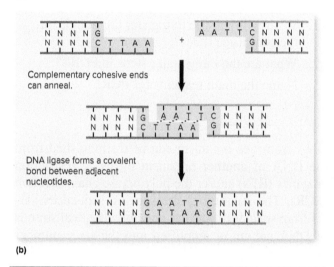

Figure 32.1 (*a*) How a restriction enzyme recognizes a sequence and cuts it. (*b*) Fragments can then anneal.

site unique to the enzyme. The restriction enzymes' names are derived from the first letter of the genus and the first two letters of the species. For example, *Eco*RI is the first restriction enzyme from *E. coli* strain R. Not only do restriction enzymes cut DNA at very specific sequences, but there are many different enzymes, each with its own recognition sequence. For example, the enzyme *Hha*I cuts at the sequence GCGC. The enzymes recognize a palindrome on double stranded DNA, which is the same sequence forward and backwards. In the two sequences of DNA shown below, each strand is cut in to two pieces because each contains GCGC; the pieces, however, differ in size between the two sequences.

Smaller (9 bp) DNA fragment   Larger (36 bp) DNA fragment

ATGGCTCAA GCGC TCACGGTAACTGCTGCATCCCGTTATACGAGCTACTT

ACCGAGTT CGCG AGT-GCCATTGACGACGTAGGGCAATATGCTCGATGA

Similarly sized DNA fragments (24 bp versus 21 bp)

ATATCGTTGAACTCCGTGTAGACT GCGC ACGTGTTACAATCCACCAAGT

TATAGCAACTTGAGGCACATCTGA CGCG TGCACAATGTTAGGTGGTTCA

Figure 32.2 A thermocycler used to perform PCR. Courtesy of Anna Oller, University of Central Missouri.

The PCR **amplifies**, or increases, the quantity of DNA strands already present. It is performed in a programmable machine called a thermocycler (**figure 32.2**), which contains a metal heat block that can change temperatures at a given time. The amplification of a specific sequence or DNA area (also known as the **target DNA**) is achieved by designing a **primer** (also known as an oligonucleotide, or oligo) using computer software programs and having a company make the primer. The primer will only bind to the complementary DNA sequence and amplify that area (**figure 32.3**). Primers vary in length from 7 to 30 bases, with shorter sequences being non-specific, whereas longer primers are generally directed to a known DNA sequence of a specific gene.

The process of PCR heats the DNA to 95°C, which is called denaturation. This causes the DNA strands to separate (**figure 32.4**). The primers bind to the DNA sequence at a lower temperature (around 50°C) during the annealing step. The annealing temperature will depend on the melting temperature of the primers. The melting temperature is provided with the primers when purchased. The extension step is performed at the optimal temperature for the polymerase (72°C) and allows *Taq* **polymerase** to add the individual nucleotides to make the complementary DNA strand. These three steps are repeated for 25–45 times, known as cycles. Once the main steps have been completed, an additional 10-minutes step at 72°C is performed to ensure all the bases were extended, and then the DNA is held at 4°C or put into the freezer until a gel can be run. The results are then visualized by **electrophoresing** the DNA in an agarose gel (exercise 33). Clean DNA can then be used

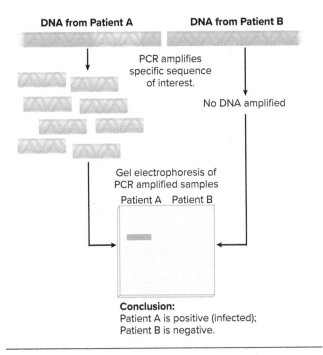

DNA from Patient A **DNA from Patient B**

PCR amplifies
specific sequence
of interest.

No DNA amplified

Gel electrophoresis of
PCR amplified samples

Patient A   Patient B

Conclusion:
Patient A is positive (infected);
Patient B is negative.

Figure 32.3 How a DNA target sequence is amplified. DNA from two different patients is obtained. The primer targeting a specific disease is used and only one patient is positive for the disease, which is visualized by running the samples on an agarose gel.

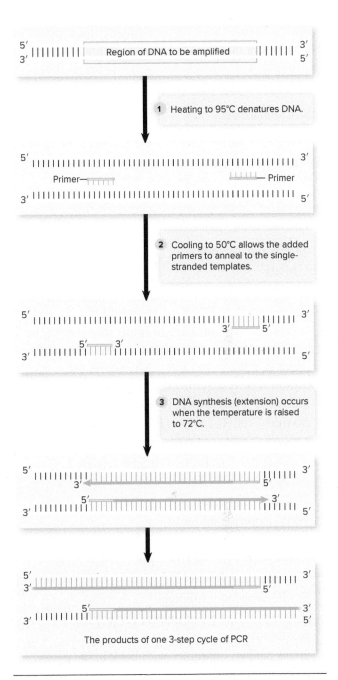

Region of DNA to be amplified

1 Heating to 95°C denatures DNA.

Primer— — Primer

2 Cooling to 50°C allows the added primers to anneal to the single-stranded templates.

3 DNA synthesis (extension) occurs when the temperature is raised to 72°C.

The products of one 3-step cycle of PCR

Figure 32.4 The three major steps involved in PCR: denaturation, annealing, and extension.

for **DNA sequencing** which determines if the DNA piece amplified is the correct DNA.

The reagents in a typical PCR reaction include *Taq* **DNA polymerase**, *Taq* DNA **buffer solution**, magnesium chloride ($MgCl_2$), water, **nucleotides** (**dNTPs**-deoxynucleoside triphosphates), DNA, and primers (oligonucleotides). The water serves as a diluent for the reaction, the buffer optimizes the pH of the solution, and $MgCl_2$ provides the ions for the reaction to work properly. The primer directs the location of amplification. The *Taq* polymerase is obtained from the hot springs bacterium *Thermus aquaticus* and recognizes the dNTPs—adenine (A), thymine (T), cytosine (C), and guanine (G)—for synthesizing the complementary DNA strand.

In this exercise, you will compare the DNA of three bacterial viruses using both restriction enzymes and PCR: phage lambda, phage φX174, and virus X (which is either lambda or φX174). The restriction enzyme *Dra*I is used to cut the DNA. It recognizes the base sequences shown below:

TTTAAA

AAATTT

Lambda contains 13 sites that can be cut by this enzyme, but φX174 has only two. You will amplify DNA using a short primer called a randomly amplified polymorphic DNA primer. Therefore, you should be able to identify virus X as either φX174 or lambda, depending on the number and size of the pieces separated by gel electrophoresis.

Note: Students should practice using micropipettors before doing this exercise and loading a practice gel (exercise 33) before the next lab due to time constraints.

DNA Isolation

The DNA of Gram-negative bacteria can be isolated using commercial kits that you buy, or you can use a standard protocol like the one listed below.

Materials

Microcentrifuge tubes (1.5 ml)

Gloves

Micropipettors

Gram-negative DNA isolation kit OR

trypticase soy or nutrient broth (depending on the bacterium) containing log phase bacteria like *E. coli, Citrobacter, Klebsiella,* etc.

1 mg/ml of lysozyme dissolved in filter sterilized STET buffer (5% (v/v)) Triton X-100

50 mM Tris-HCl, pH 8.0

50 mM EDTA, pH 8.0, 8% (w/v) sucrose

Tris-EDTA (TE) buffer (10 mM Tris-HCl pH 8, 1 mM EDTA)

70% ethanol (cold)

100% isopropanol (cold)

Water bath (boiling)

Microcentrifuge rack that can be placed into the water or a floatie

Sterile wooden toothpicks

Permanent fine point marker for writing on plastic tubes

Glass beaker, container with bleach, or autoclave bag

Vortexer (optional)

Forceps

Ice

PROCEDURE

Because we are isolating DNA in an open room, keep microcentrifuge tubes closed as much as possible to avoid contamination. Further, try not to cough, sneeze, breathe, etc., directly above/into the tubes.

Tip: Always centrifuge your tubes in the same direction, with the hinge facing the back of the rotor (away) and the lid facing the middle of the rotor. This way, any precipitate will be seen directly under the hinge so you know exactly where to look for your DNA. Do not worry if you do not see the DNA. If you see a very large pellet, it most likely is not DNA and has contaminating proteins.

1. The day before the isolation, inoculate the bacterium into a trypticase soy or nutrient broth tube and grow for 12–24 hours at 37°C.

2. Pipet the bacteria from the broth tube to a 1.5 ml microcentrifuge tube, label the tube with a marker, and centrifuge at 12,000 rpm for 1 minute to pellet the cells.

   a. If a small pellet is seen, pipet the **supernatant** (liquid above the pellet) off and dispense into a glass beaker or other item for disposal. Pipet another 1 ml of the broth into the 1.5 ml tube and centrifuge again.

3. Pipet the supernatant off into the discard (like a glass beaker).

4. Pipet 20 μl of lysozyme to the pellet, and gently pipet up and down with the liquid to break up the pellet.

5. Make sure the microcentrifuge cap is firmly locked down into place.

6. Place the centrifuge tube containing the lysozyme (with STET) into a rack or floatie and place into the boiling water for 1 minute. Make sure the tube is at least halfway submerged in the boiling water.

7. Using forceps, remove the microcentrifuge tube from the boiling water and place on ice for 2 minutes. This contains the **cell lysate**.

8. Centrifuge the tubes for 10 minutes at 12,000 rpm. You should see a large pellet. The pellet contains the cell components like proteins, membranes, etc.

9. Carefully pipet off the supernatant into a new, labeled microcentrifuge tube. The supernatant contains the DNA.

10. Pipet 200 μl of refrigerated 100% isopropanol into the microcentrifuge tube containing the supernatant.

11. **Invert** the tube 50 times. The first few inversions are when you may see the DNA precipitate out.

12. Centrifuge the tubes for 5 minutes at 12,000 rpm. You may be able to see a small pellet of DNA.

13. Pipet off the supernatant into the glass beaker.

14. Add 300 μl of refrigerated 70% ethanol to the tube. Invert the tube 2 times.

15. Centrifuge the tubes for 1 minute at 12,000 rpm.

16. Open the lid to the centrifuge tube. **Quickly and in one motion**, invert the tube onto a piece of towel. Gently tap to remove the excess ethanol.

17. Allow the tubes to dry for about 15 minutes. If you could see a DNA pellet before, it will now become clear as it dries.

18. Resuspend the DNA in 25–50 μl of TE buffer by pipetting up and down. Allow to set at room temperature overnight or at 55°C for 4 hours. This step allows the DNA to regain its conformation.

Restriction Enzyme Digestion

Materials

Lambda DNA
φX174 DNA
Unknown phage (either lambda or φX174)
Restriction enzyme *Dra*I (or another enzyme)
Sterile 0.5 ml microcentrifuge tubes or Eppendorf tubes
Micropipettors
Tris-EDTA (TE) buffer
Stop mix (includes tracking dye) like EDTA
Water bath or incubator
Microfuge/Centrifuge
Disposable gloves
Distilled water (sterile)
Ice in containers

PROCEDURE

You may use the phage DNA or your instructor may have you isolate your own DNA from a bacterium to use in the restriction enzyme digestion and PCR.

1. Obtain 3 sterile tubes and label with a marker. Add _____ μl sterile distilled water (as determined by the instructor) to a sterile

Eppendorf tube on ice, and then add 1 μl 10X TE buffer. Buffer is added before the enzyme so that the conditions are immediately optimal for the enzyme.

2. Add _____ μl lambda DNA to the tube. The amount added is usually 0.5–1.0 μg/μl. The instructor will indicate how much you should add.

3. Add the restriction enzyme *Dra*I (or another enzyme).

4. Repeat for φX174 DNA and the unknown virus DNA.

5. Each tube contains:
   10X TE buffer
   DNA
   Enzyme
   Sterile, distilled water to bring total volume to 10 μl, including enzyme.

6. Invert several times and incubate the tubes for 30 minutes at 37°C.

7. Add stop mix (usually 1 μl of 10X, or EDTA). This stops the reaction. Some stop mixes also contain a loading dye that helps the sample sink to the bottom of the well.

8. Tubes may be stored at −20°C until a gel can be run.

Polymerase Chain Reaction

Materials

DNA (Lambda and φX174 or DNA you isolated)
Primer 5'-CCGCAGCCA-3'(100 mM)
Sterile 0.5 ml microcentrifuge tubes
0.5 ml or 0.2 ml microcentrifuge PCR tubes, sterile
Thermocycler (PCR tubes must fit heat block)
Micropipettors
PCR Mastermix OR:
  Taq polymerase (5 U μl^{-1})
  Taq polymerase buffer (100 mM Tris-HCl, 500 mM KCl, 0.8% Nonidet P-40)
  dNTPs (10 mM)
  MgCl$_2$ (25 mM)
Sterile double-distilled purified water (PCR reagent grade)

Ice bath

Microfuge/Centrifuge

Permanent fine point marker for writing on plastic tubes (Sharpies will melt off)

Disposable gloves

PROCEDURE

1. Your instructor will assign your team a letter or number to help with writing small labels on the tops of tubes.

2. Wear gloves during this procedure to prevent your cells from contaminating the tubes.

3. You can perform this exercise on the bench top. You want to minimize potential air contaminants, so keeping the centrifuge tube lids shut is helpful.

4. Always pipet any liquids slowly with micropipettors, so as to not create bubbles. Bubbling the liquids will inactivate any enzymes present.

5. If any small drops of liquid have accumulated on the side of any centrifuge tube, pulse spin tubes to ensure all liquid is at the bottom of the centrifuge tube.

6. Program the thermocycler for 30 cycles according to the following parameters (or your own parameters, listed on the right side):

Program parameters:	Your own, specific parameters:
a. 95°C for 1 minute (denature)	95°C _____
b. 50°C for 2 minutes (anneal)	__°C _____
c. 72°C for 1 minute (extend)	72°C
d. Repeat a–c for 30 cycles.	
e. 72°C for 10 minutes	72°C for 10 minutes (extend)
f. 4°C indefinitely	4°C indefinitely
g. Need only 1 cycle of e and f.	

7. You need small ice cubes or crushed ice. You can also use bench top ice coolers.

8. ALL tubes MUST be kept on ice at all times!

    a. Insert each tube about half-way into the ice, so you can still read tube labels.

    b. As you finish using each tube, move it to the back of the ice bucket. This helps you keep track of tubes.

    c. Once you are done with each tube, place them in the proper discard container.

9. Label your tube on the top and side with a marker PCR-1, 2, 3, etc. Keep the centrifuge tube lids shut.

10. Add the following items into a 0.2 or 0.5 ml microcentrifuge tube in the following order.

11. The list on the left is the predetermined reaction, but your instructor may modify it. If so, write the amounts in the list on the right.

12. Tip: If your instructor has pipetted the primer and *Taq* polymerase into individual or separate tubes, gently pipet 5 μl of the water solution into the tube, add 1 μl to the pipet dial, and then pipet the solution back into the water.

    a. 17 μl Water   _____ μl Water

    b. 0.25 μl Primer   _____ μl Primer

    c. 6 μl Mastermix   _____ μl Mastermix

    OR:

    If Mastermix is added, d–g are not needed. Skip to h to add DNA.

    d. 2.5 μl *Taq* DNA buffer   _____ μl *Taq* DNA buffer

    e. 2 μl MgCl$_2$ _____ μl MgCl$_2$

    f. 1 μl dNTPs   _____ μl dNTPs

    g. 0.15 μl *Taq* DNA polymerase   _____ μl *Taq* DNA polymerase

    h. 2 μl DNA   _____ μl DNA

13. Now pipet the 25 μl into the 0.2 (or 0.5) ml thin-walled PCR tube. After adding all ingredients together, make sure to balance the centrifuge and pulse spin each PCR tube for 3–5 seconds.

14. Add the PCR tubes containing all reagents to the programmed thermocycler.

15. The PCR run will take about 3 hours to complete. The samples can then be stored at −20°C until a gel can be run.

EXERCISE

32

Laboratory Report: Identifying DNA with Restriction Enzymes and the Polymerase Chain Reaction (PCR) ASM34, 36

Questions

1. Would you have the same number of fragments from each phage if you used a different restriction enzyme? Why or why not?

2. Name and describe the purpose of each of the main reagents in an enzyme digestion.

3. What would happen if the PCR primer was not designed properly?

4. Name and explain two things that might inhibit a PCR reaction from working properly.

33

Electrophoresis ASM 34, 36

Definitions

Agarose. Highly purified form of agar used in gel electrophoresis.

Buffer. Substance in a solution that maintains the pH.

Comb. A piece of plastic containing a variable number of teeth that forms indentations, or wells, when the gel solidifies and the comb is removed. DNA is added to the wells.

Electrophoresis. Technique that uses electrical current to separate either DNA or RNA fragments or proteins.

Gel electrophoresis. A technique that uses electricity to separate DNA or RNA in a gel by fragment size.

Ladder. A standard marker with known base pair sizes added to each gel run. Sample sizes are compared to the ladder.

Loading dye. A dye that weighs down DNA so it stays in wells and gives an estimation of band location.

UV transilluminator. A piece of equipment that uses a wavelength of UV light to excite the molecules in the gel so samples can be visualized.

Objectives

1. List the major steps in performing electrophoresis.
2. Explain how electrical current can be used to move molecules.
3. Interpret the size of molecules electrophoresed.

Pre-lab Questions

1. What size molecules will be seen at the top of the gel by the wells?
2. Why must you add a loading dye to the DNA solution?
3. What is the purpose of the DNA ladder?

Getting Started

Electrophoresis is a widely used molecular biology technique based upon the principle of attraction of negatively and positively charged molecules (similar to North–South magnets). Electrophoresis uses an electrical current to separate out molecules based on their base pair size. It is used to visualize PCR or restriction enzyme digests. One end of the electrophoresis apparatus is negatively charged (electrodes appear black), whereas the other end is positively charged (electrodes appear red) (**figure 33.1**). DNA is negatively charged, so it is attracted to a positive electrical charge. This is why DNA is added to the negative end of the gel. The time DNA is allowed to migrate or how long electricity is applied is important. If the gel runs for too long, the DNA will be phoresed out of the gel and into the buffer solution. Once the DNA is in the buffer solution, it cannot be recovered.

The amount of voltage applied also makes a difference on how the gel appears. The lower (slower) the voltage, the more precise separation that occurs. Thus, if two bands are close in size, they can be separated, but if the gel is run at a higher (faster) voltage, then you may not be able to see the two different bands.

Different mediums can be used in electrophoresis such as acrylamide for separating proteins, and agarose for separating DNA or RNA. **Agarose** is a purified agar that will solidify once it has been boiled and allowed to cool. It is porous to allow molecules to move through it, and different concentrations of agar can be used to make the gel more or less porous. Small molecules move through the gel faster than larger molecules. The faster a molecule travels, the farther away from the loading wells it will move. Small molecules are found toward the bottom or end of the gel. A thicker, less porous gel is used to separate large molecules (thousands of base pairs) from one another, whereas a thinner, more porous gel is used to separate smaller sized molecules (hundreds of base pairs) from one another.

Once the agarose solution has boiled, it needs to cool for 5–10 minutes so the heat does not melt or warp the plastic trays. The plastic trays serve as

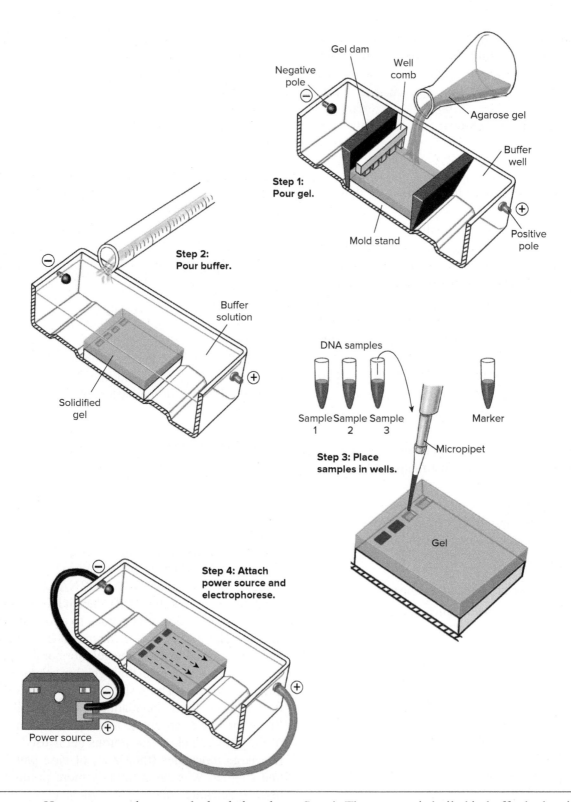

Figure 33.1 How agarose gels are made, loaded, and run. *Step* 1: The agarose is boiled in buffer by heating, cooled, and poured into the gel mold. Ethidium bromide or another stain is added so bands can be visualized. A comb with teeth is added to the negatively charged end to create wells. *Step* 2: Once the gel has cooled, the comb is pulled out, and buffer is poured over the gel. *Step* 3: The samples are loaded into the wells of the gel, which is still in the gel box. *Step* 4: The electricity is turned on to allow the molecules to separate based on charge and size.

a mold, similar to a mold you would use to make Jello™. The agarose solution is then poured into the trays and **combs** are added. Once the gel has solidified, the comb is pulled out, leaving a well. The DNA is then added to the wells. Combs may have varying numbers of teeth, usually between 5 and 10. The combs may also have different thicknesses. Thicker combs allow more volume of DNA to be added, so visualization of small quantities of DNA is enhanced. Thinner, longer combs allow for sharper bands so if two bands are close in size, they are easier to visualize and separate.

Different types of **buffers** can be used, depending upon the type and size of molecules being separated. A 1× TAE (tris-acetate-EDTA) solution is used for DNA. For RNA separation, a TBE (trisborate-EDTA) solution is used instead. Agarose is added to 1× TAE buffer and boiled. The TAE buffer is the same solution used for running the gel in the electrophoresis apparatus. DNA can be viewed in the gel a few different ways. First, the DNA can be stained with a dye (methylene blue or a fluorescent

dye) in a glass or plastic tub after the gel has been run, and then destained in distilled water. Second, a solution of ethidium bromide or a fluorescent dye can be added directly to the flask of agarose before it is poured, or it can be directly added to the electrophoresis buffer at the positive end. Ethidium bromide (and other fluorescent stains) is positively charged so it migrates to the negative end, and the DNA traps it, so when the **UV transilluminator** is turned on, the DNA can be seen in the gel. The new fluorescent dyes do not have the same safety concerns as ethidium bromide and are excellent at detecting small amounts of DNA.

The DNA samples need to be mixed with a 6× **loading dye** that contains a dye (bromophenol blue) or dyes dissolved in a solution that increases the density of the solution so the DNA stays at the bottom of the well. If a loading dye is not used, the DNA will float into the buffer solution and cannot be recovered. The dye provides an estimation of how far down the gel the molecules have traveled. DNA bands are seen, which are fragments of the DNA (**figure 33.2**).

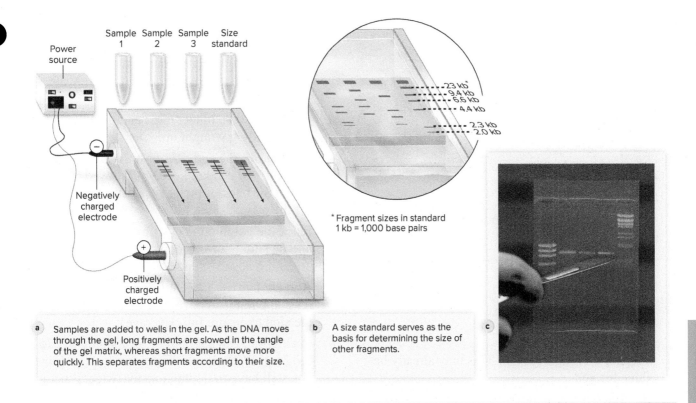

a — Samples are added to wells in the gel. As the DNA moves through the gel, long fragments are slowed in the tangle of the gel matrix, whereas short fragments move more quickly. This separates fragments according to their size.

b — A size standard serves as the basis for determining the size of other fragments.

Figure 33.2 (*a*) How electrophoresis separates molecules based on size. (*b*) A standard is loaded onto every gel so fragment size can be determined. (*c*) Visualization of DNA bands on a UV transilluminator. (*c*) Richard T. Nowitz/Science Source.

A standard must be added to the first well of the gel for each gel run. Different **ladders** can be purchased, all with varying sized DNA fragments. We use a 1 kb (kilobase) DNA ladder (standard), which means the lowest band is 1,000 base pairs (1 kb = 1,000 base pairs). The ladder is also mixed with 6× loading dye to keep it in the well. Some ladders come premixed with the loading dye.

You will perform agarose **gel electrophoresis** using your PCR and restriction enzyme digestion samples from the previous lab to estimate band fragment sizes.

Materials

Electrophoresis apparatus

Combs (10 wells)

Large binder clips to hold combs in place, 2 small gel trays (holds 50 ml)

Masking tape (if trays do not have end caps)

Agarose

1× TAE (tris-acetate-EDTA) solution

Micropipettors (1–2 µl, 5 µl, and 15 µl)

Hot-hands™ or potholders

DNA ladder

6× loading dye

Ethidium bromide or a fluorescent dye (ethidium bromide = 1 µl of a 10× solution per gel)

Gloves

UV transilluminator

DNA (from previous PCR and restriction enzyme digestion)

PROCEDURE

1. Prepare a 1% agarose gel. This may already be done for you.

   a. Add 0.5 g agarose to 50 ml 1 × TAE buffer in a 150 ml flask. Add a Kimwipe to the top of the flask to help prevent evaporation.

   b. Boil until the agarose molecules completely disappear. Watch the flask so it does not boil over. (If microwaving, boiling occurs at about 30 seconds. Open the microwave to swirl and avoid boiling over; continue boiling until the solution no longer bubbles, about 1 minute.)

   c. Remove the flask with a potholder and allow to cool for about 5 minutes on the benchtop.

   d. While the flask cools set up the gel tray by placing end caps on the tray (or taping the ends with masking tape) and placing the 10-well comb at the end with the negative (black) electrode.

   e. Put on gloves. Add 1 µl of ethidium bromide (or other dye) to the agarose solution, and pipet up and down to remove as much of the ethidium bromide as possible. Discard in the appropriate container.

   f. Swirl the liquid to distribute the dye. Pour the gel into the tray and allow to solidify. (You may pour the gels in the refrigerator to facilitate solidifying.) Do not move gels once they are poured until they have turned opaque white, signaling they are solidified.

   g. Remove the binder clips and then the comb by pulling straight up.

   h. Remove the end caps or masking tape. You will need to keep your gloved fingers on the ends of the gel or it may fall out of the tray.

2. Place the gel into the electrophoresis apparatus and be sure the wells containing your sample are at the negative end.

3. Add enough 1× TAE buffer to the electrophoresis apparatus to cover the wells (look in from the side to ensure full coverage).

4. Each group can add 2 µl of 6× loading dye to the 25 µl PCR sample. If you have a 1 µl pipet, you will have to do the transfer twice. Pipet up and down to mix the sample.

5. Each group can add 2 µl of 6× loading dye to the enzyme digestion reaction. If you have a

1 μl pipet, you have to do the transfer twice. Pipet up and down to mix the sample.

Note: One set of gels will be used for the PCR samples, and the other gels will be used for the restriction enzyme digestion. Write your sample name on the board next to the lane you used.

6. Add 5 μl of a 1 kb DNA ladder to lane 1 of each gel by aiming the pipet tip over the well and **slowly** pushing the pipet plunger to dispense the solution into the gel. (If you push the plunger too quickly, your sample may be pushed out of the well and into the buffer solution.) Do not release the plunger until you have removed the pipet tip from the solution or it will suck the buffer and some of the sample back up.

7. Add 15 μl of the PCR sample to lane 2, etc., of the appropriate gel. Each group should continue until all groups have added their samples.

8. Add 15 μl of the enzyme digestion reaction to lane 2, etc., of the appropriate gel. Each group should continue until all groups have added their samples.

9. Plug the electrical cords into the gel and into the electrical box.

10. Allow your gel to run for about an hour at 80 volts.

11. Once the gel is phoresed, turn off the electrical current and disconnect the electrical cords.

12. Wearing gloves, carefully remove the gel tray and move the gel onto the UV transilluminator. Wearing goggles, turn off the lights, and turn on the UV transilluminator to visualize the DNA bands. Depending on the dye used, prolonged exposure to the UV light may cause the bands to fade.

13. Photograph the gel or make a drawing of the bands.

14. Depending upon the expected PCR product size, a smudged band at the very bottom of each sample may be leftover reagents (color plate 77).

EXERCISE

33

Results and Questions

1. Record the gel results for the **PCR** sample in the table. Record the base pair size for band 1, band 2, band 3, etc.

	Ladder	Lane 1	Lane 2	Lane 3	Lane 4	Lane 5	Lane 6	Lane 7	Lane 8	Lane 9
Sample Identity										
Total # bands seen										
Band 1 size										
Band 2 size										
Band 3 size										
Band 4 size										
Band 5 size										

2. Record the gel results for the **restriction enzyme digestion** sample in the table. Record the base pair size for band 1, band 2, band 3, etc.

	Ladder	Lane 1	Lane 2	Lane 3	Lane 4	Lane 5	Lane 6	Lane 7	Lane 8	Lane 9
Sample Identity										
Total # bands seen										
Band 1 size										
Band 2 size										
Band 3 size										
Band 4 size										
Band 5 size										

3. What is the probable identity of phage X?

4. How were you able to estimate the size of the DNA fragments?

5. How bright were the bands in comparison to the ladder (lighter, the same, a little brighter/darker, much darker)?

6. Explain reasons why you may not have had any bands present, if you didn't see any.

34

Definitions

16S. *S* is an abbreviation for Svedberg. It is a unit of mass that is measured by the rate a particle sediments in a centrifuge. The prokaryotic ribosome is made up of two main parts: 30S and 50S. The 30S particle is made up of the 16S rRNA plus 21 polypeptide chains.

DNA nucleotide sequence. The order that bases are found in a piece of DNA.

Signature sequences. DNA sequences of about 5–10 bases long found at a particular location in the 16S rRNA. These DNA sequences are unique to Archaea, Bacteria, or Eukarya (eukaryotes).

Objectives

1. Explain the importance of the 16S rRNA sequence for the identification of organisms.

2. Interpret rRNA data using the Ribosomal Database Project.

Pre-lab Questions

1. What sequence information is used to identify organisms in this lab?

2. What are signature sequences?

3. If an organism is related to another, will the rRNA sequence be more similar or different?

Getting Started

Identifying and classifying bacteria have always been more difficult than identifying and classifying plants and animals. Bacteria have very few differences in their structure, and while they are metabolically extremely diverse, it has not been clear which of these characteristics are the most important for identification or grouping. For example, is nitrogen fixation more significant than anaerobic growth, or endospore formation more important than photosynthesis?

The ability to sequence the DNA of microorganisms has offered a new solution. The closer organisms are related to each other, the more similar their **DNA nucleotide sequences**. Certain genes common to all bacteria can be sequenced and compared. What genes should be compared? The DNA coding for a part of the ribosome, namely the **16S** ribosome portion, is a good choice. Ribosomes are critically important for protein synthesis, and any mutation is quite likely to be harmful. Some mutations, however, are neutral or perhaps even advantageous, and these mutations then become part of the permanent genome. These sequences change very slowly over time and are described as highly conserved. The 16S rRNA DNA segment is found in all organisms (slightly larger in eukaryotes) with the same function, so the sequences can easily be compared.

This approach has had exciting results. Some specific base sequences are always found in some organisms and not others. These are called **signature sequences**. Also, new relationships among bacteria can be determined by comparing sequences of organisms base by base by means of computer programs. The more the sequences diverge, the more the organisms have evolved from one another.

Perhaps even more useful, these sequences can be used to identify bacteria. The sequences of at least 85,000 organisms are in public databases, with new sequences continually added. There is a public database called the Ribosomal Database Project (RDP) in which sequences can be submitted. Another one is the National Center for Biotechnology Information (NCBI) run by the National Institutes of Health in the United States. You can search for matches to proteins, nucleotides, genomes, etc. If a new, unknown sequence is submitted to a database management computer, it will respond in seconds with the most likely identification of the species containing the sequence. As the number of rRNA sequences grows in the database, so do the possible applications of this information. For example, in a clinical setting, rRNA sequence data can now be used to identify a pathogen more quickly than by the use of standard culture techniques. Faster identification of an organism enables clinicians to treat disease-causing organisms more efficiently, often with a better outcome for the patient.

This exercise will give you a chance to submit the 16S ribosomal RNA DNA sequence of an organism to the RDP or the NCBI, and to identify the organism. Since many schools do not have the resources for determining the specific sequences, you will be given the sequences, which you can enter on the Web and immediately receive an identification. You may also look for other sequences posted on the Web or in microbiological journals, such as *Journal of Bacteriology,* published by the American Society for Microbiology. Finally, you will search for appropriate primers you could use for the microbes in this lab to amplify them using PCR.

PROCEDURE

1. Open your Internet browser (Firefox, Chrome, or Safari for example) on the computer.
2. Type in the URL rdp.cme.msu.edu/
3. Under "RDP Analysis Tools," click "Sequence Match."
4. Type the sequence in the box labeled "Cut and Paste Sequence(s)." This is easier to do if one

person reads the sequence while the other person types in the letters. Sometimes only a few hundred bases are necessary for an identification.

5. Click "Submit."
6. Read identification under "Hierarchy View."
7. Alternatively copy the sequence into the National Center for Biotechnology Institute at https://blast.ncbi.nlm.nih.gov and click search. (You will get the same results at both sites.) You will be performing a nucleotide BLAST search. Record your results in the Laboratory Report.
8. Now copy the sequence into a Primer-BLAST program at www.ncbi.nlm.nih.gov/tools /primer-blast. Once you submit the DNA sequence, a list of primers and a schematic of where they bind can be seen. Which primer set is the best one to use? Record the most appropriate primers to use in the Laboratory Report.

Hint: The organisms were studied in

1. Exercise 19
2. Exercise 19
3. Exercise 20
4. Exercise 13 (genus only)
5. Photosynthetic bacteria are not found in any exercise.

Organism 1

```
  1    tctctgatgt tagcggcgga cgggtgagta acacgtggat aacctaccta taagactggg
 61    ataacttcgg gaaaccggag ctaataccgg ataatatttt gaaccgcatg gttcaaaagt
121    gaaagacggt cttgctgtca cttatagatg gatccgcgct gcattagcta gttggtaagg
181    taacggctta ccaaggcaac gatgcatagc cgacctgaga gggtgatcgg ccacactgga
241    actgagacac ggtccagact cctacgggag gcagcagtag ggaatcttcc gcaatgggcg
301    aaagcctgac ggagcaacgc cgcgtgagtg atgaaggtct tcggatcgta aaactctgtt
361    attagggaag aacatatgtg taagtaactg tgcacatctt gacggtacct aatcagaaag
421    ccacggctaa ctacgtgcca gcagccgcgg taatacgtag gtggcaagcg ttatccggaa
481    ttattgggcg taaagcgcgc gtaggcggtt ttttaagtct gatgtgaaag cccacggctc
541    aaccgtggag ggtcattgga aactggaaaa cttgagtgca gaagaggaaa gtggaattcc
601    atgtgtagcg gttaaatgcg cagagatatg gaggaacacc agtggcgaag gcgactttct
661    ggtctgtaac tgacgctgat gtgcgaaagc gtgggaatca aacaggatta gataccctgg
721    tagtccacgc cgtaaacgat gagtgctaag tgttaggggg tttccgcccc ttagtgctgc
781    agctaacgca ttaagcactc cgcctgggga gtacgaccgc aaggttgaaa ctcaaaggaa
```

841 ttgacggggga cccgcacaag cggtggagca tgtggtttaa ttcgaagcaa cgcgaagaac
901 cttaccaaat cttgacatcc tttgacaact ctagagatag agccttcccc ttcgggggac
961 aaagtgacag gtggtgcatg gttgtcgtca gctcgtgtcg tgagatgttg ggttaagtcc
1021 cgcaacgagc gcaacccttta agcttagttg ccatcattaa gttgggcact ctaagttgac
1081 tgccggtgac aaaccggagg aaggtgggga tgacgtcaaa tcatcatgcc ccttatgatt
1141 tgggctacac acgtgctaca atggacaata caaagggcag cgaaaccgcg aggtcaagca
1201 aatcccataa agttgttctc agttcggatt gtagtctgca actcgactac atgaagctgg
1261 aatcgctagt aatcgtagat cagcatgcta cggtgaatac gttcccgggt cttgtacaca
1321 ccgcccgtca caccacgaga gtttgtaaca

Organism 2

1 taacacgtgg ataacctacc tataagactg ggataacttc gggaaaccgg agctaatacc
61 ggataatata ttgaaccgca tggttcaata gtgaaagacg gttttgctgt cacttataga
121 tggatccgcg ccgcattagc tagttggtaa ggtaacggct taccaaggca acgatgcgta
181 gccgacctga gagggtgatc ggccacactg gaactgagac acggtccaga ctcctacggg
241 aggcagcagt agggaatctt ccgcaatggg cgaaagcctg acggagcaac gccgcgtgag
301 tgatgaaggt cttcggatcg taaaactctg ttattaggga agaacaaatg tgtaagtaac
361 tatgcacgtc ttgacggtac ctaatcagaa agccacggct aactacgtgc

Organism 3

1 gcctaataca tgcaagtaga acgctgagaa ctggtgcttg caccggttca aggagttgcg
61 aacgggtgag taacgcgtag gtaacctacc tcatagcggg ggataactat tggaaacgat
121 agctaatacc gcataagaga gactaacgca tgttagtaat ttaaaagggg caattgctcc
181 actatgagat ggacctgcgt tgtattagct agttggtgag gtaaaggctc accaaggcga
241 cgatacatag ccgacctgag agggtgatcg gccacactgg gactgagaca cggcccagac
301 tcctacggga ggcagcagta gggaatcttc ggcaatgggg gcaaccctga ccgagcaacg
361 ccgcgtgagt gaagaaggtt ttcggatcgt aaagctctgt tgttagagaa gaatgatggt
421 gggagtggaa aatccaccaa gtgacggtaa ctaaccagaa agggacggct aactacgtgc
481 cagcagccgc ggtaatacgt aggtcccgag cgttgtccgg atttattggg cgtaaagcga
541 gcgcaggcgg ttttttaagt ctgaagttaa aggcattggc tcaaccaatg tacgctttgg
601 aaactggaga acttgagtgc agaaggggag agtggaattc catgtgtagc ggtgaaatgc
661 gtagatatat ggaggaacac cggtggcgaa agcggctctc tggtctgtaa ctgacgctga
721 ggctcgaaag cgtggggagc aaagaggatt agataccctg gtagtccacg ccgtaaacga
781 tgagtgctag gtgttaggcc ctttccgggg cttagtgccg gagctaacgc attaagcact
841 ccgcctgggg agtacgaccg caaggttgaa actcaaagga attgacgggg gcccgcacaa
901 gcggtggagc atgtggttta attcgaagca acgcgaagaa ccttaccagg tcttgacatc
961 ccgatgcccg ctctagagat agagttttac ttcggtacat cggtgacagg tggtgcatgg
1021 ttgtcgtcag ctcgtgtcgt gagatgttgg gttaagtccc gcaacgagcg caaccccctat
1081 tgttagttgc catcattaag ttgggcactc tag

Organism 4

1 ggtaccactc ggcccgaccg aacgcactcg cgcggatgac cggccgacct ccgcctacgc
61 aatacgctgt ggcgtgtgtc cctggtgtgg gccgccatca cgaagcgctg ctggttcgac
121 ggtgtttat gtaccccacc actcggatga gatgcgaacg acgtgaggtg gctcggtgca
181 cccgacgcca ctgattgacg cccctcgtc ccgttcggac ggaacccgac tgggttcagt
241 ccgatgccct taagtacaac agggtacttc ggtggaatgc gaacgacaat ggggccgccc

301 ggttacacgg gtggccgacg catgactccg ctgatcggtt cggcgttcgg ccgaactcga
361 ttcgatgccc ttaagtaata acgggtgttc cgatgagatg cgaacgacaa tgaggctatc
421 cggcttcgtc cgggtggctg atgcatctct tcgacgctct ccatggtgtc ggtctcactc
481 tcagtgagtg tgattcgatg cccttaagta ataacgggcg ttacgaggaa ttgcgaacga
541 caatgtggct acctggttct cccaggtggt taacgcgtgt tcctcgccgc cctggtgggc
601 aaacgtcacg ctcgattcga gcgtgattcg atgcccttaa gtaataacgg ggcgttcggg
661 gaaatgcgaa cgtcgtcttg gactgatcgg agtccgatgg gtttatgacc tgtcgaactc
721 tacggtctgg tccgaaggaa tgaggattcc acacctgcgg tccgccgtaa agatggaatc
781 tgatgttagc cttgatggtt tggtgacatc caactggcca cgacgatacg tcgtgtgcta
841 agggacacat tacgtgtccc cgccaaacca agacttgata gtcttggtcg ctgggaacca
901 tcccagcaaa ttccggttga tcctgccgga ggccattgc

Organism 5

1   agagtttgat cctggctcag agcgaacgct ggcggcaggc ttaacacatg caagtcgaac
61   gggcgtagca atacgtcagt ggcagacggg tgagtaacgc gtgggaacgt accttttggt
121   tcggaacaac acagggaaac ttgtgctaat accggataag cccttacggg gaaagattta
181   tcgccgaaag atcggcccgc gtctgattag ctagttggtg aggtaatggc tcaccaaggc
241   gacgatcagt agctggtctg agaggatgat cagccacatt gggactgaga cacggcccaa
301   actcctacgg gaggcagcag tggggaatat tggacaatgg gcgaaagcct gatccagcca
361   tgccgcgtga gtgatgaagg ccctaggytt gtaaagctct tttgtgcggg aagataatga
421   cggtaccgca agaataagcc ccggctaact tcgtgccagc agccgcggta atacgaaggg
481   ggctagcgtt gctcggaatc actgggcgta aagggtgcgt aggcgggttt ctaagtcaga
541   ggtgaaagcc tggagctcaa ctccagaact gcctttgata ctggaagtct tgagtatggc
601   agaggtgagt ggaactgcga gtgtagaggt gaaattcgta gatattcgca agaacaccag
661   tggcgaaggc ggctcactgg gccattactg acgctgaggc acgaaagcgt ggggagcaaa
721   caggattaga taccctggta gtccacgccg taaacgatga atgccagccg ttagtgggtt
781   tactcactag tggcgcagct aacgctttaa gcattccgcc tggggagtac ggtcgcaaga
841   ttaaaactca aaggaattga cgggggcccg cacaagcggt ggagcatgtg gtttaattcg
901   acgcaacgcg cagaacctta ccagcccttg acatgtccag gaccggtcgc agagacgtga
961   ccttctcttc ggagcctgga gcacaggtgc tgcatggctg tcgtcagctc gtgtcgtgag
1021 atgttgggtt aagtcccgca acgagcgcaa ccccgtcct tagttgctac catttagttg
1081 agcactctaa ggagactgcc ggtgataagc cgcgaggaag gtggggatga cgtcaagtcc
1141 tcatggccct tacgggctgg gctacacacg tgctacaatg gcggtgacaa tgggaagcta
1201 aggggtgacc cttcgcaaat ctcaaaaagc cgtctcagtt cggattgggc tctgcaactc
1261 gagcccatga agttggaatc gctagtaatc gtggatcagc atgccacggt gaatacgttc
1321 ccgggccttg tacacaccgc ccgtcacacc atgggagttg gctttacctg aagacggtgc
1381 gctaaccagc aatggggca gccggccacg gtagggtcag cgactggggt gaagtcgtaa
1441 caaggtagcc gtaggggaac ctgcggctgg atcacctcct t

EXERCISE

34

Laboratory Report: Bacterial Identification using Ribosomal RNA ^{ASM 28, 31, 34, 36}

Results

Record the organism identification, determine if it is classified as Bacteria or Archaea, and the most appropriate primer to use. Your instructor may have you attach a copy of the program results for verification.

Organism Identification	Bacteria or Archaea	Best Primer 5′—>3′
1 _____	_____	_____
2 _____	_____	_____
3 _____	_____	_____
4 _____	_____	_____
5 _____	_____	_____

Questions

1. What is an advantage of identifying an organism using the Ribosomal Database Project?

2. What is a disadvantage of using databases?

3. Can you identify a clinical application based on rRNA identification of an organism?

4. How do you know that the primers you chose are the best ones to use for the microbe?

EXERCISE

35

Microarray ASM 11, 15, 17, 19, 36

Definitions

Hybridize. The binding of two complementary strands of DNA.

Microarray. A molecular biology technique that uses a slide containing many known nucleotide sequences to determine gene expression.

Probe. A single-stranded piece of nucleic acid that has been tagged with a detectable marker.

Reverse transcription. A process that converts RNA to complementary DNA (cDNA) using the enzyme reverse transcriptase.

Objectives

1. Describe the concept of a microarray.
2. Explain how a microarray can detect gene expression.
3. Classify which genes in the microarray were expressed.

Pre-lab Questions

1. What is the purpose of a microarray?
2. List the main steps of a microarray from the beginning to detection.
3. If you see yellow on a computerized microarray spot, what does it mean?

Getting Started

Microarrays, which are also known as *DNA chips,* determine if genes are being expressed at a given point in time. Microarrays are advantageous over other molecular biology techniques in that thousands of genes can be scanned quickly, saving time. Microarrays are currently being used in determining gene expression levels (color plate 78) between healthy and diseased tissues, and different growth conditions for certain bacterial species. For example, their use could compare gene expression of *Escherichia coli*

grown in glucose to that being grown in lactose. It could also determine the sequence differences among different bacteria. Currently, microarray slides must be custom-made by a company or research lab and are fairly costly. Some scientists believe that eventually microarrays will be used to diagnose diseases such as cancer or microbial diseases and help direct their respective treatments.

The mRNA is isolated from the microbe and converted to single-stranded complementary DNA (cDNA) via **reverse transcription**. The cDNA is then fluorescently labeled. One set of cDNA is labeled with a green fluorescent marker, whereas the other one is labeled with a red fluorescent marker. The genome of the microbe or specific gene sequence must be known, as that is the sequence that will be placed in spots on the microarray slide. A **probe** is the nucleotide sequence that will bind, or **hybridize**, to its complementary sequence. If the cDNA sequence is complementary to the known sequence, then it will bind. The excess nucleotides that did not hybridize are washed away. The spots where the cDNA hybridized to the known sequences will fluoresce when excited by a laser (**figure 35.1**). The amount of hybridizing will also determine how bright the fluorescence will be. For example, if the entire DNA sequence is complementary, then it will fluoresce brighter than if only a few bases are complementary. Areas will fluoresce red or green, depending on which cDNA is bound to the probe. Areas where both cDNAs hybridized will fluoresce yellow, and areas where neither bound will appear dark. A computer scanner reads the slide and determines if a gene is expressed; it will also quantitate the amount of expression.

Although microarray slides have to be read by an analyzer, the activity today will let you visualize the results. (**Note:** The colors you see today are NOT representative of the reds and greens actually seen on a computer readout!) The slide you will use today will look at six genes. The hybridizing solution then allows for binding of complementary nucleotides.

(1) Known genome (*E. coli*) cut by restriction enzymes and placed on slides

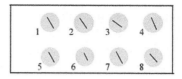

E. coli grown in glucose   *E. coli* grown in lactose

(2) Bacteria to be tested are grown in media

(3) mRNA isolated from bacteria

(4) mRNA converted to cDNA

(5) cDNA labeled with fluorescent marker

Green (G) —————————   Red (R) —————————

(6) Labeled cDNA added to slide to see if any sequences are complementary to known genome

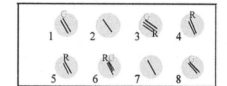

(7) Dyes fluoresce under laser excitation and are read by a computer

(8) **Results:** The glucose sequence hybridized to numbers 1 and 8 and would fluoresce green; the lactose sequence hybridized to numbers 4 and 5 and would fluoresce red; both sequences hybridized to numbers 3 and 6 and would fluoresce yellow; and neither sequence bound to numbers 2 and 7, so they would not fluoresce at all.

Figure 35.1 The concept of a DNA microarray.

Materials

Water bath or hot plate to heat bottles to 70°C

Safety pin (to dislodge clogged cDNA bottle tip)

Gloves (1 pair per group)

DNA microarray kit containing:

DNA microarray slides, 10 (reusable) (each group needs one slide)

cDNA bottles—one each of #1 (*C4BPA*), #2 (*ODC1*), #3 (*SIAT9*), #4 (*FGG*), #5 (*HBG1*), #6 (*CYP24*) (shared between groups)

Hybridization bottles, 2 (not heated, shared between groups)

Alternatively: micropipets (capacity for 30 µl) and tips

PROCEDURE

Escherichia coli was grown in lactose and in glucose broth. Six known genes were added to your microarray slide. You need to determine which sugar induced gene expression.

Each of the six dropper bottles is labeled as 1 (*C4BPA*), 2 (*ODC1*), 3 (*SIAT9*), 4 (*FGG*), 5 (*HBG1*), 6 (*CYP24*) for the six genes you will look at today.

1. You will work in groups. Each group will obtain a microarray glass slide by holding the edges of the slide.

2. Add the solution gel from bottle 1 to slide dot 1 on your slide. It should fill the circle but not overflow. Alternatively, you may pipet 30 µl of gel onto your slide. You need a new pipet tip for each dot.

3. You will repeat with the rest of the solution gels onto the slide dots (bottle 2 to dot 2, bottle 3 to dot 3, and so on). Try to make sure to get an equal amount on each dot area.

4. Wait until the dots have hardened before moving on to the next step, usually about 5 minutes.

5. The persons handling the slide and the hybridizing solution must wear gloves.

6. Add 1 or 2 drops of hybridizing solution onto each dot, being careful to not touch the dropper tip to the gel.

7. The dots will change colors. Record the colors you see on each spot in the Laboratory Report.

8. Interpret your results and what they mean. Be sure to record the shade of the color seen (light or dark pink, etc.).

Pink will indicate the *E. coli* genes expressed when grown in glucose.

Blue will indicate *E. coli* genes expressed when grown in lactose.

Purple will indicate *E. coli* genes expressed in cells grown in both sugars.

Colorless will indicate genes are NOT transcribed by cells grown in either sugar.

9. Use a paper towel to wipe off the six spots into the small trash can provided. Rinse the slide in water and dry it with another paper towel. Dispose of your gloves and paper towels in the proper container.

10. Place your cleaned slide on a paper towel to dry.

EXERCISE

35

Laboratory Report:
Microarray ASM 11, 15, 17, 19, 36

Results

Indicate the colors you saw for each gene and which sugar induced expression. Explain your results.

Color	Is the gene expressed?	If so, by which sugar?	Result/Conclusions
Dot #1	Yes/No		
#2	Yes/No		
#3	Yes/No		
#4	Yes/No		
#5	Yes/No		
#6	Yes/No		

Questions

1. What would you expect to happen if you added the wrong gene to the slide dot?

2. Would microarrays be more beneficial to science than other technologies like PCR? Why or why not?

EXERCISE

36

Enzyme-Linked Immunosorbent
Assay (ELISA) [ASM 34, 36]

Objectives

1. Explain the main steps in performing an ELISA test.
2. Interpret positive and negative ELISA test results.

Pre-lab Questions

1. What is the purpose of the chromogen?
2. What is attached to the microtiter well in an indirect ELISA?
3. What is the color of a positive ELISA test?

Getting Started

The enzyme-linked immunosorbent assay (ELISA) test is widely used for identification of plant, animal, and human pathogens, including viruses. It is used as an initial screening test for diseases such as hepatitis A, B, and C, and HIV infection. In the clinical setting, it is used to identify a variety of microbial pathogens because of test sensitivity and simplicity, often requiring only a swab or blood sample from the infected host. Of great importance to the success of this technique is the plastic microwell plate, which can attract the reactants to its surface and hold on to them. In this exercise, the ELISA technique is used to identify a sexually transmitted disease, HIV.

Various modifications of the ELISA antigen–antibody technique exist. In the direct immunosorbent assay, known as the double antibody sandwich assay (**figure 36.1a**), the microwell is coated with antibodies. In an indirect immunosorbent assay (**figure 36.1b**), the patient's antiserum is added to a microwell previously coated with antigens. If antibodies related to the antigens are present, they attach to the adsorbed antigens. After washing to remove unbound specimen components, an antibody conjugate that has coupled to the enzyme, horseradish peroxidase, is added. If binding occurs between the antigen and the antibody conjugate, a sandwich is formed, containing adsorbed antigens, patient antibodies, and the horseradish peroxidase enzyme. (Peroxidases are enzymes that catalyze the oxidation of organic substrates.) Next, the organic substrate used for this test, urea peroxide, is added. When oxidized by the peroxidase enzyme, free oxygen (O) is released. A color indicator, tetramethylbenzidine, is added, which when oxidized by the free oxygen, produces a yellow color (color plate 79). Lack of color means that the patient's antiserum does not contain antibodies to that pathogen. Once the reaction is complete, the color change is determined spectrophotometrically on a plate reader that calculates cutoff values for wells testing positive. Samples testing positive are repeated, and if they are still positive, a Western blot is performed to confirm the positive result. Specimens that yield an indeterminate result are retested in a clinical laboratory. If the retest result was still indeterminate, a second specimen would be obtained. Absorbances obtained with such reactions may be lower than expected but are still positive.

Materials

An AIDS Simutest Kit that contains the following (per group):

Antigen-coated microtiter plate containing 96 microwells

Serum from patients A through H (can be divided so each group does all samples or only a few (A–D) as time allows)

Conjugate bottle

Chromogen bottle

Pipet(s) (the ones from the kit are calibrated)

Other materials (per group)

Beaker (100 ml or larger)

Distilled or deionized water (200–400 ml), distilled water bottles work well (phosphate buffered saline (PBS) may also be used, depending on the kit purchased).

Timer

Gloves (recommended)

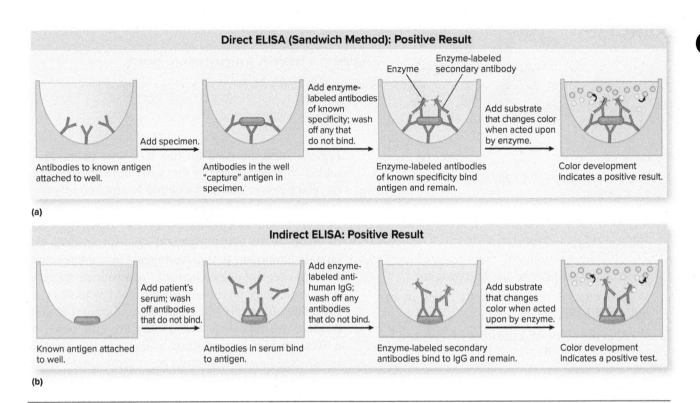

Figure 36.1 The ELISA technique. (*a*) The direct sandwich method, and (*b*) the indirect antibody method.

You will be performing a simulated ELISA test for which you can visually determine the results.

PROCEDURE

Note: This is a **SIMULATED** HIV test, and no infectious materials are present in this kit. Due to some of the chemicals used in this test, gloves are recommended but not required.

1. Each group should obtain a microtiter plate, the serum samples assigned, a conjugate bottle, a chromogen bottle, a pipet, a beaker with distilled water, and extra distilled water.

2. Using the pipet, add 6 drops of serum sample to the first two wells of the row. Sample A should be put in row A.

3. Using the water in the beaker, rinse out the pipet 4 or 5 times.

4. Add 6 drops of serum sample B to the first two wells of row B. Rinse the pipet out.

5. Repeat these steps until you have added each sample to the corresponding row.

6. Pour out the beaker water into the sink and add about 100 ml of distilled water to it.

7. Add 6 drops of distilled water to each well from wells 2–7 for each sample (A, B, etc.). **Leave wells 1 and 8 blank of distilled water.** Depending on the kit used, you may need to block with detergent, and #8 may require bovine serum albumin (BSA).

8. You will now be performing a serial dilution on the sample wells.

9. Mix well 2 of sample A by gently pipetting up and down. Take 6 drops from well 2, and transfer it to well 3. (If you removed too much liquid, add 6 drops to well 3 and then add the rest back to well 2.) Mix well 3 and transfer 6 drops into well 4. Continue this serial dilution until well 8 (6 drops from #4 into #5, 6 drops from #5 into #6, 6 drops from #6 into #7, and 6 drops from #7 into #8, and discard 6 drops from #8).

10. Rinse the pipet out 4 or 5 times in the beaker water.

11. Repeat the serial dilution steps from step 9 for sample B, sample C, and so on.

12. Allow the microtiter plate to incubate 10 minutes at room temperature.

13. Pour out the beaker water into the sink and add about 100 ml of distilled water to it.

14. Add 2 drops of conjugate to each well (1–7) in each row. Allow to incubate at room temperature for 5 minutes. (You do not need to pipet up and down.)

15. Rinse the pipet out.

16. Add 3 drops of chromogen into each well (1–7) in each row. (You do not need to pipet up and down.) A light yellow color/colorless is considered to be negative. An orange/red color is considered to be positive.

17. Record your results in the table in the Laboratory Report.

18. Once you are finished recording your results, take your microtiter plate to the sink and **gently** wash out the wells. You do not want to dislodge the pellet at the bottom of the wells. Dry the bottom of the plate with a paper towel and leave the plates upside down to dry.

19. A negative result indicates that serum antibody to HIV antigens is either absent or below the level of detection of the assay, or the specimen was obtained too early in the response.

20. A positive result implies the presence of antibody to HIV. An early acute-phase patient may only present an IgM response, whereas the chronic or convalescent patient may only present an IgG response. Some test kits will only detect IgG or IgM, whereas others will test for both.

EXERCISE

36

Laboratory Report: Enzyme-Linked
Immunosorbent Assay (ELISA) [ASM 34, 36]

Results and Questions

1. Discuss the test results and their significance.

	Well Color	Result	What Does the Result Mean?
Sample A			
Sample B			
Sample C			
Sample D			
Sample E			
Sample F			
Sample G			
Sample H			

2. What is the importance of rinsing the wells when conducting a real (not simulated) ELISA test?

3. What is the purpose of the conjugate?

4. Discuss the pros and cons of using the indirect ELISA and the double antibody sandwich ELISA. You may need to consult your text for answering this question.

37

Differential White Blood Cell Stains

Definitions

Amoeboid. To make movements or changes in shape by means of protoplasmic flow.

Basophil. A granulocyte in which the cytoplasmic granules stain dark purplish blue with methylene blue, a blue basophilic-type dye found in Wright's stain.

Eosinophil. A granulocyte in which the cytoplasmic granules stain red with eosin, a red acidophilic-type dye found in Wright's stain.

Lymphocyte. A colorless agranulocyte produced in lymphoid tissue. It has a single nucleus and very little cytoplasm. It often looks like a dark purple round dot.

Megakaryocyte. A large cell with a lobulated nucleus that is found in bone marrow. It is the cell from which platelets originate.

Monocyte. A large agranulocyte normally found in the lymph nodes, spleen, bone marrow, and loose connective tissue. It is phagocytic with sluggish movements.

Neutrophil. A mature granulocyte present in peripheral circulation. The cytoplasmic granules stain poorly or not at all with Wright's stain. The nuclei of most neutrophils are large, contain several lobes, and are described as polymorphonuclear (PMN) leukocytes.

Platelet. A small oval-to-round, pink fragment, 3 μm in diameter. Plays a role in blood clotting.

Objectives

1. Describe the blood's microscopic cell types, their origin, morphology, number, and role in fighting disease.

2. Interpret stained blood slides: Observe the cellular appearance of normal blood and perform a differential WBC count.

Pre-lab Questions

1. Name the three types of granulated white blood cells.

2. What cell is important for blood clotting?

3. Which blood cell is increased in a bacterial infection?

Getting Started

This exercise examines the cellular forms of the immune system, specifically the white blood cells (WBCs). For the most part, they can be distinguished from one another using a blood smear stained with Wright's stain, a differential stain. This stain uses a combination of an acid stain, such as eosin, and a basic stain, such as methylene blue. They are contained in an alcoholic solvent (methyl alcohol), which fixes the stains to the cell constituents, particularly because the basophilic granules are known to be water soluble. With this stain, a blood smear shows a range in color from the bright red of acid material to the deep blue of basic cell material. In between are neutral materials that exhibit a lilac color. There are also other color combinations, depending upon the pH of various cell constituents.

The two main groups of WBCs are the granulocytes (cytoplasm contains granules) and the agranulocytes (clear cytoplasm). The WBCs, or leukocytes, represent approximately 1/800 of the total blood cells. Common cell types found in the granulocytes are **neutrophils**, **eosinophils**, and **basophils**. The granulocytes are highly phagocytic and contain a complex, segmented nucleus. The granules in neutrophils stain light pink, red in eosinophils, and purple in basophils. Eosinophils often have a nucleus with two lobes, whereas the other granulocytes have four or more lobes. The agranulocytes are relatively inactive and have a simple nucleus, or kidney-shaped nucleus. The basic agranulocyte cell types

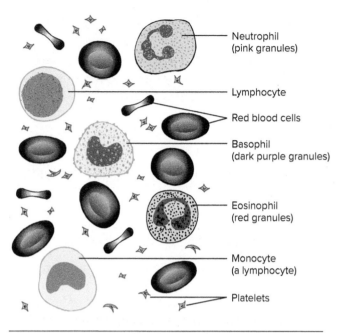

Neutrophil (pink granules)

Lymphocyte

Red blood cells

Basophil (dark purple granules)

Eosinophil (red granules)

Monocyte (a lymphocyte)

Platelets

Figure 37.1 Blood cell types present in human peripheral blood. The granular leukocyte names find their origin from the color reaction produced by the granules after staining with acidic and basic components of the staining solution. Neutrophil = neutral-colored granules; basophil = basic color; and eosinophil = acid color.

are the **lymphocytes** and **monocytes**. Lymphocytes are smaller cells with little cytoplasm, and monocytes are large cells with more cytoplasm that often contains cytoplasmic vacuoles. Another WBC type found is the **platelet** (very small, multinucleate, irregular pinched off parts of a **megakaryocyte**). Platelets are responsible for clotting the blood to prevent excessive bleeding. The appearance of these cell types in blood stained with a differential stain is illustrated in **figure 37.1**. Many of the leukocytes are **amoeboid**, capable of moving independently through the bloodstream. They also move out into the tissues, where they repel infection and remove damaged body cells from bruised or torn tissue. They are defensive cells, specialized in defending the body against infection by microorganisms and other foreign invaders.

Differential blood stains are important in disease diagnosis because certain WBCs either increase or decrease in number depending on the disease. In making such judgments, it is important to know the appearance of normal blood (color plate 80). The

microscopic field shown in the color plate includes mostly RBCs with a neutrophil, an eosinophil, a lymphocyte, and some platelets. Neutrophils are often increased in bacterial infections, eosinophils are often increased in parasitic infections (and allergies), and lymphocytes are often increased in viral infections. A Giemsa stain can be used to see *Plasmodium* species in RBCs, but they cannot be seen with only the Wright's stain. In a differential performed on human blood, you should see approximately 40–75 neutrophils, 13–43 lymphocytes, 2–12 monocytes, 1–4 eosinophils, and 1 basophil (**table 37.1**).

The red blood cells (RBCs), also called erythrocytes, make up the largest cell population. RBCs constitute an offensive weapon because they transport oxygen to various body parts and they break down carbon dioxide to a less toxic form. The RBCs in humans and all other mammals (except members of the family Camelidae, such as the camel) are biconcave, circular discs without nuclei.

RBCs are produced in the red bone marrow of certain bones. As they develop, they produce massive quantities of hemoglobin, the oxygen-transporting pigment that contains iron, which in the oxygenated form gives blood its red color. Worn-out RBCs are broken down at the rate of 2 million cells per second in the liver and spleen by phagocytic WBCs. Some of the components of the RBCs are then recycled in order for the body to maintain a constant number of these cells in the blood.

In this exercise, you will have an opportunity to stain differentially and to observe blood cell characteristics on slides.

Materials

Either animal blood or prepared commercial unstained or stained human peripheral blood smears. For student use, blood should be dispensed with either a plastic dropper or a dropping bottle capable of dispensing a small drop.

Note: In the event of spilled blood, use disposable gloves and towels to remove blood. Then disinfect the area with a 10% hypochlorite bleach dilution.

For use with whole blood

New microscope slides, 3

Plastic droppers for dispensing blood on slides, dispensing Wright's stain, and adding phosphate buffer

Hazardous waste container for droppers and slides

For use with whole blood and unstained prepared slides

Wright's stain (or Wright–Giemsa), dropping bottle, 1 per two students

Phosphate buffer, pH 6.8, dropping bottle, 1 per two students

Wash bottle containing distilled water, 1 per two students

A Coplin jar with 95% ethanol

Staining rack

Colored pencils for drawings: pink, blue, purple, or lavender

PROCEDURE

We recommend using blood from a cow, horse, pig, or dog or commercially prepared slides to minimize disease risk.

Safety Precautions: Working with blood carries risks from unknown contaminants. Disposable gloves and goggles must be worn when preparing slides.

1. Prepare three clean microscope slides free of oil and dust particles as follows:

   a. Wash slides with a detergent solution and then rinse thoroughly.

   b. Immerse slides in a jar of 95% alcohol.

   c. Air dry and polish with lens paper.

2. Place a drop of blood on one end of a clean slide (**figure 37.2a**). Repeat with a second clean slide.

3. Spread the drop of blood on the slide as follows:

   a. Place the slide on your laboratory benchtop. With your thumb and middle finger, firmly hold the sides of the slide on the end where the drop of blood is located.

   b. With your other hand, place the narrow edge of a clean slide approximately ½" in *front* of the drop at an angle of about 30° (**figure 37.2b**).

   c. Carefully push the spreader slide *back* until it comes in contact with the drop, at which point the drop will spread outward to both edges of the slide (**figure 37.2c**).

   d. Immediately, and with a firm, steady movement, push the blood slowly toward the opposite end of the slide (**figure 37.2d**).

Note: Use of these procedural restraints (a small drop; a small spreader slide angle; and a *slow*, steady spreader slide movement) should provide a thin film for study of RBCs. A good smear has the following characteristics: smooth, without serrations; even edges; and uniform distribution over the middle two-thirds of the slide.

   e. Allow slide to air dry for 5 minutes. Do not blot.

   f. For the second slide, prepare a thicker film by using a larger spreader slide angle (45°) and by pushing the blood more rapidly to the opposite end of the slide. The second slide is best for determining the differential WBC count.

Note: The unused end of the first spreader slide can be used to prepare the second slide. Discard used spreader slides in the hazardous waste container.

4. Stain the blood smears with Wright's stain as follows:

   a. Suspend the slides so that they lie *flat* on the staining rack supports.

   b. Flood or add 15 drops of Wright's stain to each blood smear. Let it stand for 3–4 minutes. This fixes the blood film.

   c. Without removing the stain, add an equal volume of phosphate buffer. Blow gently through a pipet on each side of the slide to help ensure mixing of stain and buffer solutions.

   d. Let stand until a green, metallic scum forms on the surface of the slide (usually within 2–4 minutes).

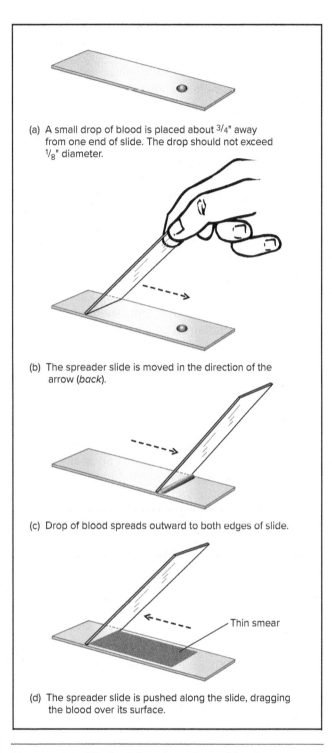

(a) A small drop of blood is placed about 3/4" away from one end of slide. The drop should not exceed 1/8" diameter.

(b) The spreader slide is moved in the direction of the arrow (*back*).

(c) Drop of blood spreads outward to both edges of slide.

Thin smear

(d) The spreader slide is pushed along the slide, dragging the blood over its surface.

Figure 37.2 Method for preparing a blood smear.

e. Wash off the stain with water. Begin washing while the stain is on the slide in order to prevent precipitation of scum-like precipitate, which cannot be removed. The initial purple appearance on the slide should be washed until it is lavender–pink.

f. Wipe off excess stain from the back of the slide and allow it to stand on end to dry (which is preferable to drying between folds of sheets of bibulous paper).

5. Examine stained blood smears:

a. Make an initial examination of the first blood smear with the low-power objective lens to find the most suitable areas for viewing with the oil immersion objective lens.

b. Next, using the oil immersion lens, make a study of the various WBC types present: basophils, eosinophils, lymphocytes, monocytes, neutrophils, and platelets. For help in this study, consult color plate 54, the Definitions section describing their staining characteristics with Wright's stain, and figure 37.1. Prepare color drawings of your findings in the Laboratory Report.

c. Conduct a differential WBC count using the second blood smear. For normal blood with a leukocyte count of 5,000–10,000 WBCs/ml, you would classify 100 leukocytes. In order to do this, you may have to examine the number and kinds of WBCs present in as many as 20 microscopic fields. Record your findings in **table 37.2** of the Laboratory Report and calculate the percentage of each WBC type.

Table 37.1 Cellular Description of Normal Blood*

Total Counts	Differential WBC Counts
RBC $4.0–5.5 \times 10^6$/L	Neutrophils 40–75%
WBC $4.5–9.0 \times 10^9$/L	Lymphocytes 25–35%
Platelets $150–450 \times 10^9$/L	Monocytes 2–10%
Basophils 0–1%	Eosinophils 0–5%

*McKenzie and Williams, *Clinical Laboratory Hematology*. Second Edition. 2010. Pearson, Boston, MA.

EXERCISE

37

Laboratory Report: Differential White Blood Cell Stains

Results

1. Color drawings of RBCs and various WBCs found in blood smears stained with Wright's stain.

RBCs   Neutrophils   Eosinophils

Basophils   Monocytes   Lymphocytes

2. Differential WBC count. In table 37.2, record the kinds of leukocytes found as you examine each microscopic field. After counting 100 WBCs, calculate their percentages from the total found for each type. Also record the number of microscopic fields examined to find 100 WBCs: _____.

Table 37.2   Kinds and Percentages of WBCs Found in Blood Smear

Neutrophils	Eosinophils	Basophils	Lymphocytes	Monocytes
Total				
Percentage				

1. What problems, if any, did you find in preparing and staining your blood smears? Indicate any differences noted between thin and thick smears.

2. Were your blood stains satisfactory? Did the stained cells resemble those in figure 37.1 and color plate 73? Were they better?

3. Compared with the normal blood percentages in **table 37.1**, how did your white blood cell count compare? Were they normal? If not, give an explanation.

4. Were there any WBC types that you did not find in your blood smear? If so, which one(s)? Why did you not find them?

5. Matching (you may wish to consult your text).
    a. Neutrophils   _____ Involved in antibody production
    b. Basophils   _____ A phagocytic cell
    c. Monocytes   _____ Increased number in parasitic infections
    d. Eosinophils   _____ Turn into macrophages
    e. Lymphocytes   _____ Release histamine

Name_____ Section_____ Points_____

Slide #_____ Circle one: G+ or G−

Directions:

You will wash your hands and put on aprons and goggles at the beginning of lab. Materials will be set out for you at your stations, not including your pens/pencils and goggles. Keep your eyes on your own area and you may not talk.

You will perform a Gram stain on the slide provided. Put your goggles on and begin staining. You may need to share staining reagents. You will need to get out your microscope and determine if the microbe is Gram positive or Gram negative. You may need to tap the slide quite a bit or blot your slide to get the slide to dry in a timely fashion.

Properly label your slide, and put a plus or a negative sign on it. Once scopes have been viewed for being in focus, you can then take your slide and paper to the front of the room. Turn the paper upside down and put the slide in the next empty slot so the paper and slide are in the same order to allow for grading efficiency. Put your microscope away. You will have _____ minutes to complete the activity, not including clean-up, which we will do at the end of the activity. Grades will be based on proper overall result, distribution of staining, your interpretation of the result you submitted, whether the microscope was focused, etc. You will tear this sheet out and hand it in with your slide for grading.

_____Microscope was not in focus, or was not in focus on the correct objective

_____Oil was on the 40×

_____Slide staining was mixed (depends on how much was mixed)

_____Extra dye was not rinsed off the slide (depends on how much dye was left on the slide)

_____Slide should have been stained Gram positive

_____Slide should have been stained Gram negative

_____Slide was not interpreted correctly

_____Slide was not properly labeled (depends on what information is missing)

If there is a discrepancy in interpretation between the slide and sheet, the Gram reaction you provided on the slide will be used.

Name_____ **Section**_____ **Points**_____

Slide #_____

You will need your microscope. You may NOT use your lab manual. You may NOT talk to anyone and keep your eyes on your own paper. You will have _____minutes to complete the quiz. Take your paper and slide up to the front when finished.

Was the slide in focus on the proper objective? +Yes −No

The purpose of this quiz is to assess your interpretation skills of Gram stains. You have a stained slide with a number on it. Always give a genus and species. Record the number of the slide above. Answer the following questions. Some questions may have multiple answers; you only need to provide one answer.

1. The microbe is considered to be (circle one): Gram Positive   Gram Negative

2. The proper term for this microbe's shape is:_____

3. The size of this bacterium is _____ μm.

4. Does this microbe have an arrangement? (circle one): Yes   No
   If yes, properly name the arrangement seen. _____

5. Properly write the scientific name of a bacterium from lab that this stain could represent.
_____

Your instructor may have you interpret an acid fast or endospore stain instead.

1. The microbe is considered to be (circle one): Endospore Positive   Endospore Negative

2. The microbe is considered to be (circle one): Acid-fast Positive   Acid-fast Negative

3. The proper term for this microbe's shape is:_____

4. The size of this bacterium is _____ μm.

5. Properly write the scientific name of a bacterium from lab that this stain could represent.

Name_____ Section_____ Points_____

Tube #_____

You will need to hand in your tube for making this streak, this sheet with your name, tube #, and lab # filled in, and your plate. You may not ask about your techniques today if the possibility of it being quizzed exists as it would be considered dishonest and you would not get any points for the quiz.

You will be graded on the quality of the isolation streak of your organism you do today.

_____ Quality of isolation overall

_____ Are there isolated colonies?

   _____ many isolated colonies to choose from

   _____ a few (<10) isolated colonies to choose from

   _____ 5

   _____ 1 or 2

   _____ It appears the loop was not flamed between streak areas

_____ Quality of the streak? (appropriate amount of the plate used?)

_____ Too little of the plate was used

_____ Too much of the plate was used

_____ Proper streak performed?

_____ Was the proper media used?

_____ Was your streak contaminated?

_____ Is the plate properly labeled?

_____ Was the plate incubated properly (at the proper temperature, upside down, etc.)?

_____ If you took an additional plate, it is considered dishonest and you will not obtain any points for the quiz.

_____ If there is no growth, zero points can be given for the streak.

Name_____ Section_____ Points_____

You may be quizzed over any of the following activities during lab. Thus, you want to make sure the items handed in for incubation are correct as they may be subject to grading. You need to hand this sheet in by the end of the lab with your name and lab # filled in. You may not ask about your techniques today, if the possibility of it being quizzed exists as it would be considered dishonest and you would not get any points for the quiz. If extra tubes, etc. are inoculated, incubated, or autoclaved, it is considered dishonest and a zero will be recorded for this quiz.

1. You may be graded on the quality of the isolation streak of your unknown you do today.

_____Quality of isolation

_____Is there isolation? (at least 5 usable colonies)

_____Is there contamination? (If an Unknown we cannot tell you.)

_____Quality of the streak

_____Appropriate amount of the plate used?

_____Use more of the plate?_____

_____Use less of the plate?_____

_____Take less from the previous section?___

_____Proper streak performed

_____Proper agar plate used

_____Was the plate properly incubated? (temperature, upside down, proper incubator, etc.)

_____Is the plate labeled appropriately? (Make sure we know WHO performed the streak.)

_____If you took an additional plate, it is considered dishonest and you will not obtain any points for the quiz. If there is no growth it is a zero. If you did not do it, it is also a zero.

2. You may be graded on the proper labeling of any tube or plate.

_____Proper labeling of the tube/plate?

_____Organism name?

_____Unknown number if Unknown?

_____Date?

_____Lab #?

_____Temperature?

_____Media? Proper media?

_____Readability?

_____Person's name who performed the inoculation?

_____Proper organism?

_____Was the tube or plate properly incubated? (temperature, upside down, proper incubator, etc.)

3. You may be graded on the proper inoculation of any tube or plate.

_____Proper inoculation on or in the tube or plate?

_____Quality of growth on the tube or plate?

_____Proper incubation temperature?

_____Proper rack or location?

_____Was the proper item/organism inoculated?

_____Contamination? (if an Unknown we do not comment.)

_____Was the tube or plate properly incubated? (temperature, upside down, proper incubator, etc.)

_____Properly labeled?

APPENDIX

Culture Media and Reagent Formulas

Dissolve media ingredients and then autoclave media at 121°C for 15 minutes at 15 psi unless otherwise noted.

Blood Agar

Per liter: Blood agar base (Difco)	40.0 g
Distilled water or deionized water	1,000 ml

Dissolve and sterilize. Cool to 45–50°C. Add 5% sterile, defibrinated sheep blood.

Bile Esculin

Per liter: Esculin	1.0 g
Ferric citrate	0.5 g
Meat extract	3.0 g
Meat peptone	5.0 g
Oxbile (bile salts)	40.0 g
Agar	15.0 g

pH 6.6 ± 0.2 at 25°C

Add components to distilled/deionized water and bring volume to 1 L. Mix thoroughly. Heat, boiling for 1 minute, then autoclave.

Brain Heart Infusion

Per liter: Beef heart infusion	250.0 g
Calf brain infusion	200.0 g
Dextrose	2.0 g
Disodium phosphate	2.5 g
NaCl	5.0 g
Proteose peptone	10.0 g

pH 7.4 ± 0.2 at 25°C

Add components to distilled/deionized water and bring volume to 1 L. Mix thoroughly. Heat, boiling for 1 minute, then autoclave.

Brewer's Anaerobic Agar

Per liter: Dextrose	10.0 g
NaCl	5.0 g
Pancreatic digest of casein	5.0 g
Proteose peptone #3	10.0 g
Resazurin	2.0 mg
Sodium formaldehyde sulfoxylate	1.0 g
Sodium thioglycollate	2.0 g
Agar	20.0 g

pH 7.4 ± 0.2 at 25°C

Add components to distilled/deionized water and bring volume to 1 L. Mix thoroughly. Heat, boiling for 1 minute, then autoclave.

Brilliant Green Lactose Bile Broth

Per liter: Ox gall, dehydrated	20.0 g
Lactose	10.0 g
Pancreatic digest of gelatin	10.0 g
Brilliant green	0.013 g

pH 7.2 ± 0.2 at 25°C

Add components to distilled/deionized water and bring volume to 1 L. Mix thoroughly. Distribute into tubes containing inverted Durham tubes, in 10.0 ml amounts for testing 1.0 ml or less of sample. Autoclave for 12 minutes. After sterilization, cool the broth rapidly. Medium is sensitive to light.

Casein/Skim Milk Agar

Per liter: Dextrose	1.0 g
Pancreatic digest of casein	5.0 g
Skim (non-fat) milk powder	50.0 g
Yeast extract	2.5 g
Agar	15.0 g

pH 7.0 ± 0.2 at 25°C

Add components to distilled/deionized water and bring volume to 1 L. Mix thoroughly. Heat, boiling for 1 minute, then autoclave. Sometimes autoclaving for 10 minutes is recommended to prevent burning of the milk.

Corn Meal Agar

Per liter: Corn meal	60.0 g
Agar	15.0 g
Tween 80*	10.0 g

Mix the corn meal to a smooth cream with water, simmer for 1 hour, filter through cheesecloth or coarse filter paper, add the agar and Tween 80 and heat until dissolved. Make up to volume, and sterilize, preferably for 30 minutes at 10 psi.

*Tween 80 addition greatly stimulates chlamydospore formation.

DNase Agar

Per liter: Deoxyribonucleic acid	2.0 g
Methyl green	0.05 g
Tryptose 20.0 g	20.0 g
NaCl	5.0 g
Agar	15.0 g

Final pH 7.3 ± 0.2

Add components to distilled/deionized water and bring volume to 1 L. Mix thoroughly. Heat, boiling for 1 minute, then autoclave. When reading results, a drop of 2N HCl may need to be added to the colony.

Endo Broth/Agar

Per liter: Peptone	10.0 g
Lactose	100.0 g
KH_2PO_4	3.5 g
Sodium sulfate	2.5 g
Agar (leave out for broth)	10.0 g

pH 7.5 (approx.)

Add components to distilled water. Add 4 ml of a 10% (weight/volume) alcoholic solution of basic fuchsin (95% ethyl alcohol). Bring to a boil to dissolve completely. Mix well before pouring.

Eosin-Methylene-Blue (EMB) Agar (Levine)

Per liter: Peptone	10.0 g
Lactose	10.0 g
K_2HPO_4	2.0 g
Eosin Y	0.4 g
Methylene blue	0.065 g
Agar	15.0 g

Add components to distilled water. Boil to dissolve completely. Sterilize at 121°C for 15 minutes. Shake medium at 60°C to oxidize the methylene blue (i.e., restore its blue color) and suspend the precipitate, which is an important part of the medium.

Gelatin Agar

Per liter: Beef extract	3.0 g
Peptone	5.0 g
Gelatin	120.0 g
Final pH	6.8 g

Add components to distilled/deionized water and bring volume to 1 L. Heat to 50°C to dissolve completely. Distribute into tubes and autoclave.

Glucose–Acetate Yeast Sporulation Agar

Per liter: Glucose	1.0 g
Yeast extract	2.0 g
Sodium acetate (with 3 H_2O)	5.0 g
Agar	15.0 g

Adjust pH to 5.5

Halobacterium Medium (ATCC Medium 213)

Per liter: NaCl	250 g
$MgSO_4 \cdot 7H_2O$	10 g
KCl	5 g
$CaCl_2 \cdot 6H_2O$	0.2 g
Yeast extract	10 g
Tryptone	2.5 g
Agar	20 g

The quantities given are for 1 L final volume. Make up two solutions, one containing yeast extract, tryptone, and agar, and the other the salts. Adjust pH of the agar solution to 7.0. Sterilize the solutions separately at 121°C for 15 minutes, then add the solutions together. Mix and dispense aseptically approximately 6 ml per 13 mm sterile tube.

Lauryl Sulfate Broth (Double Strength)
Halve to Make Single Strength

Per liter: Tryptose	40.0 g
Lactose	10.0 g
NaCl	10.0 g
K_2HPO_4	5.5 g
KH_2PO_4	5.5 g
Sodium lauryl sulfate	0.2 g

pH 6.8 (approx.)
Dispense 10 ml quantities in large test tubes containing inverted Durham tubes.

MacConkey Agar

Per liter: Peptone	17.0 g
Proteose peptone	3.0 g
Bacto bile salts no. 3	10.0 g
NaCl	5.0 g
Agar	13.5 g
Neutral red	0.03 g
Crystal violet	0.001 g

pH 7.1

Lysine Decarboxylase (LDC)/Ornithine Decarboxylase (ODC)

Per liter: Peptone	5.0 g
Yeast extract	3.0 g
Dextrose	1.0 g

L-Lysine hydrochloride 5.0 g (*substitute ornithine to make ODC*)

Bromocresol purple	0.2 g

pH 6.8 (approx.)
Final pH 6.8 ± 0.2
Add components to distilled/deionized water and bring volume to 1 L. Mix thoroughly. Heat, boiling for 1 minute to dissolve components. Distribute into tubes and autoclave. After inoculation, add a thin sterile mineral oil overlay to the tubes before incubation.

Mannitol Salt Agar for *Staphylococcus* Isolation

Per liter: Phenol red	0.025 g
D-mannitol	10.0 g
Beef extract	1.0 g
Meat peptone	5 g
Casein peptone	5 g
Agar	15 g
NaCl	75.0 g

Add components to distilled/deionized water and bring volume up to 1 L. Mix thoroughly. Heat, boiling for 1 minute to dissolve components, then autoclave.

McFarland Standard 3

1% BaCl	0.3 ml
1% H_2SO_4	9.70 ml

Mix the two ingredients into a sterile tube.

Methyl Red Voges–Proskauer Medium

Per liter: Buffered peptone	7.0 g
Dipotassium phosphate	5.0 g
Glucose	5.0 g

Final pH 6.9 ± 0.2
Distribute 5 ml/tube.
Two tubes are needed for the test, one for Voges–Proskauer and the other for methyl red.

Mineral Salts + 0.5% Glucose (or Other Sugar)

Per liter:	*1×*	*10×*
$(NH_4)_2SO_4$	1.0 g	10.0 g
K_2HPO_4	7.0 g	70.0 g
KH_2PO_4	3.0 g	30.0 g
$MgSO_4 \cdot 7 H_2O$	0.1 g	1.0 g

pH 7.0
Sterilize 20 minutes at 121°C.

Glucose agar base:	
Agar	6.0 g
Glucose	2.0 g
Distilled water	360 ml

Sterilize for 15 minutes at 118°C.

Add 40 ml of the 10× salts to 360 ml glucose agar before pouring into plates.

Motility Medium

Per liter: Dehydrated nutrient broth	8.0 g
Agar	3.0 g

Mix the dehydrated nutrient broth and agar in the water. Heat until the agar has dissolved and adjust to pH 6.8.
Dissolve 25 mg TTC in 5 ml of distilled water. Make up the dye solution just before it is added to the medium.

Triphenyl tetrazolium chloride (TTC)	5 ml of 0.5% aqueous solution

Add the diluted dye. Dispense 10 ml/tube.

Mueller–Hinton Agar

Per liter: Beef, infusion from	300.0 g
Casamino acids, tech	17.5 g
Starch	1.5 g
Agar	17.0 g

When using a commercial base:

Mueller–Hinton broth (dehydrated)	21.0 g
Agar	17.0 g

MUG (4-methylumbelliferyl-β-D-glucuronide)

Per liter: Casein peptone	20.0 g
Meat extract	2.0 g
NaCl	5.0 g
Yeast extract	1.0 g
Sorbitol	10.0 g
Ammonium ferric citrate	0.5 g
Sodium thiosulfate	2.0 g
Bromothymol blue	0.025 g
Deoxycholic acid sodium salt	1.12 g
Agar	13.0 g
4-methylumbelliferyl-β-D-glucuronide	0.1 g

Add components to distilled/deionized water and bring volume to 1 L. Mix thoroughly. Heat, boiling for 1 minute, then autoclave.

Nitrate Broth

Per liter: Beef extract	3.0 g
Peptone	5.0 g
Potassium nitrate	1.0 g
Final pH 7.0	

Add components to distilled/deionized water and bring volume to 1 L. Mix thoroughly. Distribute into tubes and autoclave.

Nutrient Broth/Agar

Per liter: Beef extract	3.0 g
Peptone	5.0 g
Agar (leave out for broth)	15.0 g
pH 6.9 (approx.)	

Nutrient + Starch Agar

Nutrient broth (See ingredients above)	1,000 ml
Agar	15.0 g
Soluble starch	2.0 g

Prepare nutrient broth. Add agar and heat to melt agar. Suspend the starch in a small quantity of cold water and add to the nutrient agar while it is still hot. pH 7.0 before and after sterilizing.

Nutrient + Starch + Glucose Agar

Add 20 g (2%) glucose to the above medium.

1% Peptone Agar Deeps (10 ml/screw cap tube)

Per liter: Peptone	10.0 g
Agar	10.0 g

Phenylalanine Deaminase

Per liter: Disodium phosphate	1.0 g
NaCl	5.0 g
Phenylalanine (D-L)	2.0 g
Yeast extract	3.0 g
Agar	12.0 g

pH 7.3 ± 0.2 at 25°C

Add components to distilled/deionized water and bring volume to 1 L. Mix thoroughly. Heat, boiling for 1 minute, then autoclave.

Sabouraud's Dextrose Broth/Agar

Per liter: Dextrose	40.0 g
Peptone	10.0 g
Agar (leave out for broth)	15.0 g
Final pH 5.6 (approx.)	

Saline (Per liter)

NaCl	8.5 g
Distilled water	1,000 ml

Simmons Citrate Agar

Per liter: Magnesium sulfate	0.2 g
Monoammonium phosphate	1.0 g
Dipotassium phosphate	1.0 g
Sodium citrate	2.0 g
NaCl	5.0 g
Agar	15.0 g
Bromthymol blue	0.08 g
pH 6.8	

Distribute in tubes or flasks and autoclave. Cool in a slanting position.

Spirit Blue

Per liter: (Base mix): Pancreatic digest of casein	10.0 g
Spirit blue dye	0.150 g
Yeast extract	5.0 g
Agar	17.0 g
pH 6.8 ± 0.2 at 25°C	

Lipase Substrate: (per 400 ml of water)

Tween 80	1.0 ml
Olive oil	100 ml

Add base mix components to distilled/deionized water and bring volume to 1 L. Mix thoroughly. Heat, boiling for 1 minute, then autoclave.

When components are cooled, add lipase substrate to autoclaved base.

Sugar Fermentation Tubes (Andrade's)

Per liter: Nutrient broth (dehydrated)	8.0 g
Glucose or other sugar	5.0 g
Andrade's indicator*	10 ml

pH 7.1 before sterilizing, pH 7.0 after sterilizing

Add 6 ml/standard tube. Add Durham tubes inverted before sterilizing—they will fill in the autoclave. Sterilize for 12–13 minutes at 118°C.

Andrade's indicator:

Acid fuchsin	1.0 g
Distilled water	750 ml
1 M NaOH	7 ml

To dissolve and neutralize the acid fuchsin solution, add 7 ml 1M NaOH. Allow to stand for several hours. If the solution is still red, add 3–4 ml more of 1M NaOH. Let stand for several hours. If solution is still red, add 1–2 ml more. Continue until solution is light red. Let stand for several hours and adjust pH to 8.0—it will still appear a light red. (When added to medium, the broth will be a very pale yellow.) This can be stored for long periods at room temperature.

Sugar Fermentation Tubes (Phenol Red)

Per liter: Beef extract	1.0 g
Proteose peptone no. 3	10.0 g
NaCl	5.0 g
Phenol red	0.018 g
Final pH 7.4	

Dissolve 16 g phenol red broth base in 100 ml of distilled water. Add 5 g of carbohydrate. Add 6 ml/standard tube. Add inverted Durham tubes. Test the first batch with known fermenters. Occasionally the carbohydrate hydrolyzes. If this occurs, lower temperatures and shorter sterilization time are recommended.

Thioglycollate Medium, Fluid

Per liter: Yeast extract	5.0 g
Casitone	15.0 g

Dextrose	5.0 g
NaCl	2.5 g
L-Cystine	0.75 g
Sodium thioglycollate	0.3 ml
Agar	0.75 g
Resazurin, certified	0.001 g

Final pH 7.1

Suspend 29.5 g of the thioglycollate medium in 1 L cold distilled water and heat to boiling to dissolve completely. Distribute into tubes and sterilize for 15 minutes at 121°C.

Triple Sugar Iron

Per liter: Beef extract	3.0 g
Dextrose	1.0 g
Ferrous sulfate	0.2 g
Lactose	10.0 g
NaCl	5.0 g
Peptone	15.0 g
Phenol red	0.024 g
Proteose peptone	5.0 g
Saccharose	10.0 g
Sodium thiosulfate	0.3 g
Yeast extract	3.0 g
Agar	12.0 g

pH 7.3 ± 0.2 at 25°C

Add components to distilled/deionized water and bring volume to 1 L. Mix thoroughly. Heat, boiling for 1 minute, then autoclave.

Tryptone Broth

Per liter: Tryptone	10.0 g
NaCl	5.0 g

pH 7.0 before sterilizing

Tryptone Agar for Base Plates

Per liter: Tryptone	10.0 g
NaCl	5.0 g
MgSO$_4$ • 7H$_2$O	1.0 g
Agar	15.0 g

pH 7.0 before sterilizing

Tryptone Soft Agar Overlay

Prepare formula for base plates as above, but use 7 g of agar.

Trypticase (Tryptic) Soy Broth (TSB)/Agar

Per liter: Trypticase peptone	17.0 g
Phytone peptone	3.0 g
NaCl	5.0 g
K$_2$HPO$_4$	2.5 g
Glucose	2.5 g
Agar (leave out for broth)	15.0 g

pH 7.3 (approx.)

For TS + streptomycin plates: Add 1.0 ml of stock streptomycin per liter of sterile, melted, cooled agar (stock streptomycin: 1.0 g streptomycin/10 ml distilled water, filter sterilized).

Trypticase Soy Yeast (TSY) Extract Agar

Per liter: Trypticase soy broth	30.0 g
Yeast extract	10.0 g
Agar	15.0 g

pH 7.2 before sterilizing

TSY + Glucose Agar

Add 2.5 g glucose/l.

TSY + Glucose + Bromocresol Purple Agar Slants

Per liter: Trypticase soy broth	30.0 g
Yeast extract	10.0 g
Glucose	7.5 g
Bromocresol purple	0.04 g
Agar	15.0 g

pH 7.3 before sterilizing, pH 7.0 after sterilizing.
Sterilize at 118°C for 15 minutes, 6 ml/tube.

TYEG Salts Agar Plates Containing 0.5%, 5%, 10%, and 20% NaCl

For basal medium composition see TYEG salts agar. The two lower salt concentrations (0.5% and 5%) can be added directly to the basal medium, but the two higher salt concentrations (10% and 20%) need to be autoclaved separately and added back to the basal medium after autoclaving.

TYEG Salts Agar Plates Containing 0%, 10%, 25%, and 50% Sucrose

For basal medium composition see TYEG salts agar. The three 10% sucrose concentrations can be added directly to the basal medium, whereas the remaining two sucrose concentrations (25% and 50%) need to be autoclaved separately and added back to the basal medium after autoclaving.

TYEG (Tryptone Yeast Extract Glucose) Salts Agar Slants

Per liter: Tryptone	10 g
Yeast extract	5 g
Glucose	2 g
K$_2$HPO$_4$	3 g
CaCl$_2$ • 6H$_2$O	0.2 g
Agar	20 g

pH 7.0 (before autoclaving). Dispense 6 ml/tube.

Urea Broth/Agar

Urea broth/agar base	29.0 g
Ingredients in this base:	
Peptone	1.0 g
Dextrose	1.0 g
NaCl	5.0 g
Monopotassium phosphate	2.0 g
Urea	20.0 g
Phenol red	0.012 g
Agar (leave out for broth)	15.0 g

Final pH 6.8

Suspend 29 g of urea agar base in 100 ml of distilled water. Filter sterilize this concentrated base. Dissolve 15 g of agar in 900 ml distilled water by boiling, and sterilize. Cool to 50–55°C and add 100 ml of the filter sterilized urea agar base under aseptic conditions. Mix and distribute in sterile tubes. Slant tubes so they will have a butt of 1′ and slant 1½′ length, and solidify in this position.

Worfel–Ferguson Slants (for *Klebsiella* urease detection)

Per liter: NaCl	2.0 g
K$_2$SO$_4$	1.0 g
MgSO$_4$ • 7H$_2$O	0.25 g
Sucrose	20.0 g
Yeast extract	2.0 g
Agar	15.0 g

No pH adjustment.

Xylose Lysine Deoxycholate Agar

Per liter: Ferric ammonium citrate	0.8 g
Lactose	7.5 g
L-Lysine	5.0 g
NaCl	5.0 g
Phenol red	0.08 g
Sodium deoxycholate	2.5 g
Sodium thiosulfate	6.8 g
Sucrose	7.5 g
Xylose	3.75 g
Agar	15.0 g

pH 7.4 ± 0.2 at 25°C
Add components to distilled/deionized water and bring volume to 1 L. Mix thoroughly. Heat, boiling for 1 minute, then autoclave.

Yeast Fermentation Broth Containing Durham Tubes

Per liter: Tryptone	10.0 g
Yeast extract	5.0 g
K_2HPO_4	5.0 g
Sugar (glucose, lactose, or maltose)	10.0 g

The method of preparation is the same for phenol red broth base.
Note: Be careful not to overautoclave, because the sugars may decompose with overheating.

Acid Alcohol for Acid-Fast Stain

Ethanol (95%)	97.0 ml
Concentrated HCl	3.0 ml

Add acid to alcohol. Do not breathe acid fumes.

Alpha-naphthol for (Voges-Proskauer)

α-naphthol	50.0 g
Absolute ethanol	1,000.0 ml

Dissolve α-naphthol in the ethanol.

Carbolfuchsin Stain

Basic fuchsin	0.3 g
Ethanol (95%)	10.0 ml
Phenol	5.0 ml
Distilled water	100.0 ml

Dissolve the basic fuchsin in ethanol. Add phenol to the water. Mix the solutions together. Allow to stand for several days. Filter before use.

Dimethyl-α-naphthylamine

Per liter: N-N-dimethyl-α-naphthylamine	6.0 g
5N acetic acid (30%)	1,000.0 ml

Dissolve the N-N-dimethyl-α-naphthylamine in the acetic acid.

DNase (Deoxyribonuclease)

Sigma DN-25 or any crude grade DNA
Add a few visible crystals to 5 ml sterile water.
Hold in ice bucket when in solution.

Ferric Chloride (12%)

Iron III (Ferric) chloride (FeCl)	12.0 g
Concentrated hydrochloric acid	2.5 ml
Deionized water (up to 100 ml total volume)	

Gram-Stain Reagents

Crystal Violet (Hucker Modification)

Solution A:	Crystal violet (Gentian violet)	40.0 g
	Ethanol (95%)	400.0 ml
Solution B:	Ammonium oxalate	1.0 g
	Distilled water	1,600.0 ml

Dissolve the crystal violet in the alcohol and let stand overnight. Strain through filter paper. Add solution B. Mix to dissolve completely.

Gram's Iodine

Iodine	1.0 g
Kl	2.0 g
Distilled water	300.0 ml

Dissolve Kl in 2 ml of water. Add iodine and dissolve. Then add 300 ml of water.

Decolorizer for Gram Stain

Use either 95% ethanol or ethanol/acetone 1:1.

Safranin O (Distilled water)

Safranin	10.0 g

Hydrogen Peroxide (3%)

Purchase 3% H_2O_2 and put into brown dropper bottles. Refrigeration is recommended.

India Ink

Smaller particles are better than larger particles.

Kovac's Reagent

Isoamyl alcohol (isopentyl)	75.0 ml
Concentrated HCl	25.0 ml
p-dimethylaminobenzaldehyde	5.0 g

Mix the alcohol and the acid. Add the aldehyde and stir to dissolve. Store in plastic or dark glass bottles.

Malachite Green (5%)

Malachite green	5.0 g
Distilled water	95.0 ml

Mix, filter, and store in brown bottles.

Methyl Red

Methyl red	0.1 g
Ethanol (95%)	300.0 ml
Distilled water	200.0 ml

Dissolve methyl red in alcohol and add water. Dispense into dropper bottles.

Methylene Blue (Loeffler's)

Solution A:	Methylene blue	0.3 g
	95% Ethanol	30.0 ml
Solution B:	0.01% (w/v) KOH	100.0 ml
	about 1 pellet KOH/	1,000 ml

Mix solutions A and B together.

Methylene Blue (Acidified)

Solution A:	Methylene blue	0.02 g
	Distilled water	100.0 ml
Solution B:	KH_2PO_4 (0.2 Molar)	99.75 ml
Solution C:	K_2HPO_4 (0.2 Molar)	0.25 ml

Mix solutions A, B, and C together (pH must be 4.6).

ONPG

Per liter: O-nitrophenyl (0.0003 M)	1.0 g
β-D-galactopyranoside	

Oxidase Reagent
N,N-dimethyl-p-phenylene diamine
hydrochloride   1.0 g
Distilled water   100.0 ml
pH 5.5
This must be prepared on the day it is used. However, it can be frozen in small vials and thawed when needed, but the test is more difficult to read.

Phosphate Buffer
Per liter: $Na_2HPO_4 \cdot 2H_2O$ 11.88 g

KH_2PO_4 9.08 g
For pH 6.8: mix 6 parts Na_2HPO_4 with 4 parts KH_2PO_4.

Potassium Hydroxide (KOH)
Per liter: Potassium hydroxide   400.0 g
Deionized water   1000.0 ml
Dissolve the potassium hydroxide into the water.

Sodium Dodecyl Sulfate (SDS) in 10× Saline Citrate
Per liter: Sodium dodecyl sulfate   5.0 g
NaCl   87.7 g
Sodium citrate   44.1 g
No pH adjustment.

Streptomycin
Prepare a stock solution of streptomycin sulfate by adding 1 g to 10 ml distilled water. Filter sterilize, and freeze in convenient amounts.

Sulfanilic Acid
Per liter: Sulfanilic acid   8.0 g
5N Acetic acid   1000.0 ml
Dissolve sulfanilic acid in the acetic acid.
To make a 5 N acetic acid solution, add 28.75 ml of glacial acetic acid to 71.25 ml of distilled water.

Voges–Proskauer Reagent
Reagent A: Alpha-naphthol   5.0 g
                    Ethyl alcohol (95%)   100 ml
                                          (bring to volume)
Reagent B: KOH   40 g
                    Distilled water   100 ml
                                          (bring to volume)
Dispense each reagent in separate dropper bottles.

Wright Stain
Per liter: Wright stain   3.0 g
Absolute methanol   1000.0 ml
Dissolve stain in methanol overnight with stirring, then filter the stain into dropper bottles.

APPENDIX

2

Living Microorganisms (Bacteria, Fungi, Protozoa, and Helminths) Chosen for Study in This Manual

Acinetobacter, aerobic Gram-negative rods or coccobacilli in pairs. Opportunistic pathogen found in soil. Naturally competent and, therefore, easily transformed.

Alcaligenes faecalis, a Gram-negative rod, obligate aerobe, non-fermentative, found in the intestines.

Amoeba proteus, a unicellular protozoan that moves by extending pseudopodia.

Aspergillus niger, a filamentous black fungus with a foot cell, columella, and conidia.

Bacillus cereus, a Gram-positive rod, forms endospores, found in soil.

Bacillus subtilis, a Gram-positive rod, forms endospores, found in soil.

Citrobacter freundii, a Gram-negative rod, produces sulfur, found in the intestinal tract and soil.

Clostridium sporogenes, a Gram-positive rod, forms endospores, obligate anaerobe, found in soil.

Corynebacterium glutamicum, a Gram-variable, irregular, club-shaped rod found in oral cavities. A diphtheroid.

Diatom, a pond alga with symmetry containing silicon dioxide in their cell walls.

Dugesia, a free-living flatworm.

Enterobacter aerogenes, a Gram-negative rod, coliform group, produces capsules found in soil and water.

Enterococcus faecalis, a Gram-positive, group D coccus, grows in chains, found in the intestinal tract, opportunistic pathogen.

Escherichia coli, a Gram-negative rod, facultative, found in the intestinal tract, coliform group, can cause diarrhea and serious kidney disease.

Escherichia coli, K-12 strain, commonly used in research. Host strain for λ phage.

Halobacterium salinarium, member of the Archaea, can live only in high salt solutions.

Klebsiella pneumoniae, Gram-negative rod, coliform group, produces capsules, opportunistic pathogen.

Kocuria (Micrococcus) roseus, Gram-positive obligate aerobe cocci arranged in packets of four or eight. Part of the normal biota of the skin. The pink to red colonies are seen as an environmental contaminant.

Micrococcus luteus, Gram-positive obligate aerobe, cocci are arranged in packets of four or eight. Part of the normal biota of the skin. The yellow colonies are frequently seen as an air contaminant.

Moraxella, Gram-negative coccus found in normal biota.

Mucor, a rapidly growing filamentous fungus commonly seen on fruits, often gray or tan in color, with sporangiospores supported by columella. A Zygomycete.

Mycobacterium phlei, an acid-fast rod with a Gram-positive type of cell wall.

Mycobacterium smegmatis, an acid-fast rod with a Gram-positive type of cell wall.

Neisseria subflava, a Gram-negative coccus that grows in pairs, oxidase positive, found in the environment.

Paramecium, a ciliated protozoan found in ponds.

Penicillium species, a filamentous fungus with metulae, sterigmata, and conidia; makes antibiotics.

Proteus, a Gram-negative rod, swarms on agar, hydrolyzes urea, can cause urinary tract infections.

Providencia stuartii, a Gram-negative rod, hydrolyzes urea, found in intestines.

Pseudomonas aeruginosa, a Gram-negative rod, obligate aerobe, motile, opportunistic pathogen, can degrade a wide variety of compounds.

Rhizopus nigricans, a filamentous fungus with stolons, coenocytic hyphae, and a sporangium containing asexual sporangiospores.

Saccharomyces cerevisiae, eukaryotic fungal yeast cell, replicates by budding. Important in bread, beer, and wine making and in fungal genetics.

Salmonella enteriditis, a Gram-negative rod, non-fermentative, produces sulfur, found in the intestines.

Saprolegnia, an oomycete often seen in ponds.

Serratia liquefaciens, a Gram-negative rod that produces gelatinase, found in soil.

Serratia marcescens, a Gram-negative rod that produces pigments, found in soil.

Shigella flexneri, a Gram-negative rod, non-fermentative, found in the intestines.

Shigella sonnei, a Gram-negative rod, non-fermentative, found in the intestines.

Sordaria firmicola, a soil fungus often used in genetic studies.

Spirillum volutans, a Gram-negative curved rod, flagella at each pole, found in pond water.

Sporosarcina ureae, a Gram-positive coccus in arrangements of eight, produces endospores, hydrolyzes urea, found in soil.

Staphylococcus aureus, a Gram-positive coccus, a component of the normal skin biota, but can cause wound infections and food poisoning.

Staphylococcus epidermidis, a Gram-positive coccus, a component of the normal skin biota.

Streptococcus bovis, a Gram-positive coccus, a non-enterococcus, found in soil.

Streptococcus durans, a Gram-positive coccus, a nonenterococcus, found in soil.

Streptococcus mutans, normal biota of the mouth, forms mucoid colonies when growing on sucrose.

Streptococcus pneumoniae, lancet-shaped Gram-positive cells arranged in pairs and short chains, pathogenic strains form capsules.

Trichoderma viride, a soil fungus in which conidia appear yellow to green with age. Conidiophores and septate hyphae are also produced. An Ascomycete.

Volvox, a pond alga with hollow spheres, often containing round daughter cells.

APPENDIX

3

Dilution Practice Problems

See exercise 8 for an explanation of making and using dilutions.

1. If a broth contained 4.3×10^2 organisms/ml, about how many colonies would you expect to count if you plated
   a. 1.0 ml?
   b. 0.1 ml?
2. Show three ways for making each:
   a. 1/100 or 10^{-2} dilution,
   b. 1/10 or 10^{-1} dilution, and
   c. 1/5 or 2×10^{-1} dilution.
3. Show two ways of obtaining a 10^{-3} dilution using 9.0 ml and 9.9 ml dilution blanks.
4. **Figure A3.1** below shows a scheme for diluting yogurt before making plate counts. An amount of 0.1 ml was plated on duplicate plates from tubes B, C, and D. The numbers in the circles represent plate counts after incubation.

   a. Which plates were in the correct range for accurate counting?
   b. What is the average of the plates?
   c. What is the total dilution of the tubes?

   A _____

   B _____

   C _____

   D _____

   d. How many organisms/ml were in the original sample of yogurt?
5. Suppose an overnight culture of *E. coli* has 2×10^9 cells/ml. How would you dilute it so that you have countable plates? Diagram the scheme.

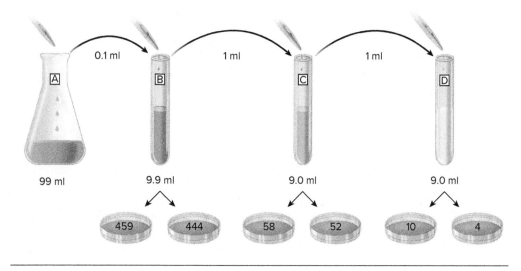

Figure A3.1

Answers

1a. 430 colonies

1b. 43 colonies

2a. 1.0 ml into 99 ml
0.1 ml into 9.9 ml
10 ml into 990 ml

2b. 1.0 ml into 9.0 ml
0.1 ml into 0.9 ml
10 ml into 90 ml

2c. 1.0 ml into 4 ml
0.1 ml into 0.4 ml
10 ml into 40 ml

3.

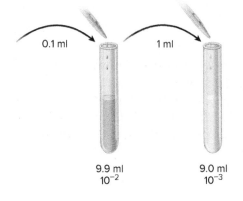

4a. Plates from tube C

4b. 55

4c. A. 10^{-2}
B. 10^{-4}
C. 10^{-5}
D. 10^{-6}

4d. The number of colonies × 1/dilution × 1/sample on plate = number of organisms/ml

$$55 \times 1/10^{-5} \times 1/0.1 = 55 \times 10^5 \times 10 = 55 \times 10^6 \text{ or } 5.5 \times 10^7 \text{ organisms/ml}$$

3.

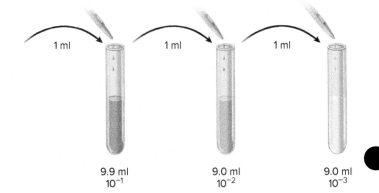

5.

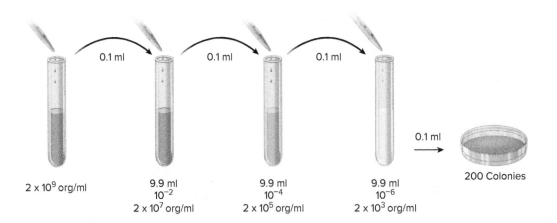

APPENDIX

Alternative Procedures

A. Spread Plate Technique

An alternative to making plate counts as described in exercise 8 is the spread plate technique (**figure A4.1**). Instead of putting the sample in melted agar, it is placed on the surface of an agar plate and spread around with a sterile rod (which resembles a hockey stick). Usually the sample size is 0.1 ml, so the following is a dilution scheme for suspension A and suspension B.

First Session

Suspension A

1. Label all water blanks with the dilution: 10^{-2}, 10^{-3}, 10^{-4}, 10^{-5}, 10^{-6}, 10^{-7}, 10^{-8}. Also label four Petri dishes (on the bottom): 10^{-5}, 10^{-6}, 10^{-7}, 10^{-8}.

2. Make serial dilutions of the bacterial suspension A:

    a. Mix the bacterial suspension by rotating between the hands, and transfer 1.0 ml of the suspension to the 99 ml water blank labeled 10^{-2}. Discard the pipet.

    b. Mix well, and transfer 1.0 ml of the 10^{-2} dilution to the 9.0 ml water blank labeled 10^{-3}. Discard the pipet.

    c. Mix well, and transfer 1.0 ml of the 10^{-3} dilution to the 9.0 ml water blank labeled 10^{-4}. Discard the pipet.

    d. Mix well, and transfer 1.0 ml of the 10^{-4} dilution to the 9.0 ml water blank labeled 10^{-5}. Discard the pipet.

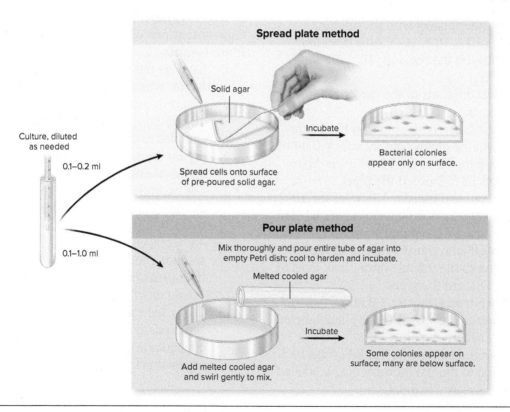

Figure A4.1 The spread plate method (top) and the pour plate method (bottom) for inoculating plates.

e. Mix well, and transfer 1.0 ml of the 10^{-5} dilution to the 9.0 ml water blank labeled 10^{-6}. Discard the pipet.

f. Mix well, and transfer 1.0 ml of the 10^{-6} dilution to the 9.0 ml water blank labeled 10^{-7}. Discard the pipet.

g. Mix well, and transfer 1.0 ml of the 10^{-7} dilution to the 9.0 ml water blank labeled 10^{-8}. Discard the pipet.

h. Mix well.

3. Starting with the most dilute suspension (10^{-8}), remove 0.1 ml and place on the surface of the agar plate. Immediately spread it around the entire surface of the plate with a sterile bent glass rod.

4. Using the same pipet and sterile glass rod (be careful not to contaminate), repeat the procedure for the 10^{-7}, 10^{-6}, and 10^{-5} dilutions.

5. Discard the pipet, and place the glass rod in a container as indicated by the instructor.

6. Invert the plates and incubate at 37°C.

Suspension B

1. Label all water blanks with the dilution: 10^{-2}, 10^{-3}, 10^{-4}, 10^{-5}, 10^{-6}. Also label four Petri dishes (on the bottom): 10^{-3}, 10^{-4}, 10^{-5}, 10^{-6}.

2. Make serial dilutions of the bacterial suspension B:

a. Mix the bacterial suspension by rotating between the hands, and transfer 1.0 ml of the suspension to the 99 ml water blank labeled 10^{-2}. Discard the pipet.

b. Mix well, and transfer 1.0 ml of the 10^{-2} dilution to the 9.0 ml water blank labeled 10^{-3}. Discard the pipet.

c. Mix well, and transfer 1.0 ml of the 10^{-3} dilution to the 9.0 ml water blank labeled 10^{-4}. Discard the pipet.

d. Mix well, and transfer 1.0 ml of the 10^{-4} dilution to the 9.0 ml water blank labeled 10^{-5}. Discard the pipet.

e. Mix well, and transfer 1.0 ml of the 10^{-5} dilution to the 9.0 ml water blank labeled 10^{-6}. Discard the pipet.

3. Starting with the most dilute suspension (10^{-6}), remove 0.1 ml and place on the surface of the agar plate. Immediately spread it around the entire surface of the plate with a sterile rod.

4. Using the same pipet and sterile glass rod (be careful not to contaminate), repeat the procedure for the 10^{-5}, 10^{-4}, and 10^{-3} dilutions.

5. Discard the pipet, and place the glass rod in a container as indicated by the instructor.

6. Invert the plates and incubate at 37°C.

Second Session for Both A and B Suspension (see page 64)

Note: In the spread plate procedure, all the samples are 0.1 ml. This appears as $\frac{1}{0.1}$ for the sample size in the formula

$$\text{Number of organisms/ml in original sample} = \text{the number of colonies on plate} \times \frac{1}{0.1} \times \frac{1}{\text{dilution}}$$

B. An Alternate Method of Calculating Concentrations of Organisms from Dilution Plate Counts

1. Convert the dilution to the dilution factor, which is the reciprocal of the dilution. For example, the dilution factor of a 10^{-1} dilution is 10. Convert all dilutions and samples to dilution factors, and then multiply by the average number of colonies on a countable plate.

2. For example: On page 65, the total dilution of the tubes in the box would be $10 \times 10 = 100$. If a 1.0 ml sample from tube B contained 48 bacteria, the concentration in the original solution (not shown) would be $48 \times 10 \times 10$ organisms/ml or 48×100 or 48×10^2 organisms/ml.

3. Using the same example, if 0.1 ml was sampled, the reciprocal of 1/.1 is 10. The concentration would be $10 \times 10 \times 10 \times 48$ organisms/ml or 48×10^3 organisms/ml.

C. Prepared Slides and Labs Used

Microbe	Category/Usage
Bacillus	Bacteria/arrangement, Gram-stain
Corynebacterium	Bacteria/arrangement, metachromatic granules
Enterobacter	Bacteria/Gram-stain
Micrococcus	Bacteria/arrangement, Gram-stain
Mycobacterium	Bacteria/arrangement, acid-fast
Proteus	Bacteria/flagella
Spirillum	Bacteria/flagella
Aspergillus	Fungi
Coprinus	Fungi
Morchella	Fungi
Mucor	Fungi
Penicillium	Fungi
Rhizopus	Fungi
Saccharomyces	Fungi
Sordaria	Fungi
Trichoderma	Fungi
Anopheles	Parasite
Ascaris	Parasite
Balantidium	Parasite
Dirofilaria	Parasite
Entamoeba	Parasite
Enterobius	Parasite
Fasciola	Parasite
Giardia	Parasite
Glossina	Parasite
Leishmania	Parasite
Opisthorchis/Clonorchis	Parasite
Plasmodium	Parasite
Schistosoma	Parasite
Taenia	Parasite
Trichinella	Parasite
Trichomonas	Parasite
Trichuris	Parasite
Trypanosoma	Parasite
Amoeba	Protozoa
Cladophora	Protozoa
Diatoms	Protozoa
Dugesia	Protozoa
Euglena	Protozoa
Paramecium	Protozoa
Planaria	Protozoa
Saprolegnia	Protozoa
Spirogyra	Protozoa
Volvox	Protozoa

APPENDIX

5

Determination of cell dimensions is often used in microbiology, where it has numerous applications. Examples include measuring changes in cell size during the growth cycle, determining the effect of various growth factors on cell size, and serving as a taxonomic assist in culture identification. Measurements are made by inserting a glass disc with inscribed graduations (**figure A5.1*a***), called an ocular micrometer, into the ocular of the microscope (**figure A5.2**).

It is not necessary to calibrate the ocular micrometer to determine the *relative* size of cells for this purpose, either a wet mount or stained preparation of cells can be examined with the ocular micrometer. By measuring the length in terms of number of ocular micrometer divisions, you might conclude that cell X is twice as long as cell Y. For such purposes, determination of cell length in absolute terms—such as number of micrometers—is not necessary.

For determining the *absolute* size of cells, it is necessary to first measure the length in micrometers (μm) between two lines of the ocular micrometer. A stage micrometer with a scale measured in micrometers is needed for μm calibration. The stage micrometer scale (**figure A5.1*b***) is such that the distance between two lines is 0.01 mm (equivalent to 10 μm). By superimposing the stage micrometer scale over the ocular micrometer scale, you can determine the absolute values in micrometers between two lines on the ocular micrometer scale. The absolute value obtained is also dependent on the objective lens used. For example, with the low-power objective lens, seven divisions on the ocular micrometer equal one division on the stage micrometer. Thus, with the low-power objective lens, one division on the ocular micrometer equals 0.01 mm/7, which equals 1.40 μm.

Procedure for Insertion, Calibration, and Use of the Ocular Micrometer

1. Place a clean stage micrometer in the mechanical slide holder of the microscope stage.
2. Using the low-power objective lens, center and focus the stage micrometer.
3. Unscrew the top lens of the ocular to be used, and then carefully place the ocular micrometer with the engraved side down on the diaphragm inside the eyepiece tube (figure A5.2). Replace the top lens of the ocular.
   Note: With some microscopes, the ocular micrometer is inserted in a retaining ring located at the base of the ocular.
4. To calibrate the ocular micrometer, rotate the ocular until the lines are superimposed over the lines of the stage micrometer.
5. Next, move the stage micrometer until the lines of the ocular and stage micrometer coincide at one end.
6. Now find a line on the ocular micrometer that coincides precisely with a line on the stage micrometer.
7. Determine the number of ocular micrometer divisions and stage micrometer divisions where the two lines coincide. An example of this step is shown in **figure A5.1*c***.

A relatively large practice microorganism for determining average cell size, with both the low- and high-power objective lenses, is a yeast cell wet mount. You must first calibrate the ocular micrometer scale for both objective lenses, using the stage micrometer. Next, determine the average cell size of a group of cells, in ocular micrometer units, using both the low- and high-power objective

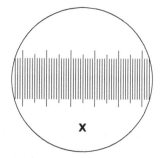

(a) Ocular micrometer
The diameter (width) of the graduations in micrometers (μm) must be determined for each objective.

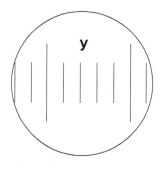

(b) Stage micrometer
The graduations are 0.01 mm (10 μm) wide.

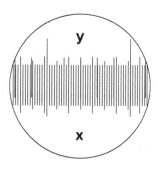

(c) Superimposing of ocular micrometer scale (x) over the stage micrometer scale (y)
Note that seven divisions of the ocular micrometer equal one division of the stage micrometer scale (0.01 mm).

One division of x = $\dfrac{0.01}{7}$

= 0.0014 mm = 1.4 μm

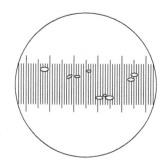

(d) Based on Figure A5.2c calculations, what is the average length in micrometers (μm) of the rod-shaped bacteria?

Figure A5.1 Calibration of the ocular micrometer.

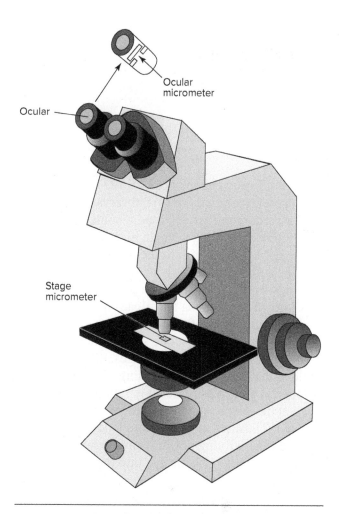

Figure A5.2 Location of the ocular and stage micrometers.

lenses. Attempt to measure the same group of cells with both lenses. You may find it easier to first measure them with the high-power objective lens, followed by the low-power objective lens. You will have mastered the technique if you obtain the same average cell size with both objective lenses.

Note: Care should be taken when inserting and removing the ocular micrometer from the ocular. Before inserting, make certain the ocular micrometer is free of dust particles by cleaning both sides with lens paper moistened with a drop of lens cleaning solution. Install and remove only in an area free of air currents. After removing the ocular micrometer, reexamine it for dust particles. If present consult your instructor.

APPENDIX

6

History

Anton van Leeuwenhoek (1632–1723), a Dutch linen draper, recorded the first observations of living microorganisms. He used a homemade microscope containing a single glass lens powerful enough to enable him to see what he described as little "animalcules" (now known as bacteria) in scrapings from his teeth and larger "animalcules" (now known as protozoa and algae) in droplets of pond water and hay infusions.

A single-lens microscope such as van Leeuwenhoek's had many disadvantages. These included distortion with increasing magnifying powers and a decreasing focal length (the distance between the specimen when in focus and the tip of the lens). Thus, when using a single lens with an increased magnifying power, van Leeuwenhoek had to practically push his eye into the lens in order to see anything.

Modern microscopes, called **compound microscopes,** have two lenses: an eyepiece or ocular lens and an objective lens (**figure A6.1**). The eyepiece lens allows comfortable viewing of the specimen from a distance. It also has some magnification capability, usually 10 times (10×)

or 20 times (20×). The purpose of the objective lens, which is located near the specimen, is to provide image magnification and clarity. Most teaching microscopes have three objective lenses with different powers of magnification (usually 10×, 45×, and 100×). Total magnification is obtained by multiplying the magnification of the ocular lens by the magnification of the objective lens. Thus, when using a 10× ocular lens with a 45× objective lens, the total magnification of the specimen image is 450 diameters.

Another giant in the early development of the microscope was a German physicist, Ernst Abbe, who developed (ca. 1883) various microscope improvements. One was the addition of a third lens, the **condenser lens,** which is located below the microscope stage (figure A6.1). By moving this lens up or down, it is possible to concentrate (intensify) the light emanating from the light source on the bottom side of the specimen slide. (The specimen is located on the top surface of the slide.)

Abbe also developed the technique of using lens immersion oil in place of water as a medium for transmission of light rays from the specimen to the lens of the **oil immersion objective lens.** Oil with a density more akin to the microscope

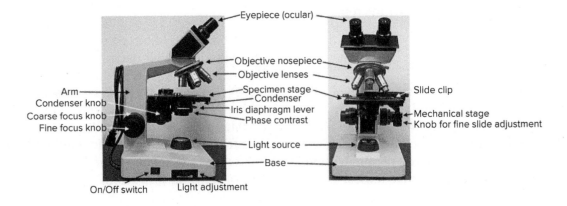

Figure A6.1 A modern bright-field microscope.

Courtesy of Anna Oller, University of Central Missouri.

lens than that of water helps to decrease the loss of transmitted light, which, in turn, increases image clarity. Finally, Abbe developed improved microscope objective lenses that were able to reduce both **chromatic** and **spherical lens aberrations.** His objective lenses had a concave lens (glass bent inward like a dish) in addition to the basic convex lens (glass bent outward). Such a combination diverges the peripheral rays of light only slightly to form an almost flat image. The earlier simple convex lenses produced distorted image shapes due to spherical lens aberrations and distorted image colors due to chromatic lens aberrations.

Spherical Lens Aberrations   These occur because light rays passing through the edge of a convex lens are bent more than light rays passing through the center. The simplest correction is the placement of a diaphragm below the lens so that only the center of the lens is used (locate the **iris diaphragm** in figure A6.1). Such aberrations can also be corrected by grinding the lenses in special ways.

Chromatic Lens Aberrations   These occur because light is refracted (bent) as well as dispersed by a lens. The blue components of light are bent more than the red components. Consequently, the blue light travels a shorter distance through the lens before converging to form a blue image. The red components travel a longer distance before converging to form a red image. When these two images are seen in front view, the central area, in which all the colors are superimposed, maintains a white appearance. The red image, which is larger than the blue image, projects beyond the central area, forming red edges outside of the central white image. Correction of a chromatic aberration is much more difficult than correction of a spherical aberration because dispersion differs among kinds of glass. Objective lenses free of spherical and chromatic aberrations, known as **apochromatic objectives,** are now available but are also considerably more expensive than **achromatic objectives.**

Some Working Principles of Light Microscopy

Microscope Objectives: The Heart of the Microscope

All the other parts of the microscope are involved in helping the objective lens attain a noteworthy image. Such an image is not necessarily the *largest,* but it is the *clearest.* A clear image helps achieve a better understanding of specimen structure. Size alone does not help achieve this end. The ability of the microscope to reveal specimen structure is termed **resolution,** whereas the ability of the microscope to increase specimen size is termed **magnification.**

Resolution, or resolving power, is also defined as the ability of an objective lens to distinguish two nearby points as distinct and separate. The maximum resolving power of the human eye when reading is 0.1 mm (100 μm). We now know that the maximum resolving power of the light microscope is approximately 0.2 μm, or 500× better than the human eye, and that it is dependent on the wavelength (λ) of light used for illumination as well as the numerical apertures (NA) of the objective and condenser lens systems. These are related in this equation:

$$\text{Resolving power } (r) = \frac{\lambda}{NA_{obj} + NA_{cond}}$$

Examining this equation, we can see that the resolving power can be increased by decreasing the wavelength and by increasing the **numerical aperture.** Blue light affords a better resolving power than red light because its wavelength is considerably shorter. However, because the range of the visible light spectrum is rather narrow, increasing the resolution by decreasing the wavelength is of limited use. Thus, the greatest boost to the resolving power is attained by increasing the numerical aperture of the condenser and objective lens systems.

By definition, the numerical aperture equals $n \sin \Theta$. The **refractive index,** *n,* refers to the medium employed between the objective lens and the upper slide surface as well as the medium employed between the lower slide surface and the condenser lens. With the low- and high-power objective lenses, the medium is air, which has a refractive index of 1, whereas with the oil immersion objective lens, the medium is oil, which has a refractive index of 1.25 or 1.56. Sin Θ is the maximum angle formed by the light rays coming from the condenser lens and passing through the specimen into the front objective lens.

Ideally, the numerical aperture of the condenser lens should be as large as the numerical aperture of the objective lens, or the latter is reduced, resulting in reduced resolution. Practically, however, the condenser lens numerical aperture is somewhat less because the condenser iris has to be closed partially in order to avoid glare. It is also important to remember that the numerical aperture of the oil immersion objective lens depends upon the use of a dispersing medium with a refractive index greater than that of air ($n = 1$). This is achieved by using oil, which must be in contact with both the condenser lens (below the slide) and the objective lens (above the slide). When immersion oil is used on only one side of the slide, the maximum numerical aperture of the oil immersion objective is 1.25—almost the same as the refractive index of air.

Note: Oil should not be placed on the surface of the condenser lens unless your microscope contains an oil-immersion-type condenser lens and your instructor authorizes its use.

Microscopes for bacteriological use are usually equipped with three objectives: 16 mm low power (10×), 4 mm high–dry power (40–45×), and 1.8 mm oil immersion (100×). The desired objective is rotated into place by means of a revolving nosepiece (figure A6.1). The millimeter number (16, 4, 1.8) refers to the **focal length** of each objective. By definition, the focal length is the distance from the principal point of focus of the objective lens to the principal point of focus of the specimen. Practically speaking, it can be said that the shorter the focal length of the objective, the shorter the **working distance** (i.e., the distance between the lens and the specimen) and the larger the opening of the condenser iris diaphragm required for proper illumination (**figure A6.2**).

The power of magnification of the three objectives—10×, 45×, and 96×—is inscribed on their sides (note that these values may vary somewhat depending upon the particular manufacturer's specifications). The total magnification is obtained by multiplying the magnification of the objective lens by the magnification of the ocular eyepiece. For example, the total magnification obtained with a 4 mm objective (45×) and a 10× ocular eyepiece is 45 × 10 = 450 diameters. The highest magnification is obtained with the oil immersion objective

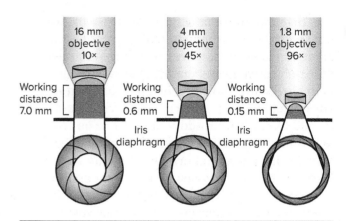

Figure A6.2 Relationship between the working distance of the objective lens and the diameter of the opening of the condenser iris diaphragm. The larger the working distance, the smaller the opening of the iris diaphragm.

lens. The bottom tip lens of this objective is very small and admits little light, which is why the iris diaphragm of the condenser lens must be wide open and the light conserved by means of immersion oil. The oil fills the space between the object and the objective lens so light is not lost (refer to figure 2.1 for visual explanation).

Note: If the microscope is **parfocal**, a slide that is in focus in one power will be in focus in other powers.

Microscope Illumination

Proper illumination is an integral part of microscopy. We cannot expect a first-class microscope to produce the best results when using a second-class illuminator. However, a first-class illuminator improves a second-class microscope almost beyond the imagination. A student microscope with only a mirror (no condenser lens) for illumination can be operated effectively by employing light from a gooseneck lamp containing a frosted or opalescent bulb. Illuminators consisting of a sheet of ground glass in front of a clear bulb are available, but they offer no advantage over a gooseneck lamp. Microscope mirrors are flat on one side and concave on the other. In the absence of a condenser lens, the concave side of the mirror should be used. Conversely, when a condenser lens is present, the flat side of the mirror should be used because condenser lenses accept only parallel rays of light and focus them on the slide.

Two or more condenser lenses are necessary for obtaining the desired numerical aperture. The Abbe condenser lens, which has a numerical aperture of 1.25, is most frequently used. The amount of light entering the objective lens is regulated by opening and closing the iris diaphragm, located between the condenser lens and the light source (figure A6.1). When the oil immersion objective lens is used, the iris diaphragm is opened farther than when the high–dry or low-power objective lenses are used. Focusing the light is controlled by raising or lowering the condenser lens using the condenser knob.

The mirror, condenser, and objective and ocular lenses must be kept clean to obtain optimal viewing. The ocular lenses are highly susceptible to etching from acids present in body sweat and should be cleaned after each use.

Definitions

Achromatic objective. A microscope objective lens in which the light emerging from the lens forms images practically free from prismatic colors.

Apochromatic objective. A microscope objective lens in which the light emerging from the lens forms images practically free from both spherical and chromatic aberrations.

Chromatic lens aberration. A distortion in the lens caused by the different refrangibilities of the colors in the visible spectrum.

Compound microscope. A microscope with more than one lens.

Condenser. A structure located below the microscope stage that contains a lens and iris diaphragm. It can be raised or lowered and is used for concentrating and focusing light from the illumination source on the specimen.

Focal length. The distance from the principal point of a lens to the principal point of focus of the specimen.

Iris diaphragm. An adjustable opening that can be used to regulate the aperture of a lens.

Magnification. The ability of a microscope to increase specimen size.

Numerical aperture. A quantity that indicates the resolving power of an objective. It is numerically equal to the product of the index of refraction of the medium in front of the objective lens (n) and the sine of the angle that the most oblique light ray entering the objective lens makes with the optical axis.

Oil immersion objective lens. High-power lens (usually 100×) that requires the immersion of the lens into a drop of oil.

Parfocal. Having a set of objectives so mounted on the microscope that they can be interchanged without having to appreciably vary the focus.

Refractive index. The ratio of the velocity of light in the first of two media to its velocity in the second medium as it passes from one medium into another medium with a different index of refraction.

Resolution. The smallest separation that two structural forms, for example, two adjacent cilia, must have in order to be distinguished optically as separate cilia.

Spherical lens aberration. An aberration caused by the spherical form of a lens that gives different focal lengths for central and marginal light rays.

Working distance. The distance between the tip of the objective lens when in focus and the slide specimen.

APPENDIX 7

The metric system enjoys widespread use throughout the world in both the sciences and non-sciences. Your studies in microbiology provide an excellent opportunity to learn how to use the metric system, particularly in the laboratory. The derivation of the word *metric* comes from the word *meter,* a measure of length.

The metric system is used in your microbiology course for media preparation, culture storage, and the measurement of cell number and size. The four basic measurements of the metric system are weight, length, volume, and temperature. The beauty of using the metric system is that all of these measurements are made in units based on multiples of 10. Not so with the English system, where measurements are made in units based on different multiples: 16 ounces in a pound, 12 inches in a foot, 2 pints in a quart, and 212°F, the temperature at which water boils, for example. The metric system weight measures in grams (g), length in meters (m), volume in milliliters (ml), and temperature in degrees Celsius (°C).

The prefixes of metric measurements indicate the multiple. The most common prefixes for metric measures used in microbiology are:

$$\text{kilo (k)} = 10^3 = 1{,}000$$

$$\text{centi (c)} = 10^{-2} = 0.01$$

$$\text{milli (m)} = 10^{-3} = 0.001$$

$$\text{micro (}\mu\text{)} = 10^{-6} = 0.000001$$

$$\text{nano (n)} = 10^{-9} = 0.000000001$$

Unit Only	Abbreviation	Equivalent
Weight		
kilogram	kg	1,000 g, 10^3 g
gram	g	1,000 mg (0.035 ounces; 454 g = 1 pound)
milligram	mg	10^{-3} g
microgram	μg	10^{-6} g
nanogram	ng	10^{-9} g
picogram	pg	10^{-12} g
Length		
kilometer	km	1,000 m (0.62 miles)
meter	m	100 cm (3.3 feet)
centimeter	cm	10^{-2} m (2.5 cm = 1 inch)
millimeter	mm	10^{-3} m
micrometer	μm	10^{-6} m (synonym = micron)
nanometer	nm	10^{-9} m
angstrom	Å	10^{-10} m
Volume		
liter	l	1,000 ml (1.1 quarts; 3.8 liters = 1 gallon)
milliliter	ml	10^{-3} liter
microliter	μl	10^{-6} liter
Temperature		
Celsius	°C	0°C = 32°F, 100°C = 212°F
		To convert from °C to °F: (°C x 9/5) + 32
		To convert from °F to °C: (°F x 5/9) − 32 or 1.8 °C = °F − 32

INDEX

Note: Page numbers followed by *f* and *t* indicate figures and tables respectively.

Plate 1 A negative stain showing *Bacillus* cells viewed on 100×. Courtesy of Anna Oller, University of Central Missouri.

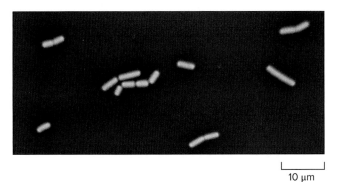

Plate 2 A Gram stain with pink Gram-negative *Escherichia coli* bacilli (*1*) and Gram-positive purple-stained *Staphylococcus aureus* cocci (*2*) viewed on 100×. Lisa Burgess/McGraw Hill

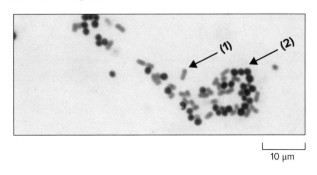

Plate 3 An acid-fast stain with methylene-blue-stained *Staphylococcus aureus* and red, acid-fast-stained *Mycobacterium phlei* cells viewed on 100×. Courtesy of Anna Oller, University of Central Missouri.

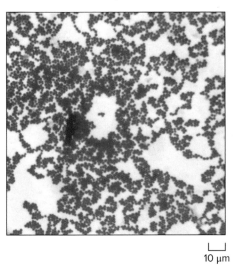

Plate 4 A capsule stain with a clear capsule surrounding the purple bacterial cell viewed on 100×. Lisa Burgess/McGraw Hill Education

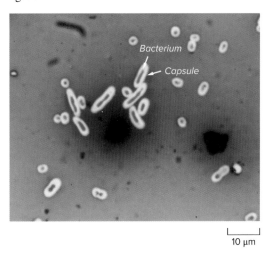

Plate 5 A positive endospore stain viewed on 100×. (*1*) Central spores in *Bacillus* species. (*2*) A free spore. Lisa Burgess/McGraw Hill Education

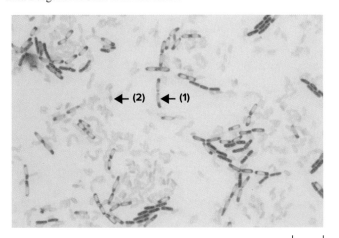

Plate 6 A lipid inclusion stain viewed on 100×. The bacterium *Bacillus* appears red, and the lipids are stained black (*1*). Courtesy of Anna Oller, University of Central Missouri.

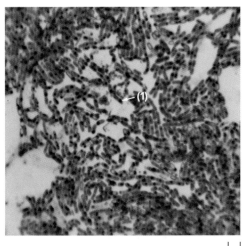

Plate 7 A flagella stain viewed on 100× showing peritrichous flagella. **Source:** Dr. William A. Clark/CDC

10 μm

Plate 9 Appearance of *Penicillium* mold growing on agar. Courtesy of Anna Oller, University of Central Missouri.

Plate 11 A *Chromobacterium violaceum* isolation streak with the most growth in the first (*1*) section, less growth in the second (*2*) section, and well-isolated colonies by the third (*3*) section of the plate. Courtesy of Anna Oller, University of Central Missouri.

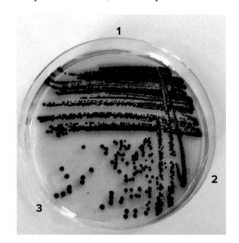

Plate 8 Appearance of various colonies growing on agar. Each number represents different colony types. Colonies labeled as 1s, 2s, etc. are most likely the same kind of microbe. *Bacillus* sp. are #1s, *Micrococcus* is #2, a yeast is #4, an actinomycete is #6, and a *Sarcina* is #11. Courtesy of Anna Oller, University of Central Missouri.

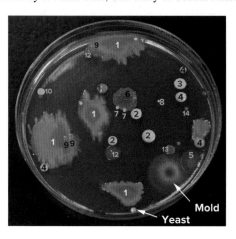

Plate 10 An isolation streak separating *Staphylococcus* and *Micrococcus* colonies. Courtesy of Anna Oller, University of Central Missouri.

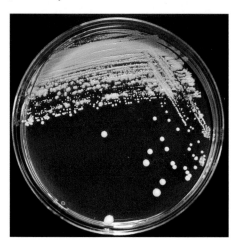

Plate 12 The spread plate technique. *Escherichia coli* was serial dilutioned, plated onto TSA, and incubated for 24 h. The most dilute solution is shown on the far left and the most concentrated solution (10^9) is shown on the far right with TNTC (too numerous to count), or >300 colonies. Courtesy of Anna Oller, University of Central Missouri.

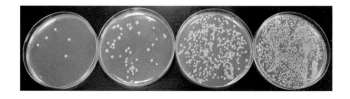

Plate 13 A semi-quantitative method using turbidity of TS broth tubes to show growth. The sharp lines of the rack can be seen in 1, so it is a 1+. The rack behind the second tube can be seen, but is less clear; thus it is a 2+. The third tube is more turbid, a 3+ and the fourth indicates complete obscureness, a 4+. Courtesy of Anna Oller, University of Central Missouri.

Plate 14 Oxygen requirements shown in thioglycollate broth tubes. (*1*) Aerobic growth at the top of the broth tube, seen with *Mycobacterium* species. (*2*) Microaerophilic bacterial growth under a thin layer of broth, seen with some *Micrococcus* species. (*3*) Facultative anaerobic growth throughout the tube, seen with *Escherichia coli*. (*4*) Anaerobic growth seen at the bottom of the tube, seen with *Clostridium* species. Courtesy of Anna Oller, University of Central Missouri.

Plate 15 An autoclave. Courtesy of Anna Oller, University of Central Missouri.

Plate 16 Autoclave tape on the left (*a*) to indicate an item was autoclaved and autoclave ampules (*b*) to verify an autoclave is working properly. The yellow ampule in (*b*) indicates bacterial growth and a failed autoclave run, whereas the purple ampule indicates a properly working autoclave. Courtesy of Anna Oller, University of Central Missouri.

Plate 17 A zone of inhibition created when a microbe is killed, or susceptible to the antibiotic. Lisa Burgess/ McGraw Hill

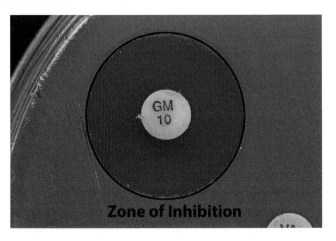

Plate 18 A minimum inhibitory concentration (MIC) of a diluted mouthwash solution used against the bacterium *Escherichia coli*. The units are milligrams (mg) per one liter. Courtesy of Anna Oller, University of Central Missouri.

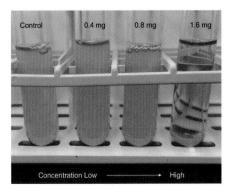

Plate 19 A lawn of growth of *Escherichia coli* subjected to antiseptics. (*1*) No clear zone, indicating resistance to iodine, (*2*) a zone of inhibition to a disinfecting wipe, indicating susceptibility, (*3*) a zone of inhibition to phenol, and (*4*) a zone of inhibition to a household cleaner. Courtesy of Anna Oller, University of Central Missouri.

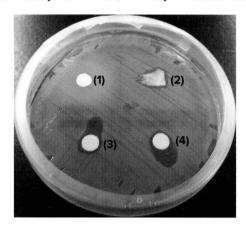

Plate 20 Plate 20 A coagulase test showing (*1*) a liquid negative result, and (*2*) a solid clump indicative of *Staphylococcus aureus*. Courtesy of Anna Oller, University of Central Missouri.

Plate 22 A mannitol salt plate with (*1*) a positive, yellow zone around *Staphylococcus aureus* and (*2*) a negative reaction seen by *Staphylococcus epidermidis*. Courtesy of Anna Oller, University of Central Missouri.

Plate 21 A latex agglutination test to test for the presence of *Staphylococcus aureus*. (*1*) A cloudy suspension of *Staphylococcus epidermidis,* whereas (*2*) *Staphylococcus aureus* shows clumping. Courtesy of Anna Oller, University of Central Missouri.

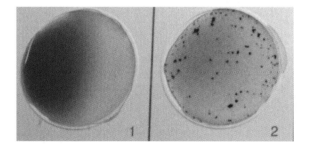

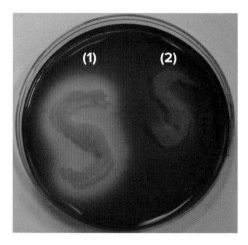

Plate 23 Beta hemolysis shown on a blood agar plate. Courtesy of Anna Oller, University of Central Missouri.

Plate 24 *Streptococcus pneumoniae* grown on a blood agar plate demonstrating alpha hemolysis and sensitivity to an optochin antibiotic disc. Courtesy of Anna Oller, University of Central Missouri.

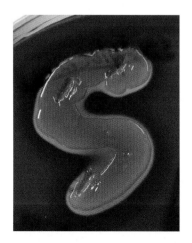

Plate 25 A nitrocefin card to test for the presence of the enzyme beta-lactamase. After adding a drop of water, bacterial colonies are added to the card. A negative reaction is seen in (*1*), while a positive reaction, a pink color, is seen in (*2*). The arrows in (*1*) are where the colony was added with a wooden applicator stick. The surrounding areas are representative of uninoculated testing areas. Courtesy of Anna Oller, University of Central Missouri.

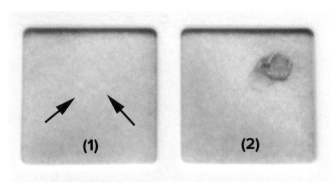

Plate 27 A catalase test showing (*1*) no bubbling (a negative test) from *Streptococcus* and (*2*) a positive bubbling test from *Staphylococcus*. Courtesy of Anna Oller, University of Central Missouri.

Plate 29 A strep throat infection caused by *Streptococcus pyogenes*, giving the characteristic strawberry tongue appearance and reddened tonsillar area. **Source:** Heinz F. Eichenwald, MD/CDC

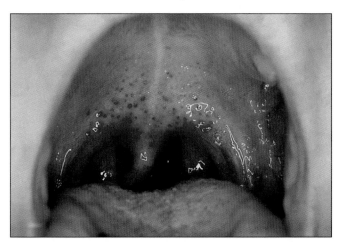

Plate 26 The oxidase test showing (*1*) a positive purple test from *Pseudomonas* and (*2*) a negative test from *Escherichia*. Courtesy of Anna Oller, University of Central Missouri.

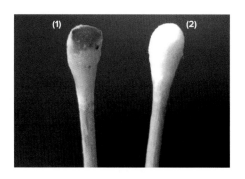

Plate 28 A throat swab sample taken from a patient with pharyngitis plated onto a blood agar plate. Notice the beta-hemolytic colonies (black arrows) and the susceptibility to the ampicillin antibiotic disc at the top (white arrow). McGraw Hill Education

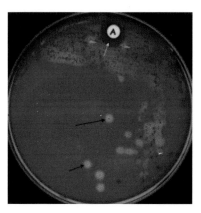

Plate 30 A rapid latex agglutination test that detects the presence of *Streptococcus pyogenes* antigens with (*1*) a negative test, and (*2*) a positive test. The test kit contains the antibodies that bind with the surface antigens on the bacterial surface. Courtesy of Anna Oller, University of Central Missouri.

Plate 31 Bile esculin slants depicting (*1*) a positive reaction given by *Enterococcus*, and (*2*) a negative reaction given by *Escherichia*. Courtesy of Anna Oller, University of Central Missouri.

Plate 32 Sodium chloride (NaCl) tubes with no growth seen in (*1*), some growth in (*2*), and heavier growth in (*3*). Notice the turbidity and appearance of rack lines. Used in conjunction with the bile esculin test, growth helps identify Group D enterococci like *Enterococcus faecalis*. Courtesy of Anna Oller, University of Central Missouri.

Plate 33 A starch plate after iodine addition depicting (*1*) a positive test with a yellow clear zone around the colony, and (*2*) a negative test with no clear zone around the yellow colony. Courtesy of Anna Oller, University of Central Missouri.

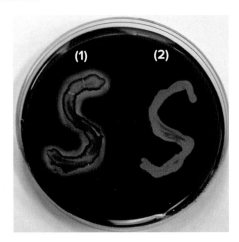

Plate 34 A casein hydrolysis, or skim milk, plate depicting (*1*) a positive clearing of media around the caseinase-producing *Bacillus* bacterium; and (*2*) a negative result given by *Escherichia coli* with no clear zone seen. Courtesy of Anna Oller, University of Central Missouri.

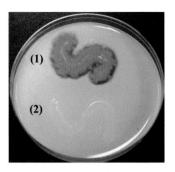

Plate 35 Gelatin hydrolysis tubes viewed after incubation and subsequent refrigeration. (*1*) The tube remained solid, a negative result for *Escherichia*, whereas (*2*) was liquefied, giving a positive result for *Serratia*. Courtesy of Annza Oller, University of Cental Missouri.

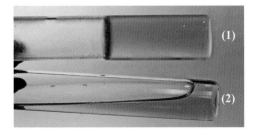

Plate 36 A spirit blue agar plate detects the production of the enzyme lipase. In (*1*) the colony and media color are lightened and oil droplets remain, a negative result. In (*2*) the colony appears dark blue with no oil droplets seen, indicating a positive test result. (1) & (2) Lisa Burgess/McGraw Hill

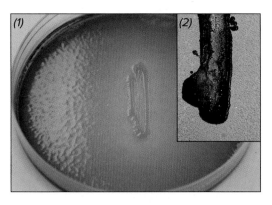

Plate 37 A DNase hydrolysis test with methyl green showing *Staphylococcus aureus* growth. If the enzyme DNase is produced, the DNA is degraded in the plate and a clear zone is seen, indicating a positive result, which is shown. Lisa Burgess/McGraw Hill

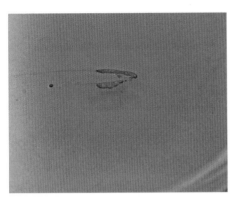

Plate 39 An indole test performed on TSA slants with Kovac's reagent added showing (*1*) a negative (yellow) result given by *Enterobacter*; and (*2*) red coloration indicative of a positive result given by *Escherichia coli.* Courtesy of Anna Oller, University of Central Missouri.

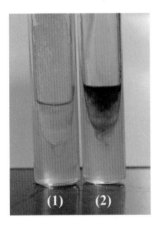

Plate 41 A nitrate test. (*1*) An uninoculated control. (*2*) Reagents A and B were added and a red color change occurred, a positive result. (*3*) Reagents A, B, and zinc were added, with no color change, which is a positive result. (*4*) Reagents A, B, and zinc were added, with a red color change, which is a negative result. Courtesy of Anna Oller, University of Central Missouri.

Plate 38 A urease test. (*1*) An uninoculated control. (*2*) A urease negative result. (*3*) A neon pink positive, given by *Proteus* or *Providencia* species. Courtesy of Anna Oller, University of Central Missouri.

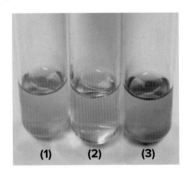

Plate 40 *Escherichia* (*1*) grown on TSA does not produce a pigment, but the *Pseudomonas* (*2*) does produce a green pigment. Courtesy of Anna Oller, University of Central Missouri.

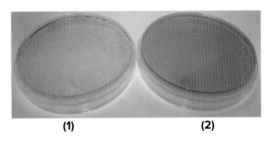

Plate 42 The methyl red test. (*1*) An uninoculated tube. After reagent has been added: A red (*2*) or pink (*3*) color indicates a positive result. A yellow (*4*) or tan (*5*) color indicates a negative result. Courtesy of Anna Oller, University of Central Missouri.

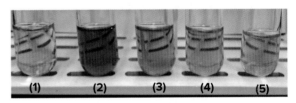

Plate 43 Voges-Proskauer tubes. (*1*) An uninoculated control. After reagents have been added: A light brown (*2*) or dark brown (*3*) color indicates a positive result. A yellow (*4*) color indicates a negative result. Courtesy of Anna Oller, University of Central Missouri.

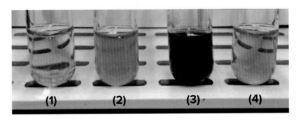

Plate 44 A phenylalanine deaminase plate after ferric chloride addition. (*1*) Yellow is a negative result. (*2*) Green is a positive result. Courtesy of Anna Oller, University of Central Missouri.

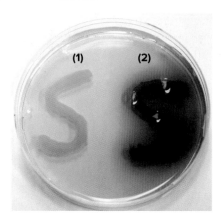

Plate 45 Sulfur indole motility tubes after Kovac's reagent has been added. (*1*) Sulfur –, indole –, motility – (*Staphylococcus*). (*2*) Sulfur –, indole +, motility + (*Escherichia*). (*3*) Sulfur +, indole –, motility + (*Salmonella*). (*4*) Sulfur +, indole +, motility + (*Proteus*). Courtesy of Anna Oller, University of Central Missouri.

Plate 46 Triple Sugar Iron (TSI) slants (*1*) Negative K/K (red/red) (*Alcaligenes*). (*2*) K/A (red/yellow) (*Shigella*). (*3*) K/@ (red/yellow plus gas) (*Enterobacter*). (*4*) A/A (yellow/yellow) (*Klebsiella*). (*5*) A/@ (yellow/yellow plus gas) *Escherichia*. (*6*) A/A⁺ (yellow/yellow plus H₂S) (*Citrobacter*). (*7*) K/A⁺ (red/yellow plus H₂S) (*Salmonella*). Courtesy of Anna Oller, University of Central Missouri.

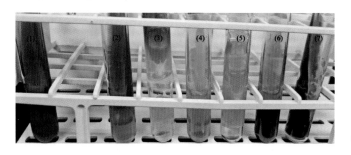

Plate 47 Sugar fermentation using Andrade's pH indicator. (*1*) Uninoculated control. (*2*) Negative control. (*3*) Acid is produced. (*4*) Acid and gas are produced. Courtesy of Anna Oller, University of Central Missouri.

Plate 48 Bromocresol purple fermentation results. *From left to right*: (*1*) Uninoculated control. (*2*) No change (negative). (*3 & 4*) Acid and gas (yellow with a gas bubble in the Durham tube). Courtesy of Anna Oller, University of Central Missouri.

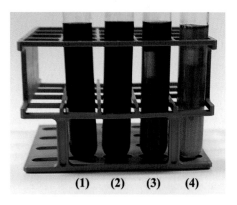

Plate 49 Glucose fermentation tubes using phenol red as the indicator. *From left to right*: (*1*) Inoculated negative control (*Pseudomonas*). (*2*) Acid produced (*Proteus*). (*3*) Acid and gas produced (*Escherichia*). Courtesy of Anna Oller, University of Central Missouri.

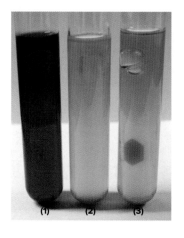

Plate 50 Citrate utilization. (*1*) A uninoculated control. (*2*) No color change is citrate negative. (*3*) A cobalt blue color is seen, which is citrate positive. Courtesy of Anna Oller, University of Central Missouri.

Plate 52 An eosin-methylene-blue (EMB) plate. (*1*) Clear *Shigella* colonies, a negative result. (*2*) A green metallic sheen indicative of *Escherichia,* a positive result. (*3*) Dark purple colonies of *Enterobacter,* a positive result. Courtesy of Anna Oller, University of Central Missouri.

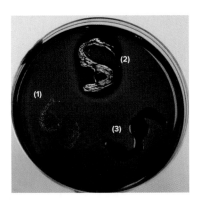

Plate 54 A lysine or ornithine decarboxylase (LDC/ODC) test after incubation. (*1*) An uninoculated control. (*2*) A yellow color indicates a negative result. (*3*) A purple color indicates a positive test result. Courtesy of Anna Oller, University of Central Missouri.

Plate 51 A MacConkey plate showing different degrees of positive results. (*1*) Yellow colonies are a negative result. (*2*) Dark pink to maroon colonies are a positive result. (*3*) A positive control, *Escherichia*. Courtesy of Anna Oller, University of Central Missouri.

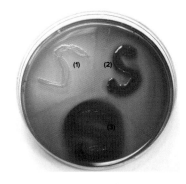

Plate 53 A Xylose Lysine Deoxycholate (XLD) plate. (*1*) Yellow colonies are a positive result (*Escherichia*). (*2*) Red colonies with black centers are a negative result (*Salmonella*). (*3*) Clear, red colonies are a negative result (*Shigella*). Courtesy of Anna Oller, University of Central Missouri.

Plate 55 A Hektoen Enteric agar plate. (*1*) Clear green colonies are a negative result (*Shigella*). (*2*) A bright orange color is a positive result (*Escherichia*). (*3*) Green colonies with black centers are a negative result (*Salmonella*). Courtesy of Anna Oller, University of Central Missouri.

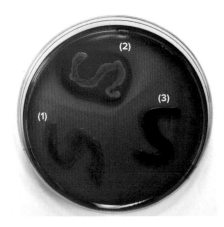

Plate 56 A *Salmonella Shigella* agar plate.
(*1*) Yellow colonies are a negative result (*Shigella*).
(*2*) Yellow colonies with black centers are a negative result (*Salmonella*). (*3*) Bright pink colonies are a positive result (*Escherichia*). Courtesy of Anna Oller, University of Central Missouri.

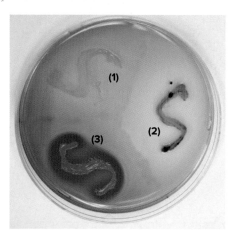

Plate 58 A Chromagar used to detect MRSA. (*1*) White colony of *Staphylococcus*. (*2*) Dark blue colony of methicillin-resistant *Staphylococcus aureus* (MRSA). Courtesy of Anna Oller, University of Central Missouri.

Plate 60 A microscopic view at 40× of *Rhizopus nigricans*. Aseptate hyphae lead to a sporangiospore on the right and two zygospores on the left. Courtesy of Anna Oller, University of Central Missouri.

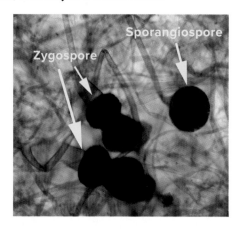

Plate 57 Addition of reagents to an EnteroPluri test, an all-in-one test. Catalin Rusnac/123RF

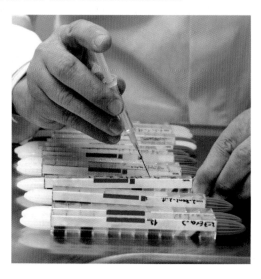

Plate 59 A rapid biochemical identification method. It determines the microbe identity and antibiotic sensitivity in a laboratory. The plate is read by a machine, and results are computed. Courtesy of Anna Oller, University of Central Missouri.

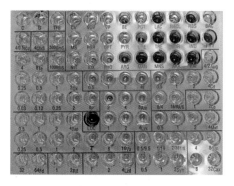

Plate 61 Ringworm caused by a dermatophyte fungus called *Trichophyton*. (*1*) Tinea capitis, affecting the scalp. (*2*) Tinea pedis, commonly called athletes foot, affecting the foot. **Source:** CDC

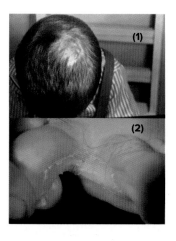

Plate 62 Labeled *Penicillium chrysogenum* reproductive structures. Viewed at 40×. Courtesy of Anna Oller, University of Central Missouri.

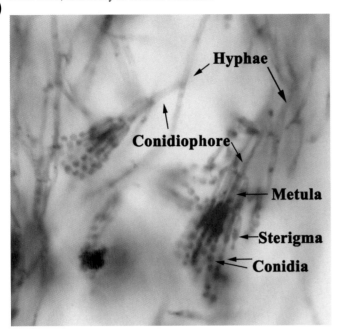

Plate 63 A stained *Morchella* cross-section containing ascospores within finger-like projections called asci. Viewed at 40×. Courtesy of Anna Oller, University of Central Missouri.

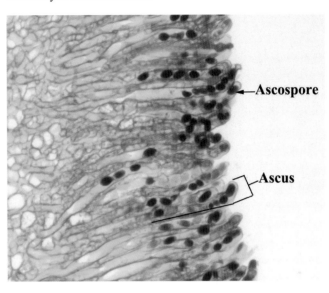

Plate 64 Basidiospores being released from the gills of a mushroom, magnified at 430×. McGraw Hill Education/Richard Gross, photographer

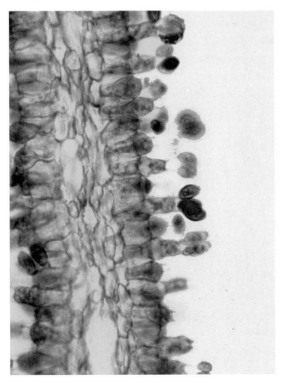

Plate 65 The yeast *Candida albicans* grown on Yeast Potato Dextrose (YPD) agar. The raised colonies have a matte appearance. Courtesy of Anna Oller, University of Central Missouri.

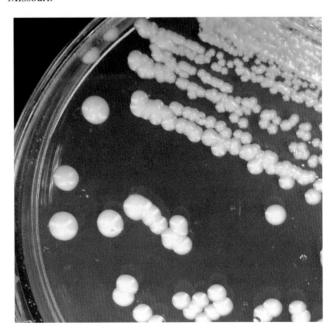

Plate 66 A lichen (*1*) viewed on 40× in which the large, red circular cells are algal cells, and the blue surrounding the algal cells are fungal cells. A lichen seen in nature (*2*) in basalt in a crustose growth form. (1) Courtesy of Anna Oller, University of Central Missouri. (2) McGraw Hill Education/Richard Gross, photographer

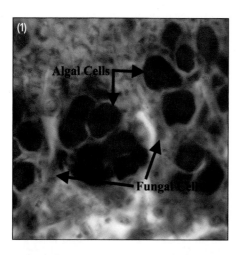

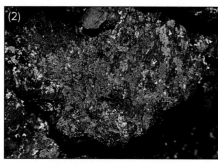

Plate 67 *Entamoeba histolytica* trophozoites viewed on 100× as (*1*) unstained and (*2*) trichrome stain. (1) Courtesy of Anna Oller, University of Central Missouri. (2) **Source:** Dr N J Wheeler, Jr./CDC

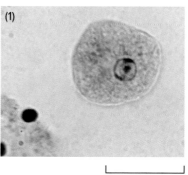

40 μm

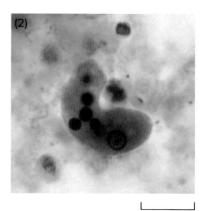

40 μm

Plate 68 *Giardia lamblia* (*intestinalis*) (*1*) trophozoites and (*2*) cysts viewed on 100×. Courtesy of Anna Oller, University of Central Missouri.

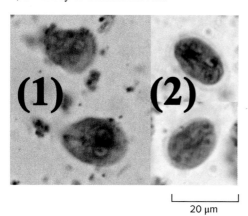

20 μm

Plate 69 Flagellated *Trypanosoma* in a blood smear viewed on 100×. Courtesy of Anna Oller, University of Central Missouri.

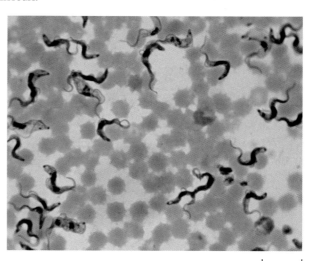

60 μm

Plate 70 *Plasmodium* ringed trophs (arrows) in a blood smear viewed on 100×. Courtesy of Anna Oller, University of Central Missouri.

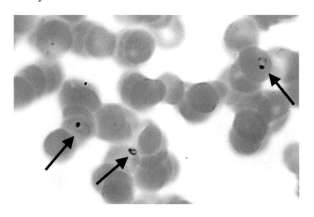

Plate 71 A *Plasmodium vivax* schizont (arrow) in a blood smear viewed on 100×. Courtesy of Anna Oller, University of Central Missouri.

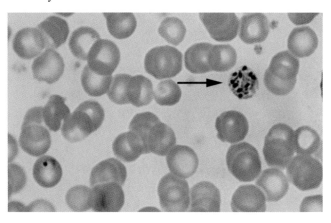

Plate 72 The liver fluke *Fasciola hepatica* viewed on 40×. Courtesy of Anna Oller, University of Central Missouri.

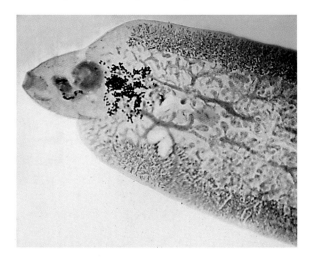

Plate 73 *Taenia* suckers and hooklets viewed on 40×. Courtesy of Anna Oller, University of Central Missouri.

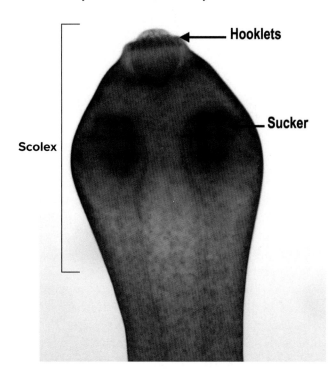

Plate 74 *Trichinella spiralis* parasites (arrows) in pork muscle tissue viewed on 4×. Courtesy of Anna Oller, University of Central Missouri.

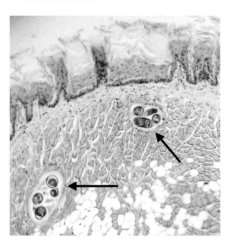

Plate 75 A prepared dog blood smear viewed on 40× showing *Dirofilaria immitis*. Courtesy of Anna Oller, University of Central Missouri.

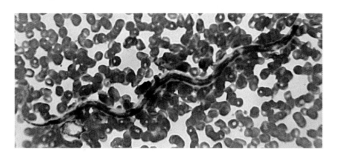

Plate 76 Zika viral particles digitally colorized and viewed via transmission electron microscope (TEM). Viral particles are stained blue with an outer envelope that can be seen around an inner core. Particles measure 40 nm in diameter. **Source:** CDC/Cynthia Goldsmith

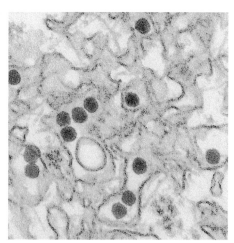

Plate 77 An electrophoresed agarose gel visualized by ethidium bromide excitation on a UV transilluminator. The 1 kb ladder (standard) is on the left, the wells are at the top, the bright band in the middle is the amplified DNA target, and the bands at the bottom are excess molecules. Courtesy of Anna Oller, University of Central Missouri.

Plate 78 (*1*) A microarray DNA chip appearance once read by a computer. (*2*) A chip used for gene sequencing. (*1*) Deco/Alamy Stock Photo. (*2*) CDC/ James Gathany

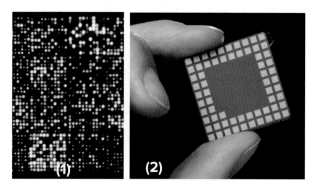

Plate 79 An ELISA 96-well plate. Courtesy of Anna Oller, University of Central Missouri.

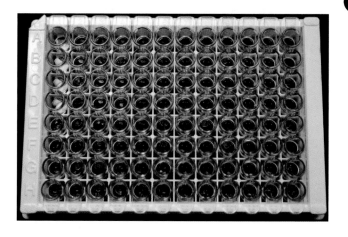

Plate 80 A blood smear stained with Wright-Giemsa and viewed on 100×. (*1*) A neutrophil. (*2*) An eosinophil. (*3*) A lymphocyte. (*4*) A platelet. (*5*) A red blood cell. Courtesy of Anna Oller, University of Central Missouri.

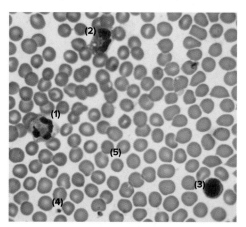